Renewable Energy: Generation and Application ICREGA'24

ICREGA'24
7th International Conference on Renewable Energy: Generation and Application, PMU campus, Al Khobar, Kingdom of Saudi Arabia, April 21st to 24th, 2024

Editors
Ala A. Hussein

Prince Mohammad Bin Fahd University, Khobar, Saudi Arabia

Peer review statement

All papers published in this volume of "Materials Research Proceedings" have been peer reviewed. The process of peer review was initiated and overseen by the above proceedings editors. All reviews were conducted by expert referees in accordance to Materials Research Forum LLC high standards.

Copyright © 2024 by authors

Content from this work may be used under the terms of the Creative Commons Attribution 3.0 license. Any further distribution of this work must maintain attribution to the author(s) and the title of the work, journal citation and DOI.

Published under License by **Materials Research Forum LLC**
Millersville, PA 17551, USA

Published as part of the proceedings series
Materials Research Proceedings
Volume 43 (2024)

ISSN 2474-3941 (Print)
ISSN 2474-395X (Online)

ISBN 978-1-64490-320-9 (Print)
ISBN 978-1-64490-321-6 (eBook)

This book contains information obtained from authentic and highly regarded sources. Reasonable efforts have been made to publish reliable data and information, but the author and publisher cannot assume responsibility for the validity of all materials or the consequences of their use. The authors and publishers have attempted to trace the copyright holders of all material reproduced in this publication and apologize to copyright holders if permission to publish in this form has not been obtained. If any copyright material has not been acknowledged please write and let us know so we may rectify in any future reprint.

Distributed worldwide by

Materials Research Forum LLC
105 Springdale Lane
Millersville, PA 17551
USA
https://www.mrforum.com

Manufactured in the United State of America
10 9 8 7 6 5 4 3 2 1

Table of Contents

Preface
Committees

Cyclic voltammetric and characterization studies of electrodeposited copper-indium for thin-film solar cell application
Leher Farooq 1

Electroluminescence image-based defective photovoltaic (solar) cell detection using a modified deep convolutional neural network
Hiren MEWADA, L. SYAMSUNDAR, Hiren Kumar THAKKAR, Miral DESAI 13

Analysis of thermal efficiency of solar flat plate collector working with hybrid nanofluids: An experimental study
Solomon MESFIN, Veeredhi VASUDEVA RAO, L. Syam SUNDAR 21

Green and environmental-friendly material for sustainable buildings
Noura Al-MAZROUEI, Amged ELHASSAN, Waleed AHMED, Ali H. Al-MARZOUQI, Essam ZANELDIN 31

Techno-economic comparison between PV and wind to produce green hydrogen in Jordan
Zaid HATAMLEH, Ahmad AL MIAARI 36

Analytical and numerical evaluation for wind turbine aerodynamic characteristics
Ashraf Abdelkhalig, Mahmoud Elgendi, Mohamed Y.E. Selim, Maryam Nooman AlMallahi, Sara Maen Asaad 44

Sustainability policies and regulation challenges in recycling EV batteries
Afnan KHALIL, Mousa HUSSEIN, Essam ZANELDIN, Waleed AHMED 51

Renewable energy sources, sustainability aspects and climate alteration: A comprehensive review
M. Amin Mir, M. Waqar Ashraf, Kim Andrews 60

Potential uses of renewable energy in construction: Advantages and challenges
Essam ZANELDIN, Waleed AHMED, Betiel PAULOS, Feruz GABIR, Makda ARAYA, El Bethel MULUYE, Deborah DEBELE3 66

Adaptive cooling framework for Photovoltaic systems: A seasonal investigation under the terrestrial conditions of Sharjah, UAE
Mena Maurice FARAG, Abdul-Kadir HAMID, Mousa HUSSEIN 73

A novel approach to address reliability concerns of wind turbines
Sorena ARTIN 82

Solar radiation forecasting using attention-based temporal convolutional network
Damilola OLAWOYIN-YUSSUF, Mohamad MOHANDES, Bo LIU, Shafiqur REHMAN 88

Application of artificial intelligence (AI) in wind energy system with a case study
Fay ALZAHRANI, Feroz SHAIK, Nayeemuddin MOHAMMED, Nasser Abdullah Shinoon AL-NA'ABI 96

Prediction of distillate output in photocatalytic solar still using artificial intelligence (AI)
Reyouf ALQAHTANI, Feroz SHAIK, Nayeemuddin MOHAMMED, Mohammad Ali KHASAWNEH, Tasneem SULTANA 104

Power systems stability of high penetration of renewable energy generations
Ahmad HARB, Hamza NAWAFLEH ... 112

Assessment of utilizing solar energy to enhance the performance of vertical aeroponic farm
Esam Jassim, Bashar Jasem, Faramarz Djavanroodi ... 118

Energy efficiency and sustainability enhancement of electric discharge machines by incorporating nano-graphite
Asaad A. ABBAS, Raed R. SHWAISH, Shukry H. AGHDEAB, Waleed AHMED ... 124

Exploring sustainable micro milling: Investigating size effects on surface roughness for renewable energy potential
Ahmet HASCELIK, Kubilay ASLANTAS, Waleed AHMED ... 132

From smart soles to green goals, interlacing sustainable innovations in the age of smart health: An exploratory search
Fahd KORAICHE, Amine DEHBI, Rachid DEHBI ... 140

Water and electricity consumption management architectures using IoT and AI: A review study
Oumaima RHALLAB, Amine DEHBI, Rachid DEHBI ... 148

A comprehensive review on computing methods for the prediction of energy cost in Kingdom of Saudi Arabia
Nayeemuddin MOHAMMED, Andi ASIZ, Mohammad Ali KHASAWNEH, Feroz SHAIK, Hiren MEWADA, Tasneem SULTANA ... 156

Vibration analysis of 3D printed PLA beam with honeycomb cell structure for renewable energy applications and sustainable solutions
Kubilay ASLANTAS, Ekrem ÖZKAYA, Waleed AHMED ... 164

Potential use of reject brine waste as a sustainable construction material
Seemab TAYYAB, Essam ZANELDIN, Waleed AHMED, Ali AL MARZOUQI ... 172

Bidding optimization for a reverse osmosis desalination plant with renewable energy in a day ahead market setting
Ebaa Al Nainoon, Ali Al Awami ... 179

Adaptive neuro-fuzzy inference system for DC power forecasting for grid-connected PV system in Sharjah
Tareq SALAMEH, Mena Maurice FARAG, Abdul Kadir HAMID, Mousa Hussein ... 188

Robust testing requirements for Li-ion battery performance analysis
Muhammad SHEIKH, Muhammad RASHID, Sheikh REHMAN ... 197

Date fruit type classification using convolutional neural networks
Abdullah ALAVI, Md Faysal AHAMED, Ali ALBELADI, Mohamed MOHANDES ... 205

Bidding optimization for hydrogen production from an electrolyzer
Nouf M. ALMUTAIRY, Ali T. ALAWAMI ... 214

Li-ion batteries life cycle from electric vehicles to energy storage
Muhammad RASHID, Muhammad SHEIKH, Sheikh REHMAN ... 223

AI-Based PV Panels Inspection using an Advanced YOLO Algorithm
Agus HAERUMAN, Sami Ul HAQ, Mohamed MOHANDES, Shafiqur REHMAN,
Sheikh Sharif Iqbal MITU 230

Impact of artificial intelligence (AI) in Martian architecture (exterior and interior)
Lindita Bande, Jose Berengueres, Aysha Alsheraifi, Anwar Ahmad, Saud Alhammadi,
Almaha Alneyadi, Amna Alkaabi, Maitha Altamimi, Yosan Asmelash 238

Sliding mode control for grid integration of point absorber type wave energy converter
Abdin Y. ELAMIN, Addy WAHYUDIE 246

Design and fabrication of low temperature flat plate collector for domestic water heating
Abdulkrim Almutairi, Abdulrahman Almarul, Riyadh Alnafessah,
Abbas Hassan Abbas Atya, Sheroz Khan, Noor Maricar 254

The utilization of IoT-based humidity monitoring method and convolutional neural networks for orchid seed germination
Muhammad Ridhan FIRDAUS, Hilal H. NUHA, Mohamed MOHANDES, Agus HAERUMAN 261

Intelligent solutions for modern agriculture: Leveraging artificial intelligence in smart farming practices
Fatima Zahrae BERRAHAL, Amine BERQIA 269

On the performance assessment of King Faisal University grid-connected solar PV facility
Mounir BOUZGUENDA 277

Solar energy powered smart water heating system
Ahmed ALTURKI, Yahya ALMAHZARI, Majid ALBLOOSHI, Faris ALGHAMDI,
Ahmed A. HUSSAIN, Ala A. HUSSEIN, Jamal NAYFEH 285

Experimental investigation on the thermal and exergy efficiency for a 2.88 kW grid connected photovoltaic/thermal system
Mena Maurice FARAG, Tareq SALAMEH, Abdul Kadir HAMID, Mousa Hussein 290

Solar PV based charging station for electric vehicles (EV)
Raghad ALGHAMDI, Batool ALSUNBUL, Raghad ALMUTAIRI, Rania ALNASSAR,
Maha ALHAJRI, Saifullah SHAFIQ, Samir EL-Nakla,
Ahmed ABUL HUSSAIN, Jamal NAYFEH 299

Role of renewable energy in decarbonisation process: Case study in KSA
Samar DERNAYKA, Saidur R. CHOWDHURY, Mohammad Ali KHASAWNEH 307

An automated and cost-efficient method for photovoltaic dust cleaning based on biaxially oriented polyamide coating material
Said Halwani, Mena Maurice Farag, Abdul-Kadir Hamid, Mousa Hussein 316

Vibration harvesting techniques for electrical power generation: A review
Omar D. MOHAMMED, Isha BUBSHAIT, Reyouf ALQAHTANI,
Kawther ALMENAYAN, Semat ALZAHER, Reem ALHUSSAIN 324

Multiport universal solar power bank
Abdullah ALTELMESSANI, Abdulqader ALJABER, Turki ALTUWAIRQI,
Ala A. HUSSEIN, Jamal NAYFEH 332

Electricity sector reforms in Saudi Arabia and their impact on demand growth and development of renewable energy
Samir El-NAKLA, Chedly Belhadj YAHYA, Jamal NAYFEH 337

Optimising solar: A techno-economic assessment and government facility compensation framework power generation
Navaid ALI, Faheem Ullah SHAIKH, Laveet KUMAR 345

A developed system design for blue energy generation
Omar D. MOHAMMED, Saud ALHARBI, Shoja ALHARBI, Waleed ALDHUWAIHI,
Nasser BINI HAMEEM 355

A review of the renewable energy technologies and innovations in geotechnical engineering
Eman J. Bani ISMAEEL, Samer RABABAH, Mohammad Ali KHASAWNEH,
Nayeemuddin MOHAMMED, Danish AHMED 360

Theoretical modeling to analyze the energy and exergy efficiencies of double air-pass solar tunnel dryer with recycled organic waste material
Dagim Kebede GARI, A. Venkata RAMAYYA, L. Syam SUNDAR 368

Renewable energy in pavement engineering and its integration with sustainable materials: A review paper
Mohammad Ali KHASAWNEH, Danish AHMED,
Fawzyah ALKHAMMAS, Zainab ALMSHAR, Maryam HUSSSAIN,
Dalya ALSHALI, Maryam ALKHURAIM, Rana, ALDAWOOD,
Hidaia ALZAYER, Nayeemuddin MOHAMMED 377

Keyword index

Preface

Renewable energy represents a pivotal shift in our approach to generating power, emphasizing sustainability and environmental responsibility. Unlike finite fossil fuels, renewable energy sources such as solar, wind, hydro, and geothermal power harness naturally replenishing elements of the Earth's ecosystem. This shift not only addresses the urgent need to mitigate climate change but also offers diverse applications across various sectors. From powering homes and businesses to driving transportation and industrial processes, renewable energy technologies are pivotal in shaping a cleaner, more resilient energy future.

Renewable energy generation involves harnessing naturally occurring and replenishing resources to produce electricity or heat. The main sources include:

- Solar Energy: Generated from sunlight through photovoltaic (PV) cells or concentrated solar power (CSP) systems. Solar energy is versatile, used in residential, commercial, and utility-scale applications for electricity generation and heating.
- Wind Energy: Captured by wind turbines that convert kinetic energy into electricity. Wind farms are increasingly common in areas with consistent wind patterns, contributing significantly to global electricity supply.
- Hydropower: Generated from flowing water in rivers, streams, and oceans. Hydroelectric dams and run-of-river systems convert water's gravitational energy into electricity, providing consistent, baseload power.
- Geothermal Energy: Extracted from heat stored beneath the Earth's surface. Geothermal power plants use steam or hot water to generate electricity and provide direct heating for buildings.
- Biomass Energy: Derived from organic materials such as wood, agricultural residues, and organic waste. Biomass can be burned directly for heat or converted into biogas or biofuels for electricity generation and heating.

Applications of renewable energy extend across various sectors:

- Electricity Generation: Renewable energy contributes to grid power, reducing reliance on fossil fuels and mitigating greenhouse gas emissions.
- Transportation: Electric vehicles (EVs) powered by renewable electricity are reducing emissions in the transportation sector.
- Heating and Cooling: Renewable energy sources like solar thermal and geothermal heat pumps provide efficient heating and cooling solutions for residential and commercial buildings.
- Industrial Processes: Renewable electricity and heat are used in industrial applications, reducing carbon footprints and enhancing sustainability.

- Rural Electrification: Off-grid renewable energy systems bring electricity to remote areas without access to traditional power grids, improving quality of life and economic opportunities.

The development and adoption of renewable energy technologies are crucial in combating climate change, enhancing energy security, and fostering sustainable development globally. In fact, renewable energy has a rich history of development and adoption worldwide, driven by technological advancements, environmental concerns, and energy security considerations. Here's an overview of its evolution and current status globally:

History of Renewable Energy:

- Early Use: Humans have utilized renewable energy sources for millennia, such as wind for sailing ships and grinding grain, and biomass for cooking and heating.
- Industrial Revolution: The 19th century saw increased use of hydropower for mechanical tasks and later for electricity generation as turbines were developed.
- 20th Century: The early 20th century saw the rise of hydropower and biomass as significant energy sources. In the mid-20th century, solar photovoltaic (PV) technology and wind turbines began development.
- Late 20th Century: The energy crises of the 1970s spurred interest in renewable energy, leading to government incentives and research in solar, wind, and geothermal energy.
- 21st Century: Rapid technological advancements, declining costs, and environmental concerns have accelerated the deployment of renewables globally.

Current Status Worldwide:

- Electricity Generation:
  - Solar PV: Solar energy capacity has grown significantly, with large-scale installations in countries like China, the United States, and India.
  - Wind Power: Wind turbines contribute a substantial portion of electricity in many countries, particularly in Europe and North America.
  - Hydropower: Remains a significant source, especially in countries like China, Brazil, and Canada.
- Policy Support: Many countries have implemented policies to promote renewable energy, including feed-in tariffs, tax incentives, and renewable portfolio standards.
- Technological Advancements: Advances in energy storage, grid integration, and efficiency improvements have bolstered renewable energy's viability and reliability.
- Global Growth: The International Renewable Energy Agency (IRENA) reports that renewable energy capacity has grown rapidly, with substantial investments in solar, wind, and other technologies.

- Challenges: Despite growth, challenges remain, including intermittency issues with solar and wind, grid integration challenges, and varying policy support across different regions.
- Future Outlook: The transition to renewables is expected to continue, driven by climate goals under the Paris Agreement, technological innovation, and economic competitiveness.

In summary, renewable energy has evolved from traditional uses to a pivotal component of global energy systems. Its history reflects a trajectory of technological innovation, policy support, and increasing global adoption, positioning it as a key solution in addressing climate change and achieving sustainable development goals worldwide. And, as technology advances and economies of scale improve, renewable energy continues to play a pivotal role in shaping a cleaner and more resilient energy landscape for the future.

Committees

HONORARY CHAIR
Dr. Issa H. Alansari, President, Prince Mohammed Bin Fahd University

CONFERENCE CHAIR
Dr. Faisal Yousif Al Anezi, Vice President, Prince Mohammed Bin Fahd University

TECHNICAL PROGRAM CHAIRS
Dr. Ala A. Hussein, Prince Mohammed Bin Fahd University, Saudi Arabia
Dr. Samir El Nakla, Prince Mohammed Bin Fahd University, Saudi Arabia
Dr. Mousa Hussein, United Arab Emirates University, United Arab Emirates

TECHNICAL PROGRAM COMMITTEE
Dr. Jamil Mosallam Bakhashwain, Prince Mohammed Bin Fahd University, Saudi Arabia
Dr. Jawad Al Asad, Prince Mohammed Bin Fahd University, Saudi Arabia
Dr. Sadiq Al Huwaidi, Prince Mohammed Bin Fahd University, Saudi Arabia
Dr. Christakis Papageorgiou, Prince Mohammed Bin Fahd University, Saudi Arabia
Dr. Hiren Mewada, Prince Mohammed Bin Fahd University, Saudi Arabia
Dr. Hussain Shareef, United Arab Emirates University, United Arab Emirates
Dr. Imad Barhumi, United Arab Emirates University, United Arab Emirates
Dr. Addy Wahyudie, United Arab Emirates University, United Arab Emirates
Dr. Nasir Saeed, United Arab Emirates University, United Arab Emirates
Dr. Waleed Ahmed, United Arab Emirates University, United Arab Emirates
Dr. Said Al-Hallaj, University of Illinois at Chicago, USA
Dr. Issa Batarseh, University of Central Florida, USA
Dr. Mohamed Farid, The University of Auckland
Dr. Yahya Meziani, Universidad de Salamanca, Spain

ORGANIZING COMMITTEE
Dr. Samir El Nakla, Prince Mohammed Bin Fahd University, Saudi Arabia
Dr. Ala Hussein, Prince Mohammed Bin Fahd University, Saudi Arabia
Dr. Christakis Papageorgiou, Prince Mohammed Bin Fahd University, Saudi Arabia
Dr. Jawad Al Asad, Prince Mohammed Bin Fahd University, Saudi Arabia
Dr. Sadiq Al Huwaidi, Prince Mohammed Bin Fahd University, Saudi Arabia
Dr. Hiren Mewada, Prince Mohammed Bin Fahd University, Saudi Arabia
Dr. Ghida Alzohbi, Prince Mohammed Bin Fahd University, Saudi Arabia
Dr. Mousa Hussein, United Arab Emirates University, United Arab Emirates
Dr. Amine El Moutaouakil, United Arab Emirates University, United Arab Emirates
Dr. Atef Abdrabou, United Arab Emirates University, United Arab Emirates
Dr. Mohammed Noor Al Tarawneh, United Arab Emirates University, United Arab Emirates
Dr. Rachid Errouissi, United Arab Emirates University, United Arab Emirates
Dr. Falah Awwad, United Arab Emirates University, United Arab Emirates
Mr. Ahmed Abul Hussain, Prince Mohammed Bin Fahd University, Saudi Arabia
Mr. Adil Humayun Khan, Prince Mohammed Bin Fahd University, Saudi Arabia
Mr. Wael Suliman, Prince Mohammed Bin Fahd University, Saudi Arabia

Cyclic voltammetric and characterization studies of electrodeposited copper-indium for thin-film solar cell application

Leher Farooq[1,a *]

[1]Department of Chemical & Petroleum Engineering, United Arab Emirates University, Abu Dhabi - 15551, UAE

[a]201570178@uaeu.ac.ae

Keywords: Electrodeposition, Copper Indium Diselenide, Thin-Film Solar Cells, Complexing Agents, Cyclic Voltammetry, Energy Dispersive X-Ray Spectroscopy, X-Ray Diffraction, Scanning Electron Microscopy

Abstract. Copper-indium layers were grown on a carbon substrate by a one-step electrodeposition technique. Four deposition electrolytes were prepared: Two unitary electrolytes comprising of 300 mM copper (II) chloride and 700 mM trisodium citrate and 300 mM indium (III) chloride and 700 mM trisodium citrate; two binary electrolytes comprising of 300 mM copper (II) chloride, 300 mM indium (III) chloride and 700 mM trisodium citrate for deposition at potentials of -0.6 V and -0.9 V, respectively. The pH of the solution was left unaltered at 4.8. X-ray diffraction studies showed the produced copper indium layers had an amorphous structure. From the scanning electron microscopy studies, it was found that copper indium layers indicated a small grain size and grain shape was irregular. The structure was not compact. Energy dispersive spectroscopy showed presence of low percentages of copper and indium in all four samples and stoichiometric molar ratios of Cu:In of 1:1 can be achieved using deposition potentials between -0.6 to -0.9 V. SEM, EDS and XRD characterization helps in the identification of the reaction mechanism, the structure and morphology of the films deposited under different conditions. These characterization studies will assist in enhancing the efficiency of the thin-films deposited for maximum conversion of sunlight to electricity using this sustainable method. In this present study, an attempt has been made to obtain near stoichiometric films for higher efficiencies.

Introduction

The direct conversion of sunlight into electricity (photovoltaics) may be the most prevailing and potential alternative to the utilization of fossil fuels. Primary energy consumption decreased by 4.5% last year, which is the first decline since 2009. Nevertheless, alternative forms to fossil fuel are in dire need to protect the environment from global warming. Copper indium diselenide (CIS) and copper indium gallium diselenide (CIGS) are two of the most favorable absorber materials for low cost photovoltaics or thin film solar cells and extensive research is being conducted in the development of these thin film solar cells [1]–[4]. Copper indium diselenide has a I-III-VI$_2$ ternary chalcopyrite structure and possesses desirable physical properties such as a direct band gap at ~1 eV and a high absorption coefficient of greater than 10^5 cm^{-1} at photon energies above the band gap. CIS thin film of about 1 μm thickness absorbs 90% of the incident sunlight with photon energy greater than its band gap. The best CIGS solar cells have the following layer structure: Mo (1 μm) coated glass/CIGS absorber/CdS buffer layer/CdS-ZnO window layer/MgF$_2$ antireflection coating/Ni-Al alloy grid [5]. These qualities make it ideal for fabrication of high efficiency polycrystalline thin film photovoltaic devices [6]. Several techniques are present for the preparation of CIS thin films including co-evaporation [7], sputtering [8]–[10], spray pyrolysis [11]–[13], electrodeposition [1], [3], [14], molecular beam epitaxy [15], etc. The highest efficiency achieved for CIS thin-film solar cell is 23.35% reported by Solar Frontier in 2019 [16]. The conductivity of this ternary semiconductor can be either n- or p-type which depends on the synthesis method and the composition of the constituent elements in the structure [17].

Several works have been published on chalcopyrite materials which have primarily used one-step electrodeposition process instead of physical vapor deposition because it allows achievement of low production cost, higher deposition speed and negligible waste of chemicals [18]. Owing to the reasoning that the co-electrodeposition process does not require very sophisticated equipment and is easy to use, this technique where the Cu-In-Ga-Se species are present in the same chemical bath is one of the suitable techniques to prepare low cost thin films [19]. PVD technique produces high quality films but is difficult to scale up. Co-electrodeposition process can be identified into two types: (1) where the elements involved have almost the same order of electrode potential, and (2) where the elements involved have different electrode potentials. In the cases where the elements involved have different electrode potentials, complexing agents such as trisodium citrate or EDTA can be utilized to reduce the difference in the deposition potentials.

Considerable amount of work has been done on the electrodeposition of CIS which includes a few studies devoted to the mechanism of CIS formation. Carbonnelle and Lamberts conducted an elementary study as well as the first step in understanding the mechanism of copper selenide electrodeposition to understanding the copper indium diselenide deposition from a ternary bath. Their results found that copper selenide is an intermediate or precursor for the formation of copper indium diselenide deposition [20].

The formation of the electrodeposited CIGS on the cathodic (working electrode) surface within an electrolysis cell typically depends on the pH level, redox (standard reduction) potential, deposition time, applied potential and the initial solution concentration along with some other factors [21]. In this paper, the final composition of the deposited film is determined using EDS analysis by studying the effects of initial concentrations of the electrolyte as well as the deposition potentials at which the reduction occurs. A detailed study through cyclic voltammetry, Energy Dispersive Spectroscopy, X-Ray Diffraction and Scanning Electron Microscopy is conducted in order to study the specific initial concentration of the electrolyte.

Materials and methods
Materials preparation

Copper (II) chloride ($CuCl_2$), indium (III) chloride ($InCl_3$) and trisodium citrate ($Na_3C_6H_5O_7$) were all purchased from Sigma-Aldrich. All chemicals were used as received and deionized water was used for the deposition bath. A small circular carbon electrode with dimensions of 2 mm in diameter was used as the substrate. The electrodeposition was carried out potentiostatically in a three-electrode cell configuration. The reference electrode was silver/silver chloride electrode and the working and counter electrodes were made up of carbon. All substrates were cleaned with acetone and rinsed with deionized water, and subsequently dried. The electrolyte bath contained 100 ml deionized water and 300 mM copper (II) chloride, 300 mM indium (III) chloride and 700 mM trisodium citrate. The pH of the solution was 4.8. With the aim of checking the stability of the electrolyte solutions prepared, they were stored for at least 20 days at room temperature and pressure and their final stability at the end of 20 days was observed before usage. The pH of the electrolytes were not altered using any acidic agents such as hydrochloric acid or sulphuric acid.

Characterization

A Solartron ModuLab XM ECS potentiostat/galvanostat was used for the film depositions and cyclic voltammetry studies. The cyclic voltammograms were measured at a scan rate of 15 mV/s and were scanned only in the negative potential region. The cathodic potentials have been fixed at -0.6 V and -0.9 V for copper and indium deposition, respectively. The deposition time has been at

15 min. All film depositions and cyclic voltammogram measurements were performed in a stagnant bath at room temperature.

The surface morphology, chemical composition and crystalline properties of the electrodeposited films were characterized by scanning electron microscopy (SEM, Jeol JSM-6010PLUS/LA), energy dispersive X-ray spectroscopy and X-ray diffraction (XRD, X' Pert PRO MRD XL XRD system from Panalytical), respectively.

Results and discussion

The deposition solutions containing electrochemically active species Cu^{2+} and In^{3+} in the presence of trisodium citrate as a complexing agent are stable for a long time. Different potentials were employed to produce Cu-rich and Cu-poor deposition layers. Cu-rich layers were produced in a binary solution containing copper (II) chloride and indium (III) chloride along with trisodium citrate at a cathodic potential of -0.6 V and Cu-poor layers were produced at a cathodic potential of -0.9 V. Both the electrolytes for the electrochemical experiments at -0.6 V and -0.9 V have the same $[Cu^{2+}]/[In^{3+}]$ initial ratios. Electrodeposition was carried out without agitation. The chemical composition of the unitary deposits of copper and indium and binary deposit of copper-indium were analyzed by EDS. The results of the chemical compositional analysis of copper at -0.6 V, indium at -0.9 V and copper-indium layer at -0.6 V and -0.9 V are shown in table 1, 2, 3 and 4, respectively. One can observe the presence of copper, indium, sodium, chloride and oxygen atoms in the deposited layer. Apart from copper and indium, the other elements present are due to improper washing of the deposited layers.

Table 1 – Chemical composition of the synthesized layer of copper at -0.6V

Formula	mass%	Atom%
O	10.50	18.87
Na	32.71	40.94
Cl	40.35	32.75
Cu	16.44	7.44
Total	100.00	100.00

Table 2 – Chemical composition of the synthesized layer of indium at -0.9V

Formula	mass%	Atom%
O	52.12	67.91
Na	28.61	25.95
Cl	6.50	3.82
In	12.77	2.32
Total	100.00	100.00

Table 3 – Chemical composition of the synthesized layer of copper-indium at -0.6V

Formula	mass%	Atom%
O	30.27	49.45
Na	26.39	30.01
Cl	14.37	10.59
Cu	18.25	7.51
In	10.72	2.44
Total	100.00	100.00

Table 4 – Chemical composition of the synthesized layer of copper-indium at -0.9V

Formula	mass%	Atom%
O	29.27	50.13
Na	18.62	22.19
Cl	26.24	20.28
Cu	6.35	2.74
In	19.52	4.66
Total	100.00	100.00

Four cyclic voltammetric experiments were conducted: copper electrodeposition at -0.6 V, indium electrodeposition at -0.9 V, copper indium electrodeposition at -0.6 V and -0.9 V. The molar ratio of Cu:In at a deposition potential of -0.6V was found to be approximately 2:1. The molar ratio of Cu:In at a deposition potential of -0.9V was found to be approximately 1:3. Thus, to obtain stoichiometric molar ratio of Cu:In = 1:1, a deposition potential between -0.6 V and -0.9 V must be chosen. The atomic ratio for Cu:In at a deposition potential of -0.6 V was found to be 3:1 and at -0.9 V is approximately 3:2. Therefore, to obtain stoichiometric atomic ratio of Cu:In = 1:1, a more negative deposition potential should be selected. Figure 1 shows the EDS diagrams and indicates that the rates of Cu and In components obtained in this study are seen to be low which matches the data obtained from the compositional analysis.

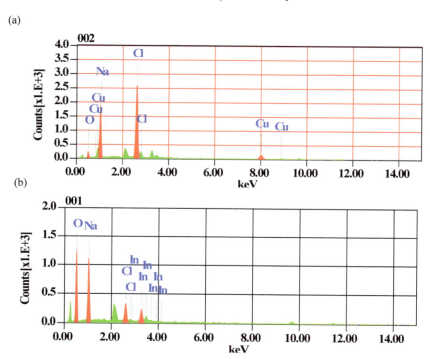

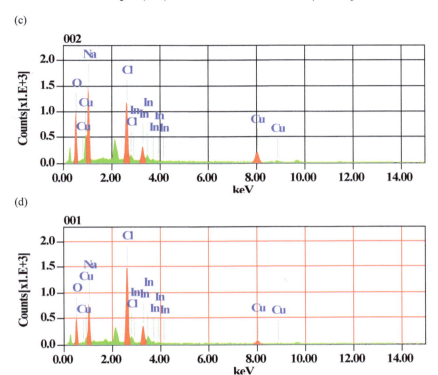

Fig. 1. EDS of the copper-indium layers.

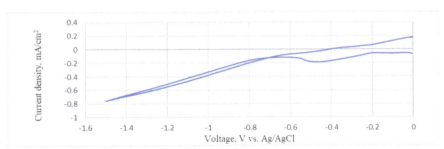

Fig. 2. 300 mM $CuCl_2$ + 300 mM $InCl_3$ without any complexing agent.
Fig. 2. shows no prominent peak for indium obtained in a binary solution of 300 mM copper (II) chloride and 300 mM indium (III) chloride without the usage of any complexing agents.

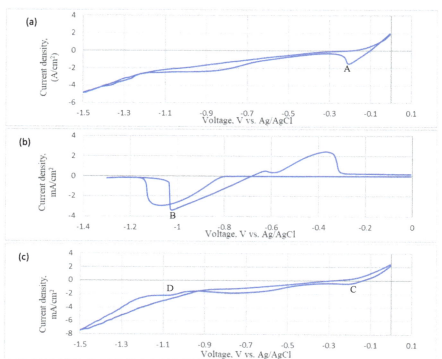

Fig. 3. Cyclic voltammograms of a Carbon electrode in (a) 300 mM CuCl$_2$ + 700 mM Trisodium Citrate, (b) 300 mM InCl$_3$ + 700 mM Trisodium Citrate and (c) 300 mM CuCl$_2$ + 300 mM InCl$_3$ + 700 mM Trisodium Citrate.

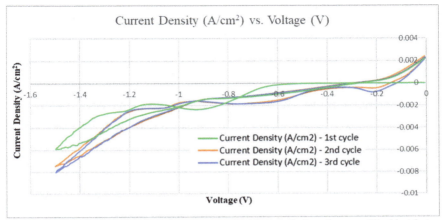

Fig. 4. Three cycles of cyclic voltammograms for binary solution of copper (II) chloride + indium (III) chloride and trisodium citrate.

On juxtaposing the successive cyclic voltammograms, slight differences are observed in the three cycles. As observed in Figure 4, for the third successive cycle, no copper peak is observed whereas an indium peak is observed but it seems to appear shifted to more negative deposition potential. Not much difference in the peaks is observed for the first and second cycles. The explanation for this could be that some copper is not oxidized back into copper ions and some nucleation sites are left covered on the carbon electrode by copper which reduces the number of nucleation sites available for copper reduction. This is known as the surface effect.

The charge transfer reaction corresponding to the peaks A and B in Figure 3 (a) and (b) at -0.2 V and -1 V potentials respectively must be related to the electrodeposition of copper and indium respectively. Cathodic peaks C and D are attributed to copper and indium electrodeposition in the binary electrolyte. These plots are for unitary solutions of copper (II) chloride and indium (III) chloride containing trisodium citrate as the complexing agent. For binary solution, respective peaks for copper and indium were found at -0.2 V and -1 V as well. For Figure 3 (a), (b) and (c), the voltammogram shows a badly defined hump at -0.8 V, -0.6 V and -0.6 V, respectively which might be attributed to the hydrogen evolution reaction. On juxtaposing Figures 2 and 3 which are cyclic voltammograms for the electrolytes in the absence and presence of the complexing agent, respectively, obtained in similar conditions, a negative shift in the peak potential of the respective element is observed by a degree of 0.2 V. This seems to suggest that some kind of complexation between the citrate anion and copper (II) and indium (III) cation occurs.

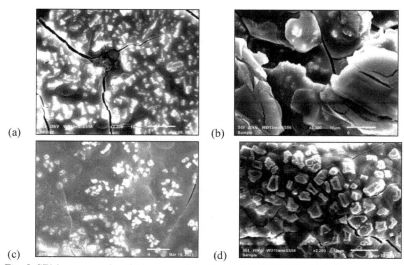

Fig. 5. SEM micrographs of (a) copper deposits at -0.6 V (*2200), (b) indium deposits at -0.9 V (*2300), (c) copper-indium deposits at -0.6 V (*1800) and (d) copper-indium deposits at -0.9 V (*2200).

High resolution scanning electron microscopy micrographs were recorded to investigate electrodeposited films of copper, indium and copper-indium. The copper films deposited at room temperature consisted mainly of grains with diameter varying about 2-4 μm as seen in Figure 5 (a). A few rods are also observed with the mean length 5 μm randomly oriented. Grains shape and size is irregular. SEM for copper film deposited at -0.6 V show crystallite structure compared to cauliflower-like structure produced in literature which could be a result of not washing the deposits after electrodeposition. Similar grain sizes and structure have been reported in literature. As can be seen from the surface morphology, the film deposit is not crack-free. Indium particles seem to be bigger than copper particles with their average size being 20 μm as compared to copper average particle size being around 2 μm as seen in Figure 5 (b). For indium, litterature has reported densely packed particles whereas those shown in Figure 5 (c) are more compact. Shape and size of the particles is more irregular. In litterature, grains had more regular circular shape [22]. The surface morphology of the Indium deposit also shows cracks on the surface. SEM depicted in Figure 5 (c) indicates smaller grain size when compared to copper alone. Again it seems that the particles are in the initial stages of nucleation. The small grains is irregular in shape and size. The size ranged from 1 to 3 μm. The surface morphology examination from the ×900 magnification shows a lot of cracks on the surface. Having cracks on the surface is not recommended and needs to be corrected by altering the deposition conditions such as deposition time, pH, concentration of electrolytes and potential. Figure 24 (a) and (b) indicates large indium particles as compared to copper deposition in the previous case. From the SEM morphology diagrams in figure 24 (c) and (d), it can be observed that copper and indium suppress each other's formation. Grain size and shape is irregular, and the average size is 5 μm. Similar to the previous SEMs for the different deposits, cracks are observed on the surface. From literature, these cracks could be avoided by improving the process conditions of the bath such as pH, deposition time, deposition voltage, bath composition, bath additives, etc. [21].

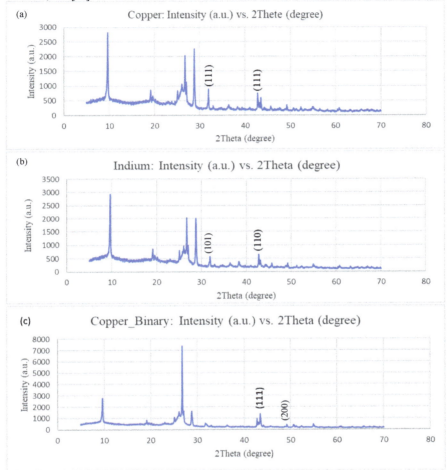

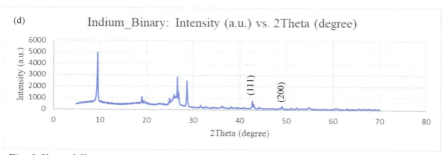

Fig. 6. X-ray diffraction patterns of the copper-indium layers at different deposition potentials.

Figure 6 depicts the X-ray diffractograms of the four deposited layers. XRD figures indicate a more amorphous structure. Figure 6 (c) and (d) show the presence of copper-indium at 2theta of 49° at a preferential plane of (200) to the growth of copper-indium. The XRD diffraction patterns cannot be compared with the JCPDS database owing to the small amount of the substance present in the deposited layer analyzed.

Conclusion
The study reports non-vacuum and easy production of copper indium layers by the co-deposition method. In this paper, the electrodeposition process for the fabrication of copper-indium has been studied which includes cyclic voltammetry, X-ray diffraction, scanning electron microscopy and energy dispersive X-ray spectroscopy. Results can be used in further studies to determine the structure, morphology and reaction mechanism of the films deposited. It can be concluded that best results are obtained at a deposition potential of -0.8 V by comparing the results of the cyclic voltammograms for the binary solutions at -0.6 V and -0.9 V. SEM characterization studies showed cracks on the surface for all four samples and irregular grain shape and size between 2-20 μm. X-ray diffraction studies indicate the presence of copper-indium at 2theta of 49° at a preferential plane of (200) to the growth of copper indium. EDS studies indicate a relatively low amount of copper and indium and that a 1:1 ratio of Cu:In will be obtained at approximately -0.8 V. This study is a good elementary study which can be used to determine which composition of the final deposited layer give maximum conversion efficiency.

References

[1] L. Farooq, A. Alraeesi, and S. Al Zahmi, "A review on the Electrodeposition of CIGS Thin-Film Solar Cells," in *Proceedings of the International Conference on Industrial Engineering and Operations Management*, 2019, pp. 158–186.

[2] V. S. Saji, I. H. Choi, and C. W. Lee, "Progress in electrodeposited absorber layer for CuIn(1-x)GaxSe2 (CIGS) solar cells," *Sol. Energy*, vol. 85, no. 11, pp. 2666–2678, 2011, https://doi.org/10.1016/j.solener.2011.08.003

[3] Y.-S. Chiu, M.-T. Hsieh, C.-M. Chang, C.-S. Chen, and T.-J. Whang, "Single-step electrodeposition of CIS thin films with the complexing agent triethanolamine," *Appl. Surf. Sci.*, vol. 299, pp. 52–57, 2014. https://doi.org/10.1016/j.apsusc.2014.01.184

[4] R. Chandran, R. Pandey, and A. Mallik, "One step electrodeposition of CuInSe2 from an acidic bath: A reduction co-deposition study," *Mater. Lett.*, vol. 160, pp. 275–277, 2015. https://doi.org/10.1016/j.matlet.2015.07.132

[5] M. E. Calixto and P. J. Sebastian, "Solartron analytic," *Sol. Energy Mater. Sol. Cells*, vol. 63, no. 4, pp. 335–345, 2000. https://doi.org/10.1016/S0927-0248(00)00053-2

[6] A. Luque and S. Hegedus, "Handbook of Photovoltaic Science and Engineering," *John Wiley Sons, Ltd*, pp. 92–100, Jan. 2003.

[7] S. Ruffenach, Y. Robin, M. Moret, R.-L. Aulombard, and O. Briot, "(112) and (220)/(204)-oriented CuInSe2 thin films grown by co-evaporation under vacuum," *Thin Solid Films*, vol. 535, pp. 143–147, 2013. https://doi.org/10.1016/j.tsf.2013.01.053

[8] M. Li *et al.*, "CIS and CIGS thin films prepared by magnetron sputtering," *Procedia Eng.*, vol. 27, pp. 12–19, 2012. https://doi.org/10.1016/j.proeng.2011.12.419

[9] S. Seeger and K. Ellmer, "Reactive magnetron sputtering of CuInS2 absorbers for thin film solar cells: Problems and prospects," *Thin Solid Films*, vol. 517, no. 10, pp. 3143–3147, 2009. https://doi.org/10.1016/j.tsf.2008.11.120

[10] H. L. Hwang, C. L. Cheng, L. M. Liu, Y. C. Liu, and C. Y. Sun, "Growth and properties of sputter-deposited CuInS2 thin films," *Thin Solid Films*, vol. 67, no. 1, pp. 83–94, 1980. https://doi.org/10.1016/0040-6090(80)90291-6

[11] M. Hashemi, S. M. Bagher Ghorashi, F. Tajabadi, and N. Taghavinia, "Aqueous spray pyrolysis of CuInSe2 thin films: Study of different indium salts in precursor solution on physical and electrical properties of sprayed thin films," *Mater. Sci. Semicond. Process.*, vol. 126, p. 105676, 2021. https://doi.org/10.1016/j.mssp.2021.105676

[12] M. Esmaeili-Zare and M. Behpour, "Influence of deposition parameters on surface morphology and application of CuInS2 thin films in solar cell and photocatalysis," *Int. J. Hydrogen Energy*, vol. 45, no. 32, pp. 16169–16182, 2020. https://doi.org/10.1016/J.IJHYDENE.2020.04.106

[13] J. Leng *et al.*, "Advances in nanostructures fabricated via spray pyrolysis and their applications in energy storage and conversion," *Chem. Soc. Rev.*, vol. 48, no. 11, pp. 3015–3072, 2019. https://doi.org/10.1039/C8CS00904J

[14] M. E. Calixto, P. J. Sebastian, R. N. Bhattacharya, and R. Noufi, "Compositional and optoelectronic properties of CIS and CIGS thin films formed by electrodeposition," *Sol. Energy Mater. Sol. Cells*, vol. 59, no. 1, pp. 75–84, 1999. https://doi.org/10.1016/S0927-0248(99)00033-1

[15] K. Abderrafi *et al.*, "Epitaxial CuInSe2 thin films grown by molecular beam epitaxy and migration enhanced epitaxy," *J. Cryst. Growth*, vol. 475, pp. 300–306, 2017. https://doi.org/10.1016/j.jcrysgro.2017.07.010

[16] "Solar Frontier Achieves World Record Thin-Film Solar Cell Efficiency of 23.35%." .

[17] S. K. Deb, "Thin-film solar cells: An overview," *Renew. Energy*, vol. 8, no. 1, pp. 375–379, 1996. https://doi.org/10.1016/0960-1481(96)88881-1

[18] O. Meglali, N. Attaf, A. Bouraiou, J. Bougdira, M. S. Aida, and G. Medjahdi, "Chemical bath composition effect on the properties of electrodeposited CuInSe2 thin films," *J. Alloys Compd.*, vol. 587, pp. 303–307, 2014. https://doi.org/10.1016/j.jallcom.2013.10.100

[19] J. Herrero and C. Guillén, "Study of the optical transitions in electrodeposited CuInSe2 thin films," *J. Appl. Phys.*, vol. 69, no. 1, pp. 429–432, Jan. 1991. https://doi.org/10.1063/1.347734

[20] K. K. Mishra and K. Rajeshwar, "A voltammetric study of the electrodeposition chemistry in the Cu + In + Se system," *J. Electroanal. Chem. Interfacial Electrochem.*, vol. 271, no. 1, pp. 279–294, 1989. https://doi.org/10.1016/0022-0728(89)80082-8

[21] A. M. Fernández and R. N. Bhattacharya, "Electrodeposition of CuIn1−xGaxSe2 precursor films: optimization of film composition and morphology," *Thin Solid Films*, vol. 474, no. 1, pp. 10–13, 2005. https://doi.org/10.1016/j.tsf.2004.02.104

[22] A. A. C. Alcanfor, L. P. M. dos Santos, D. F. Dias, A. N. Correia, and P. de Lima-Neto, "Electrodeposition of indium on copper from deep eutectic solvents based on choline chloride and ethylene glycol," *Electrochim. Acta*, 2017. https://doi.org/10.1016/j.electacta.2017.03.082

Electroluminescence image-based defective photovoltaic (solar) cell detection using a modified deep convolutional neural network

Hiren MEWADA[1,a *], L. SYAMSUNDAR[2,b], Hiren Kumar THAKKAR[3,c] and Miral DESAI[4,d]

[1]Electrical Engineering Department, Prince Mohammad Bin Fahd University, P.O. Box 1664, Al Khobar 31952, Kingdom of Saudi Arabia

[2]Mechanical Engineering Department, Prince Mohammad Bin Fahd University, P.O. Box 1664, Al Khobar 31952, Kingdom of Saudi Arabia

[3]Department of Computer Science and Engineering, School of Technology, Pandit Deendayal Energy University, Gandhinagar, Gujarat, India, 382421

[4]V. T. Patel Department of Electronics and Communication Engineering, Chandubhai S Patel Institute of Technology, Charotar University of Science and Technology, Changa, Gujarat, India

[a]hmewada@pmu.edu.sa, [b]slingala@pmu.edu.sa, [c]hiren.pdeu@gmail.com, [d]miraldesai.ec@charusat.ac.in

Keywords: Renewable Energy, Photovoltaic Solar Panels, Deep Convolution Neural Network, Image Classification

Abstract. Electroluminescence (EL) imaging of photovoltaic solar cells can detect and classify solar panel faults. This method allows technicians and manufacturers to identify defective panels that may affect performance and longevity. However, noise in EL images and solar cell silicon granularity make this process difficult. The paper presents an automated deep-learning framework to identify faulty and normal solar cells from images. Xception, a popular CNN network, is modified to reduce complexity and solve overfitting issues. Few separable convolution layers were removed from the original Xception network, and lateral dropout layers were added. The proposed deep CNN is tested on ELVP. To balance two classes, images are augmented with two rotations and dimensional shifting. Finally, the proposed model is compared to a pretrained CNN network and leading methods. The quantitative analysis showed that the model performed better than previous methods, with 94.382% accuracy, 92% precision, 95.12% recall rate, and 93.53% F1 score. Module fault identification helps with maintenance planning. Solar energy's widespread adoption and growth as a renewable and sustainable power source may result.

Introduction

Solar power has grown in popularity as a renewable energy source. Over the past decade, massive solar power plants worldwide have enabled large-scale solar energy component production. The photovoltaic (PV) module is essential to solar power. How well solar energy systems work depends on solar module efficiency. Crystalline silicon (c-Si) photovoltaic (PV) modules are the most popular due to their low cost per watt and well-established manufacturing process. Tang et al. [1], say this technology accounts for 97% of monocrystalline and polycrystalline module sales.

Solar panel fault classification is necessary for several reasons. First, a solar panel fault can reduce energy output. It streamlines maintenance planning and resource allocation. Technicians can optimize their efforts and address critical issues quickly by categorizing and prioritizing faults by severity and impact. Additionally, fault classification in electroluminescent solar panels has helped develop predictive maintenance strategies. By analyzing historical fault data and understanding patterns and trends, predictive models can predict and prevent faults. In solar panel manufacturing, fault classification is crucial to quality control. Before deployment, manufacturers can identify and classify panel defects using imaging. This ensures customers receive high-quality panels, improving satisfaction and reducing the risk of premature failure or performance degradation.

These solar panel faults can be identified using I-V curve measurements, thermal-infrared imaging (IR), and electroluminescence (EL) imaging. The I-V curve approach utilizes graphs to display PV module voltage and current output under specific radiation conditions. While the I-V curve can show module status, it cannot identify faulty cells or their locations. Using infrared imaging to monitor solar modules and cells is another popular method. An open circuit can cool a place, while a large current can heat it. IR imaging can detect dead cells, hot spots, and short circuits. However, thermal cameras' low resolution prevents them from detecting microcracks [2]. Such issues can be resolved via EL imaging. One nondestructive way to detect PV module defects is with an EL test. EL imaging can discover faulty cells and regions with ease and provide a thorough evaluation of all PV module cells. The overall status and longevity of the module can be determined at any stage[3].

Electroluminescence (EL) imaging is useful for fault detection and characterization in PV panels. As solar energy becomes more popular as a clean and sustainable power source, PV panel performance and longevity are crucial. Solar panel faults can be identified and classified noninvasively and efficiently using EL imaging, maximizing energy generation and maintenance. Images of solar cells' electroluminescent response to an electric field are captured using EL imaging. These images allow the detection of cracks, hotspots, and degradation patterns that may not be visible to the naked eye.

Some common fault types observed in EL images include cracks, i.e., visible breaks in solar cell structures, localized areas of high temperature, referred to as hotspots, and degradation patterns. This leads to reduced efficiency, resulting in diminished power output. By classifying these faults, technicians can identify the severity of the issue and take appropriate actions to rectify or mitigate its impact.

There are key challenges associated with fault classification. Manually inspecting and classifying faults in each panel is time-consuming and impractical as the number of solar panels in large-scale installations increases. Automating the classification process through the development of intelligent systems and algorithms is crucial for efficiently handling the volume of data generated by multiple panels. Thus, there is a need to create an automated classification system that can effectively categorize and differentiate between various fault types, which requires a deep understanding of the underlying physics and characteristics of each fault. Deep learning has emerged as a powerful data-driven approach for image classification applications because it can learn directly from a set of images. However, there are certain challenges associated with using deep learning for EL image classification. The limited spatial resolution, lack of color information, presence of noise and artifacts in images and limited spectral information are the major challenges associated with the use of deep learning for accurate classification.

There are many studies on EL imaging-based faulty panel identification. Overall, these methods can be categorized into traditional and machine learning algorithms. Several of the traditional methods include a logical gate-based image processing algorithm to enhance the crack regions [4], independent component analysis [5] and an anisotropic filter-based SVM approach for crack detection [6]. In this section, we review machine learning approaches. Zhang et al. [7] proposed a lightweight CNN with a size of 1.85 M learnable parameters based on the ResNet architecture for EL image classification. Based on the probability values, the overall set of images is divided into two classes, where panels are considered defective if their probability values are greater than 0.5. Their model achieved a maximum classification accuracy of 91.74%. In [8], the EL images were enhanced by equalizing the histogram of the images. Then, global information using the GCAM algorithm was integrated into an EfficientNet architecture to detect types of cracks or defects in the images. Attention-based deep learning was used in [9] to identify faulty panels from surface images of panels and achieved 98.66% accuracy. An RGB dataset of panels with dust, cement, cracks, etc., are classified as faulty types in this research. The dataset contains dusty panel images that cannot be considered defective panels. In [10], the authors prepared their dataset using an OPT-M311 camera. They used image augmentation with rotation, brightness adaptation and mirror shifting. This dataset was used to train the CNN network. Their network performed well, with 98.40% accuracy. However, validation on open-source datasets was not presented in the paper. Rahman et al. [11] tested various pretrained CNN architectures, including VGG and its variant, ResNet50 variant and the Xception network, to identify defective panels from a set of images. In their experiment, they sorted images into three categories, i.e., uncracked, cracked and unsure, which are too distorted. Although unsure images have notation, due to large distortions, they differentiate them from a set of functional and defective images. The Inception V3 network achieved 96% accuracy on monocrystalline panels, and VGG16 achieved 91.2% accuracy on polycrystalline panels. The limitation of this work is that the overall accuracy of combining both methods was not presented in the paper.

This paper presents a low-complexity, resource-efficient deep convolution network for EL image classification. We modified the Xception network to classify EL images into two classes, i.e., normal or defective. The major challenges in EL images are their low resolution and lack of spectral information. The use of a depthwise separable convolution layer in the Xception network allows efficient modeling of spatial relations within images that capture both global and local features. Therefore, we modified an Xception network by reducing its learning parameters and providing efficient classification with a small dataset. The remainder of this paper is organized as follows: Section 2 presents the proposed modified Xception network for EL image classification. Section 3 discusses the dataset and results using the proposed network. Finally, a conclusion is established based on the experimental results.

Proposed Methodology
The deep neural network plays a vital role in image categorization. A wide variety of pretrained networks are available for use in image classification. However, the types of images are the key aspect of selecting the CNN network. As explained in Section 1, electroluminescence imaging is a better choice for detecting faults within the cell of a PV module. However, the lack of spectral information and resolution makes this process challenging. A traditional CNN applies a set of filters K to the input feature map X with dimensions H × W × C, where H represents the height, W represents the width, and C represents the number of input channels. For the EL images, C is 1. Each filter has dimensions F x F x C, where F represents the filter size. The convolution operation is performed by sliding each filter across the input feature map, computing the elementwise multiplication between the filter and the corresponding spatial region of the input, and summing the results to produce an output feature map.

The Xception network modifies this convolutional operation by separating the spatial and channelwise information. It introduces two separate convolutional operations: depthwise convolution and pointwise convolution. The depthwise convolution operates on each input channel independently. It applies a set of depthwise filters, denoted as K_d, to each input channel of the feature map X. The depthwise filters have dimensions F × F × 1, where 1 represents the number of input channels. The depthwise convolution produces a set of intermediate feature maps, denoted as M, with dimensions H x W. Mathematically, the depthwise convolution can be represented as:
$$M = DepthwiseConv(X, K_d) \tag{1}$$
The pointwise convolution performs a 1x1 convolution on the intermediate feature maps M obtained from the depthwise convolution. It applies a set of pointwise filters, denoted as K_p, to combine and transform the intermediate feature maps. The pointwise filters have dimensions 1 x 1 x C', where C' represents the number of output channels. The pointwise convolution produces

the final output feature map, denoted as Y, with dimensions H x W x C'. Mathematically, it can be represented as:

$$Y = PointwiseConv(M, K_p) \qquad (2)$$

By separating the spatial and channelwise convolutions, the Xception network reduces the number of parameters and computations compared to standard convolutions. This parameter efficiency makes the Xception network computationally efficient and suitable for deep learning tasks with limited computational resources. The Xception network architecture repeats the depthwise and pointwise convolutions in multiple layers, enabling the network to learn hierarchical representations of features at different scales and complexities. Additionally, the Xception network often incorporates other common components found in deep neural networks, such as pooling layers, activation functions, and fully connected layers, to further enhance its performance. The structure of the Xception network, shown in Figure 1, is established based on the experimental results.

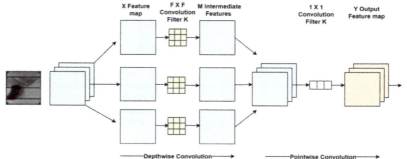

Figure 1 Xception network structure for convolution operation

The Xception network is built using stacking depth-separable convolution and comprises 14 modules, each with 36 convolutional layers. All of the layers utilize this technique except for the first 2 and the ones linked by residuals; the basic network is also built using this method. The pretrained Xception network has 170 deep layers with 22.9 million learnable parameters. The internal structure represented in Figure 2 is repeated 12 times with additional separable convolution layers in a few of the blocks. Here, this network is modified to classify grayscale EL images. In the proposed network, the initial layer is modified for grayscale images. The initial 80 layers of the pretrained Xception network are the same as those of the proposed network. The size of the network is reduced by removing the last four blocks. In addition, dropout layers are added to block 7 and the last block of the structure to reduce the feature size. These dropout layers help to solve the problem of network overfitting. The overall proposed structure has 120 layers with 14.3 M learnable parameters.

Figure 2 One block of convolution operations in the Xception network

Result Analysis and Discussion
Dataset
There is a wide variety of PV cell faults; however, not all of them will cause a significant drop in power output. The power output of the module is unaffected by some flaws, but it can be reduced over time, or the cells can disengage from the module due to others [12]. EL imaging can reveal cracks, microcracks, fractures, disconnections, silicone material flaws, finger disruptions, and unconnected cells. In this study, overall, PV cells were classified into two categories, i.e., normal and defective, if the panel had any of the above abnormalities. The ELVP dataset prepared using 18 monocrystalline and 26 polycrystalline PV panels from [13] was used in the experiments. A total of 2624 EL solar cell images, including 1508 normal images and 1116 defective images, are available in this dataset. Out of 1116 defective images, 715 images are faulty with 100% probability, and the remaining images have a lower probability of being faulty. Therefore, in the experiment, we used 1508 normal images and 716 defective images. We divided the EL images into a training set consisting of 80% (1206 normal and 573 defective) and a test set consisting of 20% for dataset partitioning. The sample images of both the normal and defective panels are shown in Figure 3.

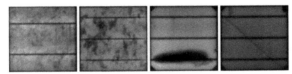

Figure 3 Sample images from the dataset (the left two images are normal panels, and the right two images are defective panels)

The small size of the dataset causes overfitting. Therefore, before using this dataset, an augmentation of images is used to enlarge the dataset. The dataset used had perfectly aligned EL images. Therefore, during the augmentation process, rotations of 90° and 180° were used. In addition, random shifting in both the X-direction and Y-direction is performed by increasing the set of training and testing images. This also helps to address the imbalance between the two classes.

Results and Discussion
The ADAM optimizer is used in a modified Xception network. The learning rate is initialized to 0.0001. The maximum number of epochs used is 10, and 64 is selected for the mini-batch size. Images are shuffled at every epoch for better network performance. We decided to quantify the classification effect and performance of our suggested model using accuracy, recall, precision, and F1 score, which are four commonly used metrics for evaluating and comparing effective models. In particular, higher values indicate better results for F1, recall, accuracy, and precision. If true positives (TPs) and true negatives (TNs) represent the number of positive results, for example, corrected detection of normal and defective panels, whereas false positives (FPs) and false negatives (FNs) represent the number of negative results, then these matrices can be calculated as follows:

$$Precision = \frac{TP}{TP+FP}, \qquad Recall = \frac{TP}{TP+FN} \qquad (3)$$

$$Accuracy = \frac{TP+TN}{TP+TN+FP+FN} \qquad F1-Score = \frac{2 \times Precision \times Recall}{Precision+Recall} \qquad (4)$$

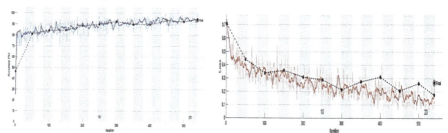

Figure 4 Accuracy and loss variation over time for the training and test datasets

The normal PV cell EL images had a uniform surface, but they had shadowed areas or impurities in the background. The backgrounds were clear and textured, but they were not defective; this put some pressure on the model to find defects. It was challenging to differentiate the surface defects of aberrant PV cells from the background in the EL image because they looked so similar to the background in the original image. The accuracy and loss analysis over the epochs for the training dataset are shown in Figure 4. Figure 4 shows that the model succeeded in achieving 93.93% accuracy with a training loss of less than 0.2.

The confusion matrix represents the TP, TN, FP and FN results of the network. Figure 5 shows the confusion matrix of the proposed network for the training dataset.

	Defective	Normal	
Defective	122 27.4%	4 0.9%	96.8% 3.2%
Normal	21 4.7%	298 67%	93.4% 6.6%
	85.3% 14.7%	98.7% 1.3%	94.4% 5.6%

Output Class / Target Class

Figure 5 Confusion matrix of the modified Xception network

In Figure 5, the first two diagonal cells show the number and percentage of correct classifications by the network. For example, 122-panel images are correctly classified as defective. This corresponds to 27.4% of all test set images. Similarly, 298 cases are correctly classified as normal. This corresponds to 67% of all test images. Twenty-one of the defective panels are incorrectly classified as normal, which corresponds to 4.7% of all images. Out of 302 normal test images, 98.7% are correct and 1.3% are incorrect. Overall, 94.4% of the predictions are correct, and 5.6% are wrong.

Some of the most successful approaches to PV defect detection in the past few years were compared to our model. We selected these approaches for testing and assessment on the same dataset as our proposed model to ensure a fair comparison; Table 1 displays the results of the method comparison.

Table 1 Comparison of the experimental results with those of other methods

Model	Accuracy	Precision	Recall	F1 Score
CNN [14]	78.38	77.86	70.10	71.84
VGG16 [14]	84.01	82.26	80.31	81.15
InceptionV3 [14]	88.96	87.73	86.72	87.20
SVM [15]	82.44	-	-	85.52
CNN [15]	88.42	-	-	88.39
L-CNN [16]	89.33	90.44	95.42	92.86
Our network	94.382	92	95.12	93.53

As shown in Table 1, pretrained networks, including CNN, VGG16 and InceptionV3, were tested in [14], and their models were compared. The presence of noise makes it challenging, and therefore, these models struggle to classify faulty panels from the set of images. In [15], the authors extracted various VGG-based CNN features, and SVM was used as a classifier. A validation of the experiment using mono-, poly- and the overall set of images was presented in the paper. However, their accuracy was limited to 88.42 max when using a CNN. In [16], a lightweight convolutional neural network was presented. They trained it from scratch, and a comparison with a support vector machine was presented in their work. Their light CNN succeeded at 89.33% accuracy for two-class classification. In contrast, the modification of the Xception network with the removal of a few separable convolution layers and the introduction of dropout layers performed well on augmented images and achieved 90% accuracy.

Conclusion
Solar cells have crystal grain boundaries due to the intrinsic silicon structure, and the presence of noise in EL images causes ambiguity in distinguishing minor cracks. In addition, EL images lack spectral information. These characterizations impose a challenge to applying conventional CNN networks to identify faulty cells from normal cells. In this paper, we used the Xception network to determine whether a panel is defective. The large number of layers in deep CNNs increases the complexity of the network; hence, the network cannot learn well from grayscale images. Therefore, the network is minimized by removing repetitive separable convolution layers. Furthermore, dropout layers are introduced in the Xception network to solve the overfitting problem, and the experimental results and a comparison with state-of-the-art methods suggest that the model's classification accuracy is improved. The experimental results are validated for binary classification only. Therefore, further validation of the network for multiclass classification will be performed in the future.

References

[1] W. Tang, Q. Yang, K. Xiong, and W. Yan, "Deep learning based automatic defect identification of photovoltaic module using electroluminescence images," *Solar Energy*, vol. 201, pp. 453–460, May 2020. https://doi.org/10.1016/j.solener.2020.03.049

[2] M. W. Akram *et al.*, "CNN based automatic detection of photovoltaic cell defects in electroluminescence images," *Energy*, vol. 189, p. 116319, Dec. 2019.

[3] U. Jahn, M. Herz, M. Köntges, D. Parlevliet, M. Paggi, and I. Tsanakas, *Review on infrared and electroluminescence imaging for PV field applications: International Energy Agency Photovoltaic Power Systems Programme: IEA PVPS Task 13, Subtask 3.3: report IEA-PVPS T13-12:2018*. Paris: International Energy Agency, 2018.

[4] M. Dhimish, V. Holmes, and P. Mather, "Novel Photovoltaic Micro Crack Detection Technique," *IEEE Trans. Device Mater. Relib.*, vol. 19, no. 2, pp. 304–312, Jun. 2019. https://doi.org/10.1109/TDMR.2019.2907019

[5] D.-M. Tsai, S.-C. Wu, and W.-Y. Chiu, "Defect Detection in Solar Modules Using ICA Basis Images," *IEEE Trans. Ind. Inf.*, vol. 9, no. 1, pp. 122–131, Feb. 2013. https://doi.org/10.1109/TII.2012.2209663

[6] S. A. Anwar and M. Z. Abdullah, "Micro-crack detection of multicrystalline solar cells featuring an improved anisotropic diffusion filter and image segmentation technique," *J Image Video Proc*, vol. 2014, no. 1, p. 15, Dec. 2014. https://doi.org/10.1186/1687-5281-2014-15

[7] J. Zhang, X. Chen, H. Wei, and K. Zhang, "A lightweight network for photovoltaic cell defect detection in electroluminescence images based on neural architecture search and knowledge distillation," *Applied Energy*, vol. 355, p. 122184, Feb. 2024.

[8] Q. Liu, M. Liu, C. Wang, and Q. M. J. Wu, "An efficient CNN-based detector for photovoltaic module cells defect detection in electroluminescence images," *Solar Energy*, vol. 267, p. 112245, Jan. 2024. https://doi.org/10.1016/j.solener.2023.112245

[9] D. Dwivedi, K. V. S. M. Babu, P. K. Yemula, P. Chakraborty, and M. Pal, "Identification of surface defects on solar PV panels and wind turbine blades using attention based deep learning model," *Engineering Applications of Artificial Intelligence*, vol. 131, p. 107836, May 2024. https://doi.org/10.1016/j.engappai.2023.107836

[10] M. Sun, S. Lv, X. Zhao, R. Li, W. Zhang, and X. Zhang, "Defect Detection of Photovoltaic Modules Based on Convolutional Neural Network," in *Machine Learning and Intelligent Communications*, vol. 226, X. Gu, G. Liu, and B. Li, Eds., in Lecture Notes of the Institute for Computer Sciences, Social Informatics and Telecommunications Engineering, vol. 226. , Cham: Springer International Publishing, 2018, pp. 122–132. https:/10.1007/978-3-319-73564-1_13

[11] Md. R. Rahman, S. Tabassum, E. Haque, M. M. Nishat, F. Faisal, and E. Hossain, "CNN-based Deep Learning Approach for Micro-crack Detection of Solar Panels," in *2021 3rd International Conference on Sustainable Technologies for Industry 4.0 (STI)*, Dhaka, Bangladesh: IEEE, Dec. 2021, pp. 1–6. doi: 10.1109/STI53101.2021.9732592

[12] M. Köntges, S. Kurtz, C. Packard, U. Jahn, K. A. Berger, and K. Kato, *Performance and reliability of photovoltaic systems: subtask 3.2: Review of failures of photovoltaic modules: IEA PVPS task 13: external final report IEA-PVPS*. Sankt Ursen: International Energy Agency, Photovoltaic Power Systems Programme, 2014.

[13] L. Pratt, J. Mattheus, and R. Klein, "A benchmark dataset for defect detection and classification in electroluminescence images of PV modules using semantic segmentation," *Systems and Soft Computing*, vol. 5, p. 200048, Dec. 2023. https://doi.org/10.1016/j.sasc.2023.200048

[14] J. Wang et al., "Deep-Learning-Based Automatic Detection of Photovoltaic Cell Defects in Electroluminescence Images," *Sensors*, vol. 23, no. 1, p. 297, Dec. 2022. https://doi.org/10.3390/s23010297

[15] S. Deitsch et al., "Automatic classification of defective photovoltaic module cells in electroluminescence images," *Solar Energy*, vol. 185, pp. 455–468, Jun. 2019. https://doi.org/10.1016/j.solener.2019.02.067

[16] M. Y. Demirci, N. Beşli, and A. Gümüşçü, "Efficient deep feature extraction and classification for identifying defective photovoltaic module cells in Electroluminescence images," *Expert Systems with Applications*, vol. 175, p. 114810, Aug. 2021. https://doi.org/10.1016/j.eswa.2021.114810

Analysis of thermal efficiency of solar flat plate collector working with hybrid nanofluids: An experimental study

Solomon MESFIN[1,a], Veeredhi VASUDEVA RAO[2,b], L. Syam SUNDAR[3,c,*]

[1]Department of Mechanical Engineering, University of Gondar, Gondar, Ethiopia

[2]Department of Mechanical Engineering, School of Engineering, CSET, University of South Africa (UNISA), South Africa

[3]Department of Mechanical Engineering, College of Engineering, Prince Mohammad Bin Fahd University, Al-Khobar, Saudi Arabia

[a]solomonmesfin5@gmail.com, [b]vasudvr@unisa.ac.za, [c]sslingala@gmail.com

Keywords: Thermal Efficiency, Flat Plate Collector; Hybrid Nanofluids; Augmentation, Solar Energy

Abstract. Thermal efficiency of solar flat-plate collector (SFPC) was analyzed experimentally through water-based mono Al_2O, CuO, and hybrid Al_2O_3-CuO nanofluids. The particle loadings used for the analysis are 0.048%, 0.096%, 0.144%, 0.192% and 0.24%, respectively. The experiments were conducted at a flow rate of 0.008 kg/s of mono, and hybrid nanofluids. The experimental outcomes indicate, the thermal efficiency of mono and hybrid nanofluids raised under the larger volume loadings in comparison with water. Results show, that the Al_2O_3-CuO hybrid nanofluid offered higher thermal efficiency values than mono Al_2O_3 and CuO nanofluids. Thermal efficiency of SFPC was found to get enhanced by 57.66%, 66.58% and 73.75% at 0.24 vol.% Al_2O_3, CuO, and Al_2O_3-CuO hybrid nanofluids, over the water data, respectively, at solar noon time of 12:00 P.M.

Introduction
The depletion of fossil resources has resulted in a global energy shortage. In the current situation, energy is critical to industrial and economic development, hence efforts must be necessary to fix the problem of using fossil fuels with alternative fuels. The ongoing movement in development toward sustainability and better environmental responsibility connected with future development, emphasizes the need for renewable energy.

Renewable energy is a rapidly growing industry, with numerous innovations and applications emerging. The concept of localized renewable energy systems has been recognized as a solution to industries and residential energy demands. The depletion of natural resources and the growing demand for traditional energy have prompted planners and decision-makers to look into other sources. As the cost of conventional energy sources rises, there is a growing interest in renewable energy sources. Recently, there has been a great deal of effort in the field of renewable energy source engineering, particularly in the generation of solar energy. Unfortunately, the spectrum of feasible applications is limited to large-output arrays, prohibitively expensive technology, and massive assemblies requiring large parcels of land and expenses because a significant portion of the development of said renewable energy generation has not been for utilities.

The sun emits solar energy into space, but only 1367 W/m^2 penetrates Earth's atmosphere [1]. The energy received from the solar mean in a certain time interval of time on a 1 m^2 area can be estimated that, the radiation during that particular time period, which is known as solar radiation or insulation. This is called as direct and diffuse solar radiation [2]. Three major criteria influencing the performance of solar energy systems in converting direct and diffuse sunlight into useable energy are a given location's geographical coordinates, topography, and climatic conditions [3].

Content from this work may be used under the terms of the Creative Commons Attribution 3.0 license. Any further distribution of this work must maintain attribution to the author(s) and the title of the work, journal citation and DOI. Published under license by Materials Research Forum LLC.

Thermal converters capture thermal energy, which is then transferred to a fluid that circulates through the panel. The solar radiation concentrators are non-concentrator type, i.e. flat plate or vacuum tubes, and concentrated type collectors. Flat non-concentrated heat collectors (which require pipes to connect the two components) are made up of numerous parts: a transparent glass cover, pipes, absorbers, and a thermal insulting layer. In contrast, the vacuum tubes' solar thermal collector is made up of absorbers, glass tubes, heat pipes, and an insulated casing [4]. In both systems, the absorber is critical to the heat transfer process, which requires high thermal conductivity. To limit the heat losses, the insulation between absorber and collector housing must be adjusted. Such factors necessitate to use of high price materials, raising the initial cost of collectors [5].

Literature review
Solar flat plate collectors (SFPC) consist of a heat-absorbing plate, serpentine tube, insulation and a heat transmitting cover. Solar energy falls on the absorber plate, and which is transmit into the working fluid in a tube based on the convection mode of heat transfer. The absorber is made from high thermal conductivity aluminum metal sheet. Even though these SFPC are used commonly, further larger in thermal efficiency of these collectors is necessary, because they are still under low thermal efficiencies [6]. Studies to improve the thermal efficiency of SFPC have included the redesigning of riser's channels [7], re-designing the top glazing in the collector and changing the heat-transfer fluid.

One of the most convenient and successful methods for increasing the improvement of SFPC is to introduce the nanofluids into the collector's tubes instead of standard heat transfer fluids such as water. Choi and Eastman introduced the term "nanofluid" for the first time [8]. The nanofluid is a colloidal dispersion of nanoparticles in base fluids (size less than 100 nm). Dispersing solid nanoparticles into the base fluid improved thermal conductivity. Nanofluids can dramatically increase thermal conductivity and heat transfer coefficient [9]. Nanofluids have a larger density and a lower specific heat, which leads to increased thermal efficiency. As a result, it minimizes the solar collector's area, weight, energy, and manufacturing costs.

Nanofluids are classified into mono and hybrid nanofluids [10]. The hybrid nanofluids are new generation and homogeneous mixture fluids which can be prepared by suspending different types (two or more than two) of nanoparticles in base fluid with developed physical and chemical bonds. Due to synergistic effect, the thermal conductivity and heat transfer of hybrid nanofluids are higher than individual nanofluids [11], Lee and Sharma [12]. Experimental results show that ethylene glycol as heat transfer fluid (HTF) in active solar flat plate heat (SFPH) system is good for cold climate countries due to its antifreeze properties.

Geovo et al. [13] considered experimental data of water diluted MgO nanofluids in SFPC for the purpose of comparison with the MATLAB software. They noticed maximum relative error of 5.36% and minimum relative error of 0.20% between experimental and software data. Choudhary et al. [14] obtained thermal efficiency increase of 16.7% for MgO-ethylene glycol/distilled water nanofluids in a FPSC under 0.2 vol% at a flow rate of 1.5 L/min. Verma et al. [15] used different nanofluids in a FPSC and analyzed energetic and exergetic parameters and obtained 23.47% increase of thermal efficiency at 0.75 vol% and at a flow rate of 0.025 kg/s for MNCNT/water as the working fluid. Moghadam et al. [16] considered water mixed CuO nanofluids in a FPSC and seen 21.8% of increase of thermal efficiency at vol. of 0.4 and at 3 kg/min. Yousefi et al. [17] found 28.3% increase of collector thermal efficiency by using Al_2O_3/water nanofluids in a FPSC. In another analysis, Yousefi et al. [18] analyzed the collector thermal efficiency by changing the nanofluids pH value, and they mentioned that, nanofluid pH also one of the influencing parameters for the collector efficiency. Belkassmi et al. [19] obtained efficiencies of 4.45%, 4.28%, and 4.22% at 2.0 lit/min, based on the experimental data of water dispersed Cu, CuO, and Al_2O_3 nanofluids in FPSC, respectively.

With the utilization of mono nanofluids, the collector thermal efficiency is enhanced. Similarly, researchers have concentrated the use of hybrid nanofluids in FPC. Elshazly et al. [20] seen an enhanced thermal efficiency of 26% by using hybrid nanofluids of MWCNT/Al$_2$O$_3$ (50:50%) in a FPSC at 1.5 lit/min.

This work is to estimate experimentally the thermal efficiency of water diluted Al$_2$O$_3$-CuO nanofluids flow in a FPSC with various volumetric concentrations. In addition to the above, the study is also focused on to investigation of efficiency of water-based copper oxide (CuO) and aluminum oxide (Al$_2$O$_3$) nanofluids alone and compare with the Al$_2$O$_3$-CuO water based nanofluid at constant mass flow rate to get the required amount of heat for water heating applications.

Experimental study
Preparation of nanofluids

The nanofluids are prepared by mixing the nanoparticles with water. Table 1 is the physical properties of Al$_2$O$_3$, CuO and water, and Table 2 is the weights of CuO and Al$_2$O$_3$ nanoparticles used for water for developement of various nanofluids. Step-by-step procedure of nanofluids preparation is mentioned in Fig. 1.

Table 1: The physical properties of Al$_2$O$_3$, CuO and water.

Nanoparticle/ base fluid	ρ, kg/m^3	C_p, (J/kgK)	k, (W/mK)	Color	Diameter, (nm)
Al$_2$O$_3$	3900	785.2	30	White	50
CuO	6510	540	33	Black	27
Water	1000	4179	0.613	---	---

Table 2: Weights of CuO, and Al$_2$O$_3$ nanoparticles used for water for the preparation of various nanofluids.

Mass of nanoparticles (g)	Particle volume loading (%)				
	$\phi = 0.048\%$	$\phi = 0.096\%$	$\phi = 0.144\%$	$\phi = 0.192\%$	$\phi = 0.24\%$
CuO (g)	25	50	75	100	125
Al$_2$O$_3$ (g)	15	30	45	60	75
Al$_2$O$_3$-CuO, (g)	20	40	60	80	100
Water (lit)	8	8	8	8	8

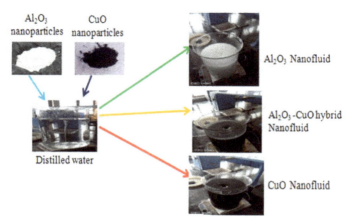

Fig. 1: Preparation of water-based mono and hybrid nanofluids.

Flat plate collector
The experimental set-up of flat plate collector is shown in Fig. 2, and which is used for water, mono, and hybrid nanofluids and its thermal efficiency is estimated. The solar collector was placed in the Gondar town, Ethiopia, which is located on 12.6° N latitude and 37.47° E longitude in the northern hemisphere with an elevation of 2133 meters above sea level. For maximum captured radiation, the flat plate solar collector was installed at 27.6° tilt angle. The set-up mainly consists of an absorber plate to absorb incident solar radiation, a single glass cover to protect collector heat loss, serpentine tube for fluid passage through the solar collector, storage tank to store working fluids for experimentation and used as heat exchanger, a pump capable to deliver the fluid to the serpentine pips, by-pass valve for returning of fluids after adjusting the control valve, adjustable valves to control flow rate one at the main flow loop and the other at the by-bass line, flow meter to measure the fluid flow rate, cold water storage tank, table to support water tank, and collector support to carry the flat plate solar collector. During the experimental test, the glass temperature, plate temperature, inlet and outlet temperatures and mass flow rates of the working fluids are measured to obtain the thermal efficiency of the flat plate solar collector.

Fig. 2: Photograph of flat plate collector.

Data analysis
Collector thermal efficiency analysis
Thermal performance of solar collector is evalauted through the instantaneous collector efficiency, which requires the amount of solar radition attracted by the collector. The useful heat energy of the working fluid is determined by the following equations.

$$Q_u = \dot{m}C_{pf}(T_o - T_i) \quad (1)$$
$$\eta_{th} = \frac{Q_u}{A_C I_T} \quad (2)$$
$$\eta_{th} = \frac{\dot{m}C_{pf}(T_o - T_i)}{A_C I_T} \quad (3)$$

Where, C_p is specific heat (KJ/kgK), T_o is outlet temperature (K), T_i is inlet temperature (K), η_{th} is the thermal efficiency, and Q_u useful heat energy or the incident solar energy, A_c is absorber area (m^2), and I_T is the incident solar energy (W/m^2).

Results and discussion
Temperature distribution
In Fig. 3, it is seen that the highest outlet temperature at solar noon for 0.24% volumetric concentrations of Al_2O_3, CuO and Al_2O_3-CuO hybrid water based nanofluids are 62.4 °C, 69 °C, and 86.7 °C, respectively. From the above experimental readings it is noticed that the Al_2O_3-CuO hybrid nanofluids outlet temperatures of flat plate solar water heating is higher than individual nanofluids and CuO nanofluid outlet temperatures is also higher than Al_2O_3 and base fluid. Due to larger random collisions of nanoparticles in base fluid, the outlet temperature of the nanofluid is raised. Therefore, increasing the volume particle concentrations of nanofluids and hybridizing the nanoparticles raises the exit temperature of the fluid by increasing the absorptivity of SFPC.

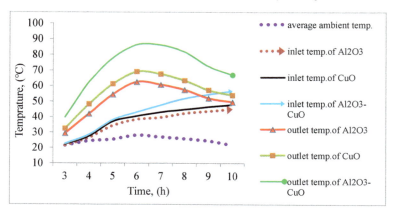

Fig. 3: Temperature records at ϕ = 0.24% particle concentration nanofluids with time.

Heat gain in the collector

Fig. 4 gives compression of useful heat gained by the distilled water and 0.048% nanoparticle concentration of single and hybrid water-based nanofluids at constant mass flow rate of 0.008 kg/s. Maximum useful energy of distilled water and 0.048% particle volume concentration of Al_2O_3, CuO and Al_2O_3-CuO hybrid water based nanofluids are 574.15 W/m², 627.76 W/m², 691.82 W/m², and 799.72 W/m² respectively. This shows that useful heat energy of distilled water was less with the same mass flow rate compared to 0.048% Al_2O_3/water nanofluid and the useful heat energy of CuO nanofluid is greater over Al_2O_3 nanofluid under a fixed particle volume loading and fluid flow rate, also, useful heat energy of Al_2O_3-CuO/water hybrid nanofluid was very high with the same mass flow rate and particle volume concentration compare to CuO/water nanofluid.

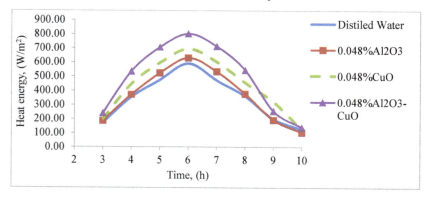

Fig. 4: Hourly variation of heat energy of water and nanofluids (ϕ = 0.048%) with time.

From Fig. 5 it is seen that the maximum useful heat energy with 0.096% particle volume concentration and constant mass flow rate of Al$_2$O$_3$, CuO and Al$_2$O$_3$-CuO hybrid water based nanofluids were 650.28 W/m^2, 738.58 W/m^2 and 834.73 W/m^2, respectively.

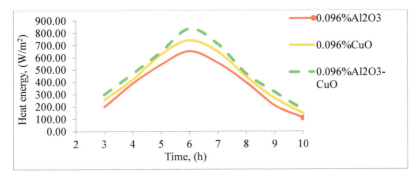

Fig. 5: Hourly variation of heat energy for (ϕ = 0.096%) nanofluids with time.

From Fig. 6, it can be seen that the maximum thermal efficiency of flat plate solar collector was 57.66%, 66.58% and 73.75% for 0.24% particle concentration of Al$_2$O$_3$, CuO and Al$_2$O$_3$-CuO hybrid water based nanofluids respectively at fixed 0.008kg/s mass flow rate. The experimental results indicated that due to the increased interactions of nanoparticles in base fluid, the efficiency of flat plate solar collector increased with percentage volume concentration of all nanofluids. Furthermore, because of the rise in internal energy between particles and reduction in agglomeration, hybrid nanofluid has higher collector efficiency than the isolations. The result is also understood that CuO/water nanofluid has greater collector efficiency than Al$_2$O$_3$/water nanofluid due to the higher thermal conductivity properties of copper oxide nanoparticles.

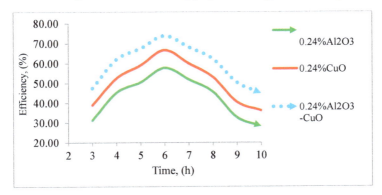

Fig. 6: Hourly variation thermal efficiency comparisons of nanofluids at (ϕ = 0.24%).

Figure 7 gives the data related to the thermal eficiency of mono, and hybrid nanofluids in a FPSC. Experimental results showed that at constant flow rate, the efficiency of flat plate solar collector was improved in all nanofluids with the increase in volumetric concentrations of nanoparticles in base fluids. The lowest thermal efficiency of the solar collector was observed about 48.32% for distilled water. The maximum thermal efficiency improvements of flat plate solar collector with 0.24 percentage nanoparticle concentration of Al_2O_3, CuO, and Al_2O_3-CuO hybrid nanofluids were 9.34%, 18.26% and 25.43% respectively compared to base fluid i.e., distilled water. Moreover, from the results it is noticed that hybrid nanofluid is more efficient than the individual nanofluids. The results are also shown that, the thermal efficiency of CuO nanofluid is higher than Al_2O_3 nanofluid and base fluid. Therefore, hybridizing nanoparticles and increasing nanoparticle concentrations from 0.048 to 0.24% improved the thermal efficiency of SPFC.

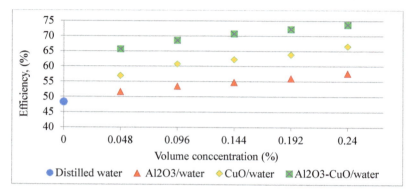

Fig. 7: Thermal efficiencies of working fluids as function of nanoparticle concentration.

Conclusions

In this research the thermal efficiency of flat plate solar collector with Al_2O_3/water, CuO/water and Al_2O_3-CuO/water hybrid nanofluid, the effect of particle volume concentrations and the two-step preparation methods were studied. Efficiencies of FPSC with each working fluid at various nanoparticle concentrations and constant flow rate 0.008kg/s were compared. The thermal efficiency of flat plate solar collectors is show to increase with the increase in nanoparticle concentration of the nanofluids. The maximum thermal efficiency of the flat plate solar collector for water and 0.24% Al_2O_3, CuO and Al_2O_3-CuO hybrid nanofluids was 48.32%, 57.66%, 66.58% and 73.75% respectively at constant flow rate 0.008 kg/s. From this, mixing of Al_2O_3 nanoparticles in water enhances the collecor efficiecy by 9.34% as compared to pure water. The SFPC with CuO nanofluid is better than Al_2O_3 nanofluid by 8.92%. Thermal efficiency SPFC wokirng with Al_2O_3-CuO hybrid nanofluid is higher than CuO/water nanofluid by 7.17%. Therefore from the 0.24 vol. % data, volume concentration of Al_2O_3-CuO/water hybrid nanofluid and 0.008kg/s, the thermal efficiency of SPFC raises upto 73.75%, which was 25.43% higher than the pure water.

References

[1] F. Yang, J. Liu, Q. Sun, L. Cheng, R. Wennersten, Simulation analysis of household solar assistant radiant floor heating system in cold area, Energy Procedia 158 (2019) 631-636. doi: 10.1016/j.egypro.2019.01.166

[2] M. Ammar, A. Mokni, H. Mhiri, P. Bournot, Parametric investigation on the performance of natural convection flat plate solar air collector with additional transparent insulation material parallel slats (TIM-PS), Solar Energy 231 (2021) 379-401. https://doi.org/10.1016/j.solener.2021.11.053

[3] A.C.M. Ango, M. Medale, C. Abid, Optimization of the design of a polymer flat plate solar collector, Solar Energy 87(2013) 64-75. https://doi.org/10.1016/j.solener.2012.10.006

[4] M. Hosseinzadeh, A. Salari, M. Sardarabadi, M. Passandideh-Fard, Optimization and parametric analysis of a nanofluid based photovoltaic thermal system: 3D numerical model with experimental validation, Energy Convers. Manag. 160 (2018) 93-108. https://doi.org/10.1016/j.enconman.2018.01.006

[5] Z. Jiandong, T. Hanzhong, C. Susu, Numerical simulation for structural parameters of flat-plate solar collector, Solar Energy 117 (2015) 192-202. https://doi.org/10.1016/j.solener.2015.04.027

[6] Z. Wang, W. Yang, F. Qiu, X. Zhang, X. Zhao, Solar water heating: From theory, application, marketing and research, Renew. Sustain. Energy Rev. 41 (2015) 68-84. https://doi.org/10.1016/j.rser.2014.08.026

[7] Y. Deng, Y. Zhao, W. Wang, Z. Quan, L. Wang, D. Yu, Experimental investigation of performance for the novel flat plate solar collector with micro-channel heat pipe array, Appl. Therm. Eng. 54 (2013) 440-449. https://doi.org/10.1016/j.applthermaleng.2013.02.001

[8] S.U.S. Choi, Enhancing thermal conductivity of fluids with nanoparticles, Am. Soc. Mech. Eng. Fluids Eng. Div. FED, 231 (1995) 99-105.

[9] T. Yousefi, E. Shojaeizadeh, F. Veysi, S. Zinadini, An experimental investigation on the effect of pH variation of MWCNT-H2O nanofluid on the efficiency of a flat-plate solar collector, Sol. Energy, 86 (2012) 771-779. https://doi.org/10.1016/j.solener.2011.12.003

[10] M.H. Ahmadi, A. Mirlohi, M. Alhuyi Nazari, R. Ghasempour, A review of thermal conductivity of various nanofluids, J. Mol. Liq. 265 (2017) 181–188. https://doi.org/10.1016/j.molliq.2018.05.124

[11] H. Chen, Y. Ding, Y. He, C. Tan, Rheological behaviour of ethylene glycol based titania nanofluids, Chem. Phys. Lett. 444 (2007) 333-337. https://doi.org/10.1016/j.cplett.2007.07.046

[12] D.W. Lee, A. Sharma, Thermal performances of the active and passive water heating systems based on annual operation, Sol. Energy 81 (2007) 207-215. https://doi.org/10.1016/j.solener.2006.03.015

[13] L. Geovo, G.D. Ri, R. Kumar, S.K. Verma, J.J. Roberts, A.Z. Mendiburu, Theoretical model for flat plate solar collectors operating with nanofluids: Case study for Porto Alegre, Brazil, Energy 263 Part B (2023) 125698, https://doi.org/10.1016/j.energy.2022.125698

[14] S. Choudhary, A. Sachdeva, P. Kumar, Investigation of the stability of MgO nanofluid and its effect on the thermal performance of flat plate solar collector, Renewable Energy 147 Part 1 (2020) 1801-1814.

[15] S.K. Verma, A.K. Tiwari, S. Tiwari, D.S. Chauhan, Performance analysis of hybrid nanofluids in flat plate solar collector as an advanced working fluid, Solar Energy 167 (2018) 231-241.

[16] A.J. Moghadam, M. Farzane-Gord, M. Sajadi, M. Hoseyn-Zadeh, Effects of CuO/water nanofluid on the efficiency of a flat-plate solar collector. Experimental Thermal and Fluid Science, 58 (2014) 9-14.

[17] Yousefi, T., Veysi, F., Shojaeizadeh, E., & Zinadini, S. (2012). An experimental investigation on the effect of Al2O3–H2O nanofluid on the efficiency of flat-plate solar collectors. Renewable Energy, 39(1), 293-298.

[18] Yousefi, T., Veisy, F., Shojaeizadeh, E., & Zinadini, S. (2012). An experimental

investigation on the effect of MWCNT-H2O nanofluid on the efficiency of flat-plate solar collectors. Experimental Thermal and Fluid Science, 39, 207-212.

[19] Belkassmi, Y., Gueraoui, K., El maimouni, L. *et al.* Numerical Investigation and Optimization of a Flat Plate Solar Collector Operating with Cu/CuO/Al$_2$O$_3$–Water Nanofluids. Trans. Tianjin Univ. 27, 64–76 (2021). https://doi.org/10.1007/s12209-020-00272-6.

[20] E. Elshazly, A.A. Abdel-Rehim, I. El-Mahallawi, 4E study of experimental thermal performance enhancement of flat plate solar collectors using MWCNT, Al$_2$O$_3$, and hybrid MWCNT/ Al$_2$O$_3$ nanofluids, Results in Engineering 16 (2022) 100723.

Green and environmental-friendly material for sustainable buildings

Noura Al-MAZROUEI[1,a]*, Amged ELHASSAN[2,b], Waleed AHMED[3,c]*, Ali H. Al-MARZOUQI[1,d], Essam ZANELDIN[4,e]

[1] Chemical and Petroleum Engineering Department, COE, UAE University, Al Ain, UAE

[2] Mechanical and Aerospace Engineering Department, COE, UAE University, Al Ain, UAE

[3] Engineering Requirements Unit, COE, UAE University, Al Ain, UAE

[4] Civil and Environmental Engineering Department, COE, UAE University, Al Ain, UAE

[a]201311706@uaeu.ac.ae, [b]201450104@uaeu.ac.ae, [c]w.ahmed@uaeu.ac.ae, [d]hassana@uaeu.ac.ae, [e]essamz@uaeu.ac.ae

Keywords: Natural Binder, Sustainable Material, Composite, Construction

Abstract. In construction, incorporating waste materials such as bio-binders and fine aggregates plays a vital role in promoting environmentally friendly building methods. Utilizing these materials for a second time reduces waste generation, contributing to conserving valuable natural resources. This underscores the importance of sustainable construction practices. This study outlines the experimental results concerning the mechanical attributes when employing a combination of micro sand silica and a bio-binder, specifically Abelmoschus esculentus. The investigated mechanical properties in this study encompass modulus, strength, and toughness. The experimentation involved mixing Abelmoschus esculentus with varying weight percentages and three distinct micro-size particles and then compressing them into cylindrical samples. Abelmoschus esculentus demonstrated favorable adhesion properties with sand silica particles, and the findings suggest a noteworthy impact on the mechanical properties upon its addition. Overall, it was observed that the optimal mechanical properties were attained with a 15% weight ratio of Abelmoschus esculentus bio-binder at a particle size of 250 μm.

Introduction

Eco-friendly building materials could be created by combining aggregates with bio-binders derived from natural resources, such as agricultural waste. This eco-friendly strategy reduces landfill trash in addition to carbon emissions. This environmentally conscious alternative to conventional cement manufacturing has the potential to make a substantial impact on mitigating global warming. The conversion of biomass stands out as a promising alternative energy solution because of its minimal greenhouse gas emissions and the generation of substitutes for petroleum. This renders it a valuable, abundant, cost-effective renewable energy source [1]. The adaptability of biomass utilization, enabling its direct use as a fuel source and transformation into diverse forms of energy, stands out as a significant advantage. In 2030, biomass is expected to ncrease for the world energy supply, making it the leading renewable energy source, according to IRENA research on the subject [2]. In light of the current imperative for environmental sustainability amid escalating energy demands, biomass emerges as an alternative fuel capable of displacing fossil fuels and promoting sustainability. Wood, agricultural and animal waste, energy crops, and industrial waste are examples of potential biomass sources for renewable energy [3]. One way to limit the use of natural resources and reduce energy usage is to produce eco-friendly construction materials from agricultural waste [4]. Furthermore, one effective way to use agricultural waste is to use plant leftovers as a bio-binder. This knowledge is anticipated to lead to a broader acceptability of using agricultural wastes in construction applications [5]. Different research has called building materials with minimal carbon emissions, sustainability, and many uses "agricultural concrete" [6]. Building waste, in particular, has a significant percentage of heavy metals, including massive solid garbage. Consequently, increased soil heavy metal concentration risks soil quality due to various biochemical processes [7]. Furthermore, the accumulation of waste generated from construction projects leads to the decomposition of specific organic substances, emitting harmful gasses that add to environmental pollution. Furthermore, air quality can be negatively impacted by pathogens and particles from garbage spreading through the atmosphere [8]. Consequently, choosing agricultural materials like Abelmoschus esculentus can help diminish the generation of construction waste, exercising control over releasing dangerous compounds into the environment and damaging pollutants. Since silica dioxide (SiO_2) makes up the majority of sand, it is frequently included as a filler in studies. When creating composite materials, one or several filler substances are combined with a matrix material, such as ceramic or polymer, sand is typically used as a filler. Strength, stiffness, toughness, and thermal stability are among the mechanical and physical properties of the composite that are improved by the use of sand as a filler. Sand is a valuable filler material in many industrial and scientific applications because of its availability, affordability, and compatibility with various matrix substances. Composite materials' mechanical and physical characteristics can be improved by adding sand as a filler, making them more appropriate for a broader range of uses. Despite its crystalline form, SiO_2, considered physiologically benign, finds uses in the pharmaceutical and agricultural industries with no known health hazards. In general, silica has qualities that allow it to work well with various polymeric materials, improving their overall qualities. Furthermore, a lot of filler is made of silica to improve the mechanical performance of polymeric substances [9]. SiO_2 nanoparticles are easily synthesized and are reasonably priced, making them useful in chromatography, chemical sensors, catalysts, and ceramics. Abelmoschus esculentus gum, commercially known as Abelmoschus esculentus, demonstrated noteworthy results as a tablet binder, suggesting its viability as an alternative for compositions [10]. Abelmoschus esculentus, a plant commonly cultivated in Africa and Asia, offers a natural, cost-effective, and non-toxic extract, showing promise across various industries. This plant's various components have long been used in medicine traditionally as fight cancer medication, antibacterial, and antidiabetic [11]. This reseasrch aims to evaluate the effects of various Abelmoschus esculentus weight percentages on the mechanical characteristics of a composite containing sand silica particles of various sizes. This study aims to evaluate this composite's potential as an ecologically friendly building material. Furthermore, TGA, XRD, and SEM investigations were performed to characterize the produced materials.

Materials and Process

The research examined the compressive strength of multiple samples produced by mixing silica sand with Abelmoschus esculentus at several ratios (five, ten, and 1fiftenn percent). The silica sand samples were placed on an automatic sieve shaker to remove impurities from their surfaces and allowed to sit for ten minutes. A range of sieves with square mesh sizes of 250 µm, 425 µm, and 850 µm were used in this shaker. A volume of 15×10^4 Liter was achieved by adding deionized water. After agitating the mixture for five minutes at room temperature, the mixture was poured into a precisely cylindrical mold with a 0.5-inch diameter and 1-inch height made of stainless steel base don ISO 604 [12]. A mold releaser was applied to prevent adhesion to the mold surfaces, and the samples were cured for 30 minutes at 176 °F with a 49 N/mm² load using a heated mechanical press. After adopting a drying process by an oven, the samples were placed in glass incubators for cooling, and their compressive mechanical properties were examined using a general Tensile Testing Machine at a 0.4 mm/min load rate.

Results and Discussion

As depicted in Figure 1, adding Abelmoschus esculentus at weight percentages of five, ten, and fifteen percent addition by weight led to an approximately thirty-five percent increase in compressive strength due to the increased number of granules of silica sand in a medium size. The silica sand mixture's compressive strength was increased by the addition of Abelmoschus esculentus, which also enhanced sand particle adherence. This enhancement can be attributed to the high concentration of galactose, rhamnose, and galacturonic acid in Abelmoschus esculentus, which enhance adhesive qualities and have crosslinking characteristics. Furthermore, for all weight percentages of additional Abelmoschus esculentus, it was depicted that the strength dropped by forty percent between 250 µm and 425 microns. The adhesion between particles can be affected by the size of the particles and the applied constant load. Unexpectedly, 850-micron silica sand particles had the lowest compressive strength, even with Abelmoschus esculentus injected in trace amounts. Evidence shows that the silica particle size noticeably affects the composite's overall strength. A noteworthy discovery was made when it was discovered that adding ten percent by weight of Abelmoschus esculentus to 850 µm silica sand may boost compressive strength by twenty-one percent. There appears to be a correlation between the amount of Abelmoschus esculentus added and the silica sand mix particle size in the elastic modulus section. An average of twenty seven percent less Abelmoschus esculentus is added to the mixture as the particle size increases from 425 to 850 microns. Including Abelmoschus esculentus raises the elastic modulus of silica sand because it strengthens the links between the particles in the composite. Particle size, in particular, does not affect a composite's Young's modulus, especially when working with micron-sized particles, as this instance shows [13]. At the nanoscale, the composite's Young's modulus may be enhanced by reducing the size of the particles [14]. The 850 µm silica sand had the lowest elastic modulus of all sizes of silica particles, mainly due to the large dust particles that affected its mechanical properties. By contrast, regardless of the amount of Abelmoschus esculentus added to the combination, the silica particle size of 250 microns demonstrated the highest elastic modulus.

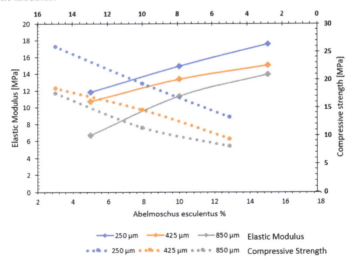

Figure 1 Elastic Modulus and Compressive Strength

Conclusion

The modulus and compressive strength of Abelmoschus esculentus, as well as the mechanical characteristics of micro silica sand, were thoroughly examined in the experiment findings. The particle size emerged as a crucial factor influencing the strength of particulate composites. Specifically, the composite exhibited the highest compressive mechanical properties at fiftenn percent by weight of Abelmoschus esculentus with a particle size of 250 microns. The maximum recorded compressive strength reached 26 MPa, while the peak elastic modulus reached 178 MPa. The potential of silica sand and Abelmoschus esculentus composites in creating environmentally friendly building materials is highlighted by this study.

References

[1] Huang, Y., Wang, Y. D., Rezvani, S., McIlveen-Wright, D. R., Anderson, M., Mondol, J., Zacharopolous, A., & Hewitt, N. J. (2013). A techno-economic assessment of biomass fuelled trigeneration system integrated with organic rankine cycle. Applied Thermal Engineering, 53(2), 325–331. https://doi.org/10.1016/j.applthermaleng.2012.03.041

[2] Muniyappan, D., Pereira Junior, A. O., M, A. V., & Ramanathan, A. (2023). Synergistic recovery of renewable hydrocarbon resources via microwave co-pyrolysis of biomass residue and plastic waste over spent toner catalyst towards sustainable solid waste management. Energy, 278. https://doi.org/10.1016/j.energy.2023.127652

[3] Di Fraia, S.; Fabozzi, S.; Macaluso, A.; Vanoli, L. Energy potential of residual biomass from agro-industry in a Mediterranean region of southern Italy (Campania). *J. Clean. Prod.* 2020, *277*, 124085. https://doi.org/10.1016/j.jclepro.2020.124085

[4] Bumanis, G.; Vitola, L.; Pundiene, I.; Sinka, M.; Bajare, D. Gypsum, Geopolymers, and Starch—Alternative Binders for Bio-Based Building Materials: A Review and Life-Cycle Assessment. Sustainability 2020, 12, 5666. https://doi.org/10.3390/su12145666

[5] Sayed, M. A., El-Gamal, S. M. A., Mohsen, A., Ramadan, M., Wetwet, M. M., Deghiedy, N. M., Swilem, A. E., & Hazem, M. M. (2024). Towards a green climate: production of slag-red brick waste-based geopolymer mingled with wo3 nanoparticles with bio-mechanical achievements. Construction and Building Materials, 413 https://doi.org/10.1016/j.conbuildmat.2024.134909

[6] Ingrassia, L.P.; Cardone, F.; Canestrari, F.; Lu, X. Experimental investigation on the bond strength between sustainable road bio-binders and aggregate substrates. *Mater. Struct.* 2019, *52*, 80. https://doi.org/10.1617/s11527-019-1381-6

[7] Ding, Z.; Gong, W.; Li, S.; Wu, Z. System Dynamics versus Agent-Based Modeling: A Review of Complexity Simulation in Construction Waste Management. Sustainability 2018, 10, 2484. https://doi.org/10.3390/su10072484

[8] Othman, A. A. E., & Abdelrahim, S. M. (2019). Achieving sustainability through reducing construction waste during the design process. Journal of Engineering, Design and Technology, 18(2), 362–377. https://doi.org/10.1108/JEDT-03-2019-0064

[9] Ahmed, W.; Siraj, S.; Al-Marzouqi, A.H. 3D printing PLA waste to produce ceramic based particulate reinforced composite using abundant silica-sand: Mechanical properties characterization. Polymers 2020, 12, 2579. https://doi.org/10.3390/polym12112579

[10] Zaharuddin, N. D., Noordin, M. I., & Kadivar, A. (2014). The use of hibiscus esculentus (okra) gum in sustaining the release of propranolol hydrochloride in a solid oral dosage form. Biomed Research International, 2014, 735891–735891. https://doi.org/10.1155/2014/735891

[11] Elkhalifa, A. E. O., Alshammari, E., Adnan, M., Alcantara, J. C., Awadelkareem, A. M., Eltoum, N. E., Mehmood, K., Panda, B. P., & Ashraf, S. A. (2021). Okra (abelmoschus esculentus) as a potential dietary medicine with nutraceutical importance for sustainable health applications. Molecules (Basel, Switzerland), 26(3). https://doi.org/10.3390/molecules26030696

[12] International Organization for Standardization. ISO 604: Plastics — Determination of compressive properties. Geneva, Switzerland: ISO, 2022.

[13] Cho, J.; Joshi, M.S.; Sun, C.T. Effect of inclusion size on mechanical properties of polymeric composites with micro and nano particles. Compos. Sci. Technol. 2006, 66, 1941–1952. https://doi.org/10.1016/j.compscitech.2005.12.028

[14] Fan, S.; Cheng, Z. A micropolar model for elastic properties in functionally graded materials. Adv. Mech. Eng. 2018. https://doi.org/10.1177/1687814018789520

Techno-economic comparison between PV and wind to produce green hydrogen in Jordan

Zaid HATAMLEH[1,a*], Ahmad AL MIAARI[1,b]

[1]Mechanical Engineering Department, King Fahd University of Petroleum and Minerals, Dhahran, Saudi Arabia

[a]zaidhatamleh08@outlook.com, [b]ahmad.miaari@gmail.com

Keywords: Hydrogen, Green Hydrogen, Renewable Energy, Photovoltaic Energy, Wind Energy, Levelized Cost of Hydrogen (LCOH)

Abstract. The world has begun to move towards searching for the best ways and means to be able to produce hydrogen gas in a healthy and environmentally friendly manner. Especially after manufacturing cars that run on hydrogen, many home appliances in the future would operate using hydrogen. This research paper aims to provide a detailed study and comparison for the potential of photovoltaic energy and wind energy in hydrogen gas production using electrolyzes technology for home applications. In this study, a mathematical model is proposed to predict hydrogen production by means of the two renewable energy sources. Furthermore, the mathematical model computes the electrical energy produced from the fuel cells using the produced hydrogen gas and evaluate its levelized cost. The study was conducted based on the technical specifications based on Jordanian codes and conditions. Results showed that photovoltaic energy system is the best solution compared to other proposed systems which can produce 30,140.5252 kg of hydrogen and produce 1,264,551 kWh/year with the lowest hydrogen levelized cost of 13.262 $/kg.

Introduction

Green hydrogen, a clean energy carrier, is crucial to decarbonization, but its cost is currently high[1]. This is mainly due to the cost of electrolyzers, which split water into hydrogen and oxygen [2]. However, electrolyzer costs are expected to decline significantly in the coming years, making green hydrogen more competitive[2]. Innovation is one of the key factors in reducing electrolyzer costs. This includes developing cheaper electrodes and catalyst materials, and increasing production volumes [2]. By 2030, green hydrogen could be cost-competitive with blue hydrogen in many countries [3]. Another important consideration is the use of renewable energy sources specifically for hydrogen production [4]. This would help reduce the overall cost of green hydrogen. Other challenges include taking into account weather fluctuations and hourly electricity consumption in the system design [5].As well as assessing the environmental impact of the entire production chain, not just the electrolyzers [6]. In addition to incorporating energy storage to manage intermittent renewable energy sources [7]. Research is ongoing to address these challenges and further reduce the cost of green hydrogen. This includes optimizing system design, developing advanced energy storage solutions, and making realistic comparisons of solar and wind energy for hydrogen production [8], [9].

Green hydrogen has the potential to play a major role in decarbonizing the energy sector. Continued research and development efforts are essential to make this a reality. The success of this technology highly depends on the hydrogen production cost per one kilogram [10]. The main idea of this study is to make a real comparison between solar energy and wind energy in hydrogen production and its cost based on Jordanian conditions and policies. The study is done by using mathematical models and using Homer program. This study aims to give a valuable insight that helps policymakers in decision making for selecting and implementing the optimal renewable hydrogen powered home[11].

Methodology
Based on what was mentioned, mathematical calculations and simulations of two different systems will be performed. The first system will be hydrogen production through photovoltaic panels. The second system is hydrogen production through wind turbines.

1) electrical load estimation
Electricity consumption bills for 12 months are used to determine the load of a single house. The energy consumption of different appliances and their operating hours are proposed in table 1.
The total annual home energy consumption for a single house in Jordan is calculated using equation 1.

Table 1 Annual Electricity Consumption for the Electrical Appliances

DATA	Summer Hours	Winter Hours	Units	Rating (W)	Summer Consumption/day	Winter Consumption/day	Annual Energy Consumption (kWh)
Small LED Spots	3	5	15	11.5	0.52	0.86	251.9
LED Smart TV	5	5	2	150	1.5	1.5	547.5
Air Conditioner	2	1	1	3000	6	3	1642.5
LED Bulbs	3	5	40	15	1.8	3	876
Phone Charger	3	3	5	25	0.38	0.38	136.9
Laptop Charger	5	5	5	65	1.63	1.63	593.1
Printer	0.5	0.5	1	1.27	0	0	0.23
Suction Fan	0.4	0.5	4	60	0.1	0.12	39.42
Speed Water Heater	0.1	0.4	1	5500	0.55	2.2	501.9
Washing Machine	1.5	1.2	1	1500	2.25	1.8	739.1
Fridge	6	6	2	600	7.20	7.2	2628
Water Pump	0.7	0.4	1	550	0.39	0.22	110.4
Water Cooler	6	6	1	5	0.03	0.03	10.9
Water Heater	0	5	1	1500	0	7.50	1368.8
Food Processor	0.1	0.1	1	250	0.03	0.03	9.1
Grill	0.1	0.4	1	1400	0.14	0.56	127.8
Water Filter (RO)	4	2	1	750	3.00	1.50	821.3

Air Fresher (Suction Duct)	0.2	0.2	1	200	0.04	0.04	14.6
Drying Machine	0	1	1	2700	0	2.7	492.8
Stand Blender	0.1	0.1	1	250	0.03	0.03	9.1

$$Energy\ (kW/h) = Power\ (kW) * Time\ (h) \tag{1}$$

The total annual electricity consumption for each home in Jordan based on the above loads is estimated to be $10921\ \frac{kWh}{Year\cdot Home}$, resulting in a total of 1,092,100 kWh/year for a hundred homes.

2) hydrogen production
The amount of hydrogen production that can satisfy the annual electricity consumption for 100 homes can be calculated from equation 2 [12].

$$Hydrogen\ (kg) = \frac{Electricity\ (kWh)}{Heating\ Value\ \left(\frac{MJ}{kg}\right)*Fuel\ Cell\ Efficiency\ (\%)*Electrochemical\ Conversion\ (\%)*Conversion\ Factor\ \left(\frac{MJ}{kWh}\right)} \tag{2}$$

A solid oxide hydrogen fuel cell from Bloom energy company was chosen for this particle study, due to its high efficiency, fuel flexibility and low emissions. The selected fuel cell has a high efficiency and electrochemical conversion of 52 % and 80% respectively. The total amount of hydrogen produced by the selected fuel cell to satisfy the annual electricity of 100 homes is calculated to be 18817.1 kg/year when using hydrogen with a heating value of 119.96 MJ/kg [13].

3) photovoltaic energy system's design
Photovoltaics are an important technology for generating electricity using solar energy[14]. This technology can be integrated in many applications such as greenhouses[15], and battery charging[16]. The photovoltaic system capacity highly depends on the sunny hours. Based on the Energy and Minerals Regulatory Commission (EMRC) in Jordan, the useful sunny hours during the year in Jordan is around 1540 h/year. The power needed to run a 100 house in Jordan can be calculated from equation 3.

$$Power = \frac{Energy}{Time} \tag{3}$$

To generate electricity with a capacity of 709 kW that satisfies the demand. A commercial photovoltaic panel Trina Vertex with a capacity of 555 W is chosen for this application. Trina is considered a well-known and trustable brand that is widely found in Jordanian market. A total of 1273 panels are needed for the one hundred houses.

The price and cost of the selected solar panels are established by the Consolidated Energy and Economic Engineering Company. Table 2 shows the cost of the suggested PV panels, where the cost of each Watt of (Trina Vertex 555W) is 0.27 $/Watt.

Table 2 PV Panel Cost

Capital Cost ($/KW)	Replacement ($/KW)	O&M ($/Year)	System Life Time (Years)	Derating Factor (%)
270	270	10	25	84.8

For the AC side of the whole PV system SMA inverters were used with 1.5 (DC/AC) ratio to ensure the highest number of operating hours for the solar inverters at their maximum power point tracking.

Regarding the cost of the solar inverters, SMA solar inverters are priced also by the Consolidated Energy and Economic Engineering Company and can be presented in table 3.

Table 3 Solar Inverter Cost

Capital Cost ($/KW)	Replacement ($/KW)	O&M ($/Year)	System Life Time (Years)
174	174	5	25

4) Wind energy system's design
The design of wind energy system higly depends on the wind speed. According to Wind Atlas software the average wind speed in Amman-Jordan is 7.12 m/s. The available wind power potential can be calcualted using equation 4.

$$\dot{W}_{Available} = 0.5 * \rho * A * V^3 \qquad (4)$$

Where ρ is the density of the air at specific temperature, A is the wind turbine swept area, and V is the average wind speed.

For this specific study Vestas V82-1.65 is selected. The wind turbine is manufactured by Vestas company with a swept area of 5281.01 m² and power preformance presented in figure 1.

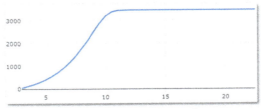

Fig. 1 Vestas V82-1.65 Power Curve

The selected wind turbine was priced by Al-Fujeij Wind Energy Company and summarized in table 4.

Table 4 Wind Turbine Specification and Cost

System Capacity (One Turbine)	Hub Height (m)	Capital Cost ($/Turbine)	Replacement ($/Turbine)	O&M ($/Year)	System Life Time (Years)
1650 kW	100	1,980,000	1,980,000	20,000	20

Results and Discussions
Energy and economic comparison for the two proposed renewable energy systems is done using Homer software [11]. Various results concerning the system potential, costs, and technical specifications are reached and discussed. The first section delves into the results of integrating the system with the PV system. while the other section discusses the results of integrating the system with wind system.

1) PV system

The PV/hydrogen proposed system consists of five main components which are the electrolyzer, PV, inverter, grid, and hydrogen tank. These components relate to each other as shown in figure 2.

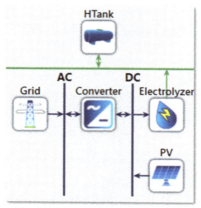

Fig. 2 PV System Schematic Diagram

The results from the mathematical models in the previous section are considered as an input for Homer software [11]. The main idea of these calculations is to reduce the size of the system as much as possible to fit the loads without creating any excess production beyond the energy-consuming facility's need. Important results after running the software can be summarized in table 5.

Table 5 Calculated PV System Output Data

Data	Unit	Value
Net Present Cost	($)	312,062
Levelized Cost of Energy (LCOE)	($/kWh)	0.0104
Payback Period	(years)	1.49
CO_2 Emissions	(kg/year)	388,581
Renewable Fraction	(%)	64.9
Annual PV Production	(KWh/year)	1,264,551
Grid Purchases	(KWh/year)	614,844
Total Hydrogen Production	(kg)	30,140.5252
Hydrogen Produced using PV (Green Hydrogen)	(kg)	21,788.2137
Levelized Cost of Hydrogen	($/Kg)	13.262

Table 5 represents the simulation results of the first experiment of this study, the amount of electrical energy produced from photovoltaic panels is considered appropriate and sufficient to cover the needs of electrical energy consumption in operating the electrolyzers for the purposes of producing green hydrogen while considering the losses during energy transmission, distribution, and losses inside the electrolyzers.

2) Wind Turbines
The wind/hydrogen proposed system consists of five main components which are the electrolyzer, wind turbine, inverter, grid, and hydrogen tank. These components relate to each other as shown in figure 3.

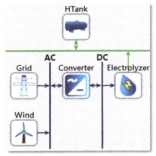

Fig. 3 Wind System Schematic Diagram

A wind farm is suggested to be in Al-Muwaqqar area in Amman, which is known for its low population density. The clean energy from the farm can be used to operate the electrolyzers to produce green hydrogen. The technical specifications and costs of the wind turbines mentioned in previous sections are also considered as an input for Homer software [11]. Different results are obtained and can be presented in table 6.

Table 6 Monthly Electric Production for the Wind System

Data	Unit	Value
Net Present Cost	($)	984,785,100
Levelized Cost of Energy (LCOE)	($/kWh)	0.1379
Payback Period	(years)	1.94
CO_2 Emissions	(kg/year)	63,533
Renewable Fraction	(%)	100
Annual Wind Production	(KWh/year)	552,150,642
Grid Purchases	(KWh/year)	100,527
Total Hydrogen Production	(kg)	9,515,287.625
Hydrogen Produced using Wind (Green Hydrogen)	(kg)	9,513,555.562
Levelized Cost of Hydrogen	($/Kg)	26.7368

It is noted from table 6 that the option of using wind turbines is considered very expensive in terms of the initial cost. Moreover, wind turbines located in Jordanian capital is not considered a good

Approach, as the wind speed inside the region does not exceed 10 m/s which is considered low compared to other locations. The turbine will not reach its natural production capacity. The high cost of producing electricity also leads to a significant increase in the cost of hydrogen.

In terms of levelized cost and hydrogen production the PV/hydrogen system is considered a better option compared to wind/Hydrogen system, where the levelized cost of the first system is 13.5 $/ kg less than of that of wind/hydrogen system.

Conclusion

A techno-economic investigation for hydrogen powered homes in Amman-Jordan based on solar energy and wind energy is presented in this study. The study utilizes mathematical models and Homer software to explore electricity, hydrogen production, and evaluate hydrogen levelized cost. The study is conducted based on Jordanian technical specifications and conditions. Based on the results it is concluded that PV integrated with hydrogen system is a better option than wind turbine integrated with hydrogen system. This is due to the ability of the system to meet the electrical load correctly without causing any disturbances in the network, while also producing 30,140.5252 kg of hydrogen at a low hydrogen levelized cost and payback period of 13.262 $/kg and 1.49 years respectively. The CO_2 emitted from the PV/hydrogen system is considered greater than from Wind turbine/ hydrogen system which was 63,533 kg/year. This high amount of CO_2 emissions is due to the electricity purchase from the grid.

References

[1] M. T. Muñoz Díaz, H. Chávez Oróstica, and J. Guajardo, "Economic Analysis: Green Hydrogen Production Systems," *Processes*, vol. 11, no. 5, p. 1390, May 2023. https://doi.org/10.3390/pr11051390

[2] O. Schmidt, A. Gambhir, I. Staffell, A. Hawkes, J. Nelson, and S. Few, "Future cost and performance of water electrolysis: An expert elicitation study," *Int. J. Hydrogen Energy*, vol. 42, no. 52, pp. 30470–30492, Dec. 2017. https://doi.org/10.1016/j.ijhydene.2017.10.045

[3] M. Shahabuddin, M. A. Rhamdhani, and G. A. Brooks, "Technoeconomic Analysis for Green Hydrogen in Terms of Production, Compression, Transportation and Storage Considering the Australian Perspective," *Processes*, vol. 11, no. 7, p. 2196, Jul. 2023. https://doi.org/10.3390/pr11072196

[4] M. S. Herdem et al., "A brief overview of solar and wind-based green hydrogen production systems: Trends and standardization," *Int. J. Hydrogen Energy*, vol. 51, pp. 340–353, Jan. 2024. https://doi.org/10.1016/j.ijhydene.2023.05.172

[5] I. Ourya, N. Nabil, S. Abderafi, N. Boutammachte, and S. Rachidi, "Assessment of green hydrogen production in Morocco, using hybrid renewable sources (PV and wind)," *Int. J. Hydrogen Energy*, vol. 48, no. 96, pp. 37428–37442, Dec. 2023. https://doi.org/10.1016/j.ijhydene.2022.12.362

[6] L. B. B. Maciel, L. Viola, W. de Queiróz Lamas, and J. L. Silveira, "Environmental studies of green hydrogen production by electrolytic process: A comparison of the use of electricity from solar PV, wind energy, and hydroelectric plants," *Int. J. Hydrogen Energy*, vol. 48, no. 93, pp. 36584–36604, Dec. 2023. https://doi.org/10.1016/j.ijhydene.2023.05.334

[7] L. Al-Ghussain, A. D. Ahmad, A. M. Abubaker, K. Hovi, M. A. Hassan, and A. Annuk, "Techno-economic feasibility of hybrid PV/wind/battery/thermal storage trigeneration system: Toward 100% energy independency and green hydrogen production," *Energy Reports*, vol. 9, pp. 752–772, Dec. 2023. https://doi.org/10.1016/j.egyr.2022.12.034

[8] Y. Ren et al., "Modelling and capacity allocation optimization of a combined pumped storage/wind/photovoltaic/hydrogen production system based on the consumption of surplus wind and photovoltaics and reduction of hydrogen production cost," *Energy Convers. Manag.*, vol. 296, p. 117662, Nov. 2023. https://doi.org/10.1016/j.enconman.2023.117662

[9] M. Nasser and H. Hassan, "Thermo-economic performance maps of green hydrogen production via water electrolysis powered by ranges of solar and wind energies," *Sustain. Energy Technol. Assessments*, vol. 60, p. 103424, Dec. 2023. https://doi.org/10.1016/j.seta.2023.103424

[10] O. A. Dabar, M. O. Awaleh, M. M. Waberi, and A.-B. I. Adan, "Wind resource assessment and techno-economic analysis of wind energy and green hydrogen production in the Republic of Djibouti," *Energy Reports*, vol. 8, pp. 8996–9016, Nov. 2022. https://doi.org/10.1016/j.egyr.2022.07.013

[11] HOMER Pro, "Microgrid Software for Designing Optimized Hybrid Microgrids n.d." https://homerenergy.com/products/pro/index.html

[12] K. W. Harrison, R. Remick, G. D. Martin, and A. Hoskin, "Hydrogen Production: Fundamentals and Cas Study Summaries e Preprint," 2010. [Online]. Available: http://www.osti.gov/bridge

[13] *The Hydrogen Economy*. Washington, D.C.: National Academies Press, 2004. https://doi.org/10.17226/10922

[14] A. Al Miaari and H. M. Ali, "Technical method in passive cooling for photovoltaic panels using phase change material," *Case Stud. Therm. Eng.*, vol. 49, p. 103283, Sep. 2023. https://doi.org/10.1016/j.csite.2023.103283

[15] A. Al Miaari, A. El Khatib, and H. M. Ali, "Design and thermal performance of an innovative greenhouse," *Sustain. Energy Technol. Assessments*, vol. 57, p. 103285, Jun. 2023. https://doi.org/10.1016/j.seta.2023.103285

[16] A. Al Miaari and H. M. Ali, "Batteries temperature prediction and thermal management using machine learning: An overview," *Energy Reports*, vol. 10, pp. 2277–2305, Nov. 2023. https://doi.org/10.1016/j.egyr.2023.08.043

Analytical and numerical evaluation for wind turbine aerodynamic characteristics

Ashraf Abdelkhalig[1,2,a], Mahmoud Elgendi[1,3,b*], Mohamed Y.E. Selim[1,3,c], Maryam Nooman AlMallahi[1,d], Sara Maen Asaad[4,e]

[1] Department of Mechanical and Aerospace Engineering, College of Engineering, United Arab Emirates University, Al Ain, UAE

[2] Aeronautical Engineering Department, College of Engineering, Sudan University of Science and Technology, Khartoum, Sudan

[3] National Water and Energy Center, United Arab Emirates University, Al Ain, P.O. Box 15551, United Arab Emirates

[4] Sustainable Energy and Power Systems Research Center, Research Institute of Science and Engineering (RISE) Engineering Sharjah, UAE

[a] ashraf891@hotmail.com, [b] mahgendi@uaeu.ac.ae, [c] mohamed.selim@uaeu.ac.ae, [d] 700039403@uaeu.ac.ae, [e] smaen@sharjah.ac.ae

Keywords: Renewable Energy, Wind Turbine, Aerodynamics

Abstract. Energy lies at the core of ultramodern society, empowering everything from heating, lighting, computers, and food products to manufacturing and transport. A rising realization of the harmful climatic belongings of anthropogenic greenhouse gas emissions is boosting governmental pressure to alleviate or avoid CO_2 discharge into the ambiance. Wind turbines are one of the most likely initial applications for renewable sources. The major challenge in the wind turbine field is designing a machine that performs efficiently, boosting its reliability and producing power. The necessity for computational and experimental proceedings for probing aeroelastic stability has increased with the increase in output power and the size of the turbines. Due to the complexity and high costs of experimental investigations, several modeling methods have been practical solutions for design and analysis objectives. In this context, this paper presents an evaluation study for the aerodynamic performance of wind turbines - by concentrating on analyses of aerodynamic workforces that act on the rotor, employing Blade Element Momentum (BEM) and with the usage of the Computational Fluid Dynamic (CFD) solver. The computed results show a reasonable agreement with the previous results found in the literature. This indicates that it is possible to predict the characteristics of wind turbines from analytical and numerical approaches with plausible reliability.

Introduction

The conversion of wind power into beneficial power has placed the foundations for one of the most significant technological progress of the 20th century. Wind turbines—elaborated to harness and utilize wind power to generate electricity—are the technology behind one of the speedy promoting industries for power production. They are currently an ordinary sight worldwide in the countryside and urban regions [1].

For successful and outsize wind energy employment, the cost of wind turbines must be reduced to be competitive with the instant options. The conduct of a wind turbine is formed by a complicated relationship of elements and sub-systems [2]. The main parts are the rotor, tower, hub, and nacelle. Extrapolating the interactive actions between the parts provides the basics for trusty design computations, optimized machine arrangements, and reduced wind electricity expenses [30]. In the aspect of rotor aerodynamics, many phenomena (e.g., atmospheric boundary layer flow) still need to be fully understood. Consequently, some methods are used to analyze the aerodynamic performance, such as wind tunnel tests or field measurements, analytical models, and Computational Fluid Dynamics (CFD) [3].

Many researchers [4-7] provided extensive surveys of the literature on the analytical and semi-empirical models (e.g., Blade Element Momentum (BEM) model). However, CFD is a vital tool for flow simulation in different cases [8–17]. With the evolution of computing implementations, using the CFD approach makes it possible to resolve wind turbine rotors fully. In this context, Ferziger and Peric [19], Jorge et al. [20], and Jiyuan et al. [21] explored the dynamic capability of CFD. They pointed out the descriptions of fundamental theories, basic techniques, and practical guidelines.

The present work aims to examine the performance characteristics of the HAWT rotor from an aerodynamics perspective and, in general, to validate the capabilities of BEM and CFD techniques applied in the wind energy field.

Method of Analysis
Analytical Study
The wind turbine performance can be predicted analytically by applying the BEM theory. In this approach, the blade is split into several separated parts along with the spread of the blade. For every part, a force equilibrium is utilized concerning two-dimensional lift and drag with the thrust and torque delivered by the part. Simultaneously, an equilibrium of axial and angular momentum applies to it. This outputs several equations that can be resolved iteratively [22]. The equations of the BEM theory given by [6] are utilized to compute the output power of the NREL turbine (see Table 1), and the details about the blade and measurement conventions can be found in [23].

Table 1. Characteristics of NREL Phase VI wind turbine [23].

Blades number	2
Blade profile	S809
Rotor radius	5.029 m
Rotational speed	72 rpm
Turbine power	19.8 kW
Power regulation	Stall

Numerical Study
Here, we investigate the aerodynamic characteristics of the S809 airfoil, represented in the blade profile. A commercial, finite volume-based solver has been used to implement this analysis. Generally, three main configurations point out commercial CFD codes corresponding to three stages of problem-solving- pre-processor, solver, and post-processor [24]. The computational domain for 2D airfoil analysis is shown in Fig 1. During the creation of the mesh around the airfoil, great care must be taken in the vicinity close to the airfoil surface to consider the boundary layer flow that might be formed, as illustrated in Fig 2. Also, the k-ε model is applied as a turbulence model.

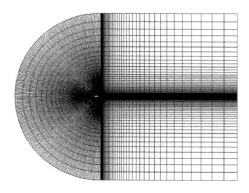

Fig. 1. Mesh generated for the airfoil section

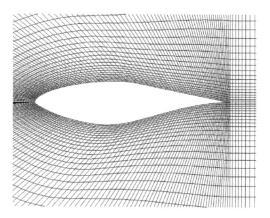

Fig. 2. Concentrated mesh generated around the airfoil.

Results and Discussion
BEM code results
The BEM code splits the blade into ten elements to determine the power generated over a range of wind speeds for each element. This has been done at wind speeds ranging from 5 to 15 m/s. Fig. 3 shows the comparison of present code results (BEM) with measured data (Exp.) [25] and other BEM predictions [26, 27] for the NREL Phase VI rotor.

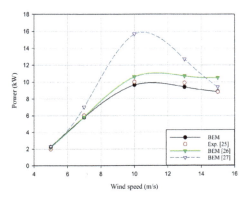

Fig. 3. Power predicted at different wind speeds.

It is illustrated by Fig. 3 that the power computed from the present BEM code compared well with experimental results at all wind velocities, except at 13m/s, where an under-prediction of 13% is realized.

2D Airfoil analysis results
Fig. 4 shows the computed pressure coefficient (Cp) distribution in the present analysis (CFD) at zero angles of attack (AOA) and 106 Reynolds number, compared with experimental data (Exp.) [28] and another computational study CFD [28] for the same airfoil and operating conditions.

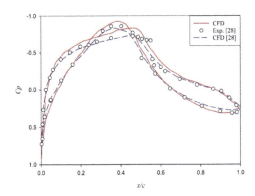

Fig. 4. Pressure coefficient for S809 airfoil.

The distribution of Cp with referenced data was validated on both airfoil surfaces. In addition, table 2 compares the numerical and experimental lift coefficient (Cl) and drag coefficient (Cd), calculated at 2x106 Reynolds number and different attack angles. Lift coefficient results are very close to the experimental data at all AOA (within 8%), while the predicted drag coefficients are up to 40% higher than the experiment results. This over-prediction of drag could be reasonable due to the laminar flow over the airfoil's forward half.

Fig. 5 (a) displays the pressure contours over the airfoil for the zero-degree angle of attack. The maximum pressure is generated at the airfoil leading edge. Also, negative pressure is created at both airfoil surfaces (i.e., top and bottom). The variation between these pressures is the source of the lift force. With the increase of AOA (Fig. 5 (b)), the negative pressure at the upper surface increases, while the negative pressure at the lower surface decreases, so the lift increases (and consequently, the lift coefficient, as indicated in Table 2).

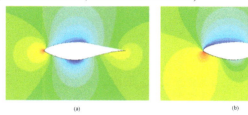

Fig. 5. Pressure contours of S809 at 2x106 Reynolds number for (a) 0 and (b) 5.13 AOA

Table 2. Comparisons between CFD and experimental Cl and Cd [28].

AOA (deg)	C_l Exp. [28]	CFD	% Error	CFD [28]	% Error [28]	C_d Exp. [28]	CFD	% Error	CFD [28]	% Error [28]
0.00	0.1469	0.152482	4	0.1324	-10	0.0070	0.012092	42	0.0108	54
1.02	0.2716	0.267285	-2	0.2492	-8	0.0072	0.012463	42	0.0110	53
5.13	0.7609	0.70615	-8	0.7123	-6	0.0070	0.018062	61	0.0124	77

Conclusion

The main goal of this paper is to carry out a characteristic aerodynamic evaluation of the blade of a HAWT. For this, the NREL Phase VI blade is analyzed analytically using the BEM method and numerically using the CFD. The results were quite satisfactory and can represent a well-grounded basis for coming research in this field. Fundamentally, the effects of changing the geometric and aerodynamic factors on the performance of wind turbines could be understood through BEM theory. More importantly, the reliability of CFD for calculating performances on a HAWT blade was confirmed. As evident from this work, the numerical investigations involved the assumption of a fully turbulent flow using the k-ε model. Nonetheless, a more advanced model with different setups needs to be considered to obtain optimal results.

References

[1] Bakırcı M and Yılmaz S 2018 Eng. Sci. Technol, an International J. 21 1128–42. https://doi.org/10.1016/j.jestch.2018.05.006

[2] Abdelrahim A and Seory A 2017 Proc. 10th Biennial International Workshop Advances in Energy Studies BIWAES2017 (Naples, Italy)

[3] Sørensen J and Shen W 2002 J. Fluid Engineering 124 393–9. https://doi.org/10.1115/1.1471361

[4] Gordon L 2011 Aerodynamics of Horizontal Axis Wind Turbines, In: Advances in Wind Energy Conversion Technology, ed Mathew S and Philip G, Springer.

[5] Pramod J 2011 Wind Energy Engineering (London: McGraw-Hill)

[6] Hansen M 2008 *Aerodynamics of Wind Turbines* 2nd ed (London: Earthscan)
[7] Manwell J, McGowan J and Rogers A 2002 *Wind Energy Explained* (John Wiley & Sons Ltd). https://doi.org/10.1002/0470846127
[8] Mekhail T, Dahab O, Sadik M, El-Gendi M and Abdel-Mohsen H 2015 *Open J. Fluid Dyn.* **05** 224–37. https://doi.org/10.4236/ojfd.2015.53025
[9] El-Gendi M, Ibrahim M, Mor K and Nakamura Y 2010 *Trans. Jpn. Soc. Aeronaut. Space Sci.* **53** 122–9. https://doi.org/10.2322/tjsass.53.122.
[10] El-Gendi M, Ibrahim M, Mor K and Nakamura Y 2009 *Trans. Jpn. Soc. Aeronaut. Space Sci.* **52** 206–12. https://doi.org/10.2322/tjsass.52.206
[11] El-Gendi M, Do K, Ibrahim M, Mor K and Nakamura Y 2010 *Trans. Jpn. Soc. Aeronaut. Space Sci.* **53** 171–9. https://doi.org/10.2322/tjsass.53.171
[12] El-Gendi M, Lee S, Joh C, Lee G, Son C, Chung W 2013 *Trans. Jpn. Soc. Aeronaut. Space Sci.* **56** 82–9. https://doi.org/10.2322/tjsass.56.82
[13] El-Gendi M 2018 *Int. J. Therm. Sci.y* **125** 369–80. https://doi.org/10.1016/j.ijthermalsci.2017.12.012
[14] El-Gendi M and Aly A 2017 *Int. J. Numer. Methods Heat Fluid Flow* **27** 2508–27. https://doi.org/10.1108/HFF-10-2016-0376
[15] El-Gendi M, Do K, Ibrahim M, Mor K and Nakamura Y 2010 *Trans. Jpn. Soc. Aeronaut. Space Sci.* **53** 171–9. https://doi.org/10.2322/tjsass.53.171
[16] El-Gendi M, Lee S and Son C 2011 *J. Korean Soc. Mar. Eng.* **35** 938–45. https://doi.org/10.5916/jkosme.2011.35.7.938
[17] Cheng S, Elgendi M, Lu F and Chamorro L 2021 *Energies* **14** 7204. https://doi.org/10.3390/en14217204
[18] Abdelkhalig A, Elgendi M and Sliem M 2022 *In: 2022 AUA Acad. Conf. Sustain. Energy Green Technol., IOP-EES*
[19] Ferziger J and Peric M 1999 *Computational Methods for Fluid Dynamics* (Berlin: Springer-Verlag). https://doi.org/10.1007/978-3-642-98037-4
[20] Jorge C, Achim H, Florian M, Thomas E and Antoine L 2001 *Advanced CFD Analysis of using Aerodynamics Using CFX* (AEA Technology GmbH)
[21] Jiyuan T, Guan H and Chaoqun L 2008 *Computational Fluid Dynamics: A Practical Approach* (Elsevier Inc.)
[22] Grant I 2005 *Wind Turbine Blade Analysis using the Blade Element Momentum Method Version 1.0* School of Engineering (Durham: Durham University)
[23] Hand M, Simms D, Fingersh L, Jager D, Cotrell J Schreck S and Larwood S 2001 *NREL Technical Report/TP-500-29955* (Golden, Colorado)
[24] Versteeg H and Malalasakera W 1995 *An Introduction to Computational Fluid Dynamics: The Finite-Volume Method* (New York: Longman Scientific & Technical)
[25] Leclerc C and Masson C 2005 *J. Sol. Energy Eng. Trans.-ASME* **127** 200–8. https://doi.org/10.1115/1.1889466
[26] Maria M and Jacobson M 2009 *Energies* **2** 816–38. https://doi.org/10.3390/en20400816
[27] Yelmule M and Anjuri E 2013 *Int. J. Renew. Energy. Res.* **3** 261–9
[28] Wolfe W and Ochs S 1997 *AIAA-97-0973*

[29] Elgendi, M., AlMallahi, M., Abdelkhalig, A. and Selim, M.Y., 2023. A review of wind turbines in complex terrain. International Journal of Thermofluids, p.100289. https://doi.org/10.1016/j.ijft.2023.100289

[30] Abdelkhalig, A., Elgendi, M., & Selim, M. Y. (2022, August). Review on validation techniques of blade element momentum method implemented in wind turbines. In IOP Conference Series: Earth and Environmental Science (Vol. 1074, No. 1, p. 012008). IOP Publishing. https://doi.org/10.1088/1755-1315/1074/1/012008

Sustainability policies and regulation challenges in recycling EV batteries

Afnan KHALIL[1,a *], Mousa HUSSEIN[2,b], Essam ZANELDIN[3,c], Waleed AHMED[4,d]

[1] Chemical and Petroleum Engineering Department, COE, UAE University, Al Ain, UAE

[2] Electrical and Communication Engineering Department, COE, UAE University, Al Ain, UAE

[3] Civil and Environmental Engineering Department, COE, UAE University, Al Ain, UAE

[4] Engineering Requirements Unit, COE, UAE University, Al Ain, UAE

[a]700043891@uaeu.ac.ae, [b]mihussein@uaeu.ac.ae, [c]essamz@uaeu.ac.ae, [d]w.ahmed@uaeu.ac.ae

Keywords: EV, Recycling, Policies, Regulations, Batteries

Abstract. The increasing use of electric vehicles has brought the critical issue of recycling the batteries of electric cars to the forefront. This paper explores the challenges posed by current recycling policies, emphasizing the gaps in regulations and the pressing need for effective authorities involvement. The complications surrounding the recycling policies of electric car batteries are explored, shedding light on the disadvantages that may restrict successful implementation. The paper underscores the importance of addressing these challenges to ensure sustainable and responsible management of electric car batteries, emphasizing the shared responsibility between authorities, manufacturers, and other stakeholders. Examining existing policies and identifying areas for improvement will contribute to the ongoing discourse on developing comprehensive and effective electric car battery recycling frameworks and a sustainable and environmentally responsible approach to the end-of-life management of electric vehicle batteries. Recommendations on how to address this crucial issue are also presented.

Introduction
Sustainability incorporates maintaining or preserving a process over an extended period of time. In commerce and policy, sustainability is driven by protecting natural and physical resources to ensure their availability for an extended duration [1]. However, the actual core of sustainable strategies goes beyond resource conservation since it prioritizes an in-depth consideration of how specific policies or corporate practices will impact not only the durability of resources but also the well-being of people, the resilience of ecosystems, and the overall stability of the economy in the future [2]. Sustainability is grounded in a reflective understanding that the Earth faces the risk of irreversible damage if substantial changes are not instituted in its management. It highlights the resolution of responsible and forward-thinking practices to safeguard the planet for future generations. As we inspect the context of fuel-powered vehicles through the lens of sustainability, a absolutely reality emerges they employ a considerable environmental impact. The emissions from conventional vehicles, including pollutants like smog, carbon monoxide, and other harmful substances, pose a significant threat to human beings and the environment [3]. What exacerbates this concern is that these emissions derive from street vehicles, directly exposing people to contaminated air that is inhaled into their lungs. This proximity increases the health risks, making vehicle emissions a pressing and immediate concern, unlike pollutants released at higher altitudes from industrial smokestacks. In fact, the sustainability discourse demands a critical evaluation of our choices, particularly in the context of transportation. Recognizing the contrary effects of conventional vehicles on human health and the environment underlines the imperative to transition to more sustainable alternatives, such as electric vehicles, as a crucial step toward mitigating the

harmful impact on our planet [4]. This shift aligns with the broader principles of sustainability, emphasizing responsible resource management and the preservation of ecosystems for the long-term well-being of our planet and its inhabitants. In pursuing a sustainable future in the automotive industry, shifting from traditional internal combustion engine vehicles to electric vehicles (EVs) is a pivotal transformation in the automotive landscape [5]. Figure 1 illustrates the design of a modern electrical car.

Fig. 1. The design of an electric car [6].

Beyond reducing emissions, this transition promises a comprehensive reimagining of our mobility and environmental responsibility approach. At the core of the move to electric vehicles is a significant reduction in greenhouse gas emissions, as EVs operate with zero tailpipe emissions. This shift reflects a commitment to a cleaner atmosphere and a more sustainable global ecosystem [7]. In densely populated urban areas, where air quality is a growing concern, adopting electric vehicles contributes to a healthier living environment. The zero-emission operation of EVs takes precautions against the pollutants that compromise air quality and impact human health and the environment [8]. Electric vehicles redefine transportation efficiency by significantly reducing the overall energy consumption compared to traditional combustion engines. This brings cost savings for users and aligns with the principles of resource efficiency, promoting responsible energy use and conservation. As the world acknowledges the finite nature of fossil fuel resources, the transition to electric vehicles strategically lessens dependence on exhaustible fuels. This move toward energy diversification supports a more sustainable energy mix, integrating renewable sources like solar and wind into our transportation infrastructure [9]. The shift to electric vehicles catalyzes technological innovation, particularly in battery technology, energy storage, and charging infrastructure. This combined effect fosters a cycle of innovation, propelling advancements in sustainable practices and contributing to a more sustainable technological landscape [10]. The desire for electric vehicles is booming as more people seek eco-friendly transportation options and governments push for cleaner mobility solutions. This rising demand for EVs reflects a global shift towards sustainable transportation fueled by lower operating costs, technological advancements, and environmental consciousness. Figure 2 shows the increase in the EV demand.

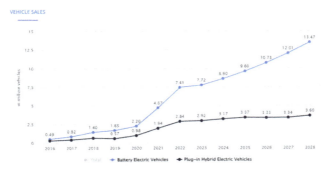

Fig. 2 Demand statistics on EV. [11]

In general, embracing electric vehicles transcends the conventional boundaries of transportation revolutions. It signifies a conscious choice towards a future where sustainability is not just an aspiration but an integral part of our collective journey. As the automotive industry accelerates towards electric mobility, incorporating these advantages holds the potential to usher in a transformative era of environmental harmony and sustainable living [12]. Despite all the benefits of this transformation from traditional vehicles to electric vehicles, the recycling of electric vehicles batteries remains a high concern. The increasing demand for electric vehicle (EV) batteries is driven by the production of new EV cars and the need for spare parts, reflecting the growing adoption of electric mobility worldwide. As more EVs hit the roads and manufacturers expand their product lines, the demand for reliable and efficient batteries continues to surge, highlighting the critical role of battery technology in the transition to sustainable transportation. Figure 3 depicts the demand and supply of lithium for batteries by sector.

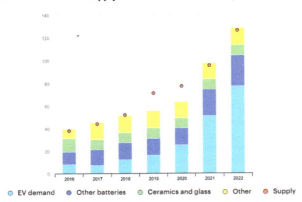

Fig. 3 Overall supply and demand of lithium for batteries by sector, 2016-2022 [13].

This article aims to understand the difficulties in recycling electric vehicle batteries by looking at different angles, particularly on how policies play a significant role. Our goal is to break down and study the complex process of recycling EV batteries, highlighting the challenges from different viewpoints and exploring how policies significantly impact the sustainable handling of these crucial components.

Challenges, Regulations, Responsibilities, and Disadvantages
In this review, our goal is to tackle the challenges, regulations, responsibilities, and disadvantages associated with the shift to electric vehicles. This involves a comprehensive understanding of the extent of the problem by exploring these essential aspects. The complicated manufacturing details of electric vehicle batteries and the absence of a well-established recycling infrastructure provide several difficulties for the recycling industry [14]. The following subsections highlight some significant obstacles to recycling policies for batteries used in electric vehicles:

Challenges in Recycling Policies
In the field of chemical complexity, lithium, cobalt, nickel, and other rare earth elements are among the valuable and potentially dangerous compounds commonly found in electric vehicle batteries. These materials must be extracted and separated for recycling, a complicated operation requiring specialized technologies. Considering that recycled materials must adhere to strict requirements to be used in creating new batteries, the complexity stems from the necessity to separate and recover these materials without sacrificing their quality [15]. Lithium-ion battery recycling may not be possible using conventional techniques like shredding and melting because of the possibility of losing essential materials and associated safety risks. Moreover, the infrastructure for recycling lithium-ion batteries is less developed than that of classic lead-acid batteries, which have clear and well-established recycling procedures. A robust collecting and transportation system and the expansion of recycling facilities are required to handle the growing quantity of end-of-life electric car batteries [16]. Considering transportation logistics as a critical point, batteries must be delivered to recycling centers properly and safely. The absence of a dedicated infrastructure for the transportation of batteries for electric vehicles might lead to higher expenses and more complicated logistics. Transportation must be safe and legal to prevent incidents and harmful emissions [17]. Besides the policy and regulation gaps, there might be exceptions to the rules and laws concerning the recycling of electric vehicle batteries in specific areas. Clear rules, rewards, and restrictions can be established to encourage producers and customers to engage in recycling initiatives in extended procedure responsibility [18]. Sustainable battery recycling depends on end-of-life extended producer responsibility (EPR) plans, which enable proper collection and disposal of items at the end of their useful lives by assuring responsible manufacturing procedures and methods. However, gaps in accountability and recycling programs may result from the lack of widespread adoption of EPR regulations for batteries used in electric vehicles [19].

Gaps in Recycling Regulations
One major obstacle to efficiency, safety, and the creation of an international framework for sustainable practices in the recycling of electric vehicles batteries is the absence of standards. Significant elements of the lack of standards are various battery layouts, where battery designs used by manufacturers of electric vehicles frequently differ in terms of cell shapes, packaging, and thermal management mechanisms. Developing consistent recycling procedures and technology is challenging due to the absence of uniformity in these design elements. In contrast, when we come to uncertain responsibilities, there may not have been a clear definition of who is responsible for what at each stage of the battery life cycle, including recycling. Roles and duties must be clearly defined to prevent gaps in the performance of recycling procedures. However, clarity regarding the obligations of manufacturers, consumers, and recycling facilities for the disposal of electric car batteries at the end of their useful lives is lacking in many countries. Controlling the disposal of electric vehicle (EV) batteries in landfills is crucial to prevent environmental contamination and maximize resource recovery. Implementing strict regulations, promoting battery recycling

programs, and incentivizing proper disposal methods are essential steps in managing the end-of-life cycle of EV batteries and mitigating potential ecological risks. Figure 4 illustrates the expected future landfill caused by EVs batteries disposals.

Fig. 4 Future landfill of EV battery disposals [6]

Well-defined policies are necessary to create accountability and encourage ethical recycling methods. Policies and guidelines for achieving goals without opposing or distancing specific technologies are known as technology-neutral guidelines. Technology-neutral policies are intended to establish fair and equal opportunities for various recycling techniques and technologies in electric car battery recycling. Among the crucial issues that should be addressed is the need to promote innovation. Technology-neutral policies promote constant creativity in the field of recycling electric vehicles batteries without favoring any particular technology over other technologies. With this strategy, the industry is free to experiment with and implement modern eco-friendly techniques without being restricted by outdated rules. In addition to the approach based on objectives, technology-neutral regulations place a greater value on the intended results and environmental goals than on recommending techniques or tools. This makes it possible to be adaptable and flexible as recycling technologies develop over time.

Authority Responsibilities and Commitments
Authorities are essential when developing and implementing efficient rules for recycling electric car batteries. In this area, authorities are primarily responsible for legislation and regulations, which are vital in controlling and directing numerous aspects of society, including sectors like the recycling of electric vehicle batteries. Regulations and legality are essential for maintaining safety, ecological sustainability, and end-of-life management of electric vehicle batteries. Legislative frameworks may require public awareness campaigns and educational programs to educate customers, companies, and other stakeholders on the significance of responsibly disposing of and recycling batteries [20]. This brings us to the matter of public awareness and education requirements. Promoting active involvement in recycling programs and developing a sustainable culture requires education. Regarding recycling electric vehicles batteries, regulations and laws seek to establish an organized, secure, and long-lasting system that handles environmental issues, encourages creativity, and guarantees the proper handling of batteries that have reached the end of their useful lives. To keep up with changes in consumption habits and technological advances, these frameworks need to be updated regularly. Furthermore, tracking and enforcing procedures are essential elements of successful legislation in the context of recycling electric car batteries. These features guarantee that set guidelines and regulations are followed, encouraging environmental sustainability, safety, and appropriate disposal of spent batteries. However, inspecting procedures are required, and supervision includes creating and applying inspection

protocols. The guidelines define the standards for evaluating consistency, influencing safety rules, ecological consequences, and conformity to recycling techniques. Inspections can occur during transportation, recycling locations, and factories. Additionally, safety inspections are required in battery recycling situations where safety is a top priority. Regulatory agencies conduct safety checks to evaluate whether companies follow safety procedures when gathering, moving, and handling wasted batteries [21]. The purpose of these examinations is to stop hazardous material releases, fires, and accidents.

Disadvantages of Recycling Policies
There are significant drawbacks to recycling regulations for electric vehicles batteries, even though they are essential for managing the end-of-life phase of these batteries and resolving environmental concerns. It's critical to recognize these difficulties to guide continuing efforts toward progress. Creating and applying innovative recycling technology for electric vehicle batteries can be quite costly. Compared to alternative approaches, including raw material extraction, recycling may not be as financially practical due to the expensive costs involved in the process and the requirement for specialized equipment and qualified employees. Also, it may fall under the field of energy efficiency of recycling. The energy intensity of recycling is the energy needed to gather, process, and repurpose wasted materials into new goods. The recycling process for electric car batteries can be expensive in terms of energy, affecting the sustainability of recycling programs from an environmental and financial perspective. There are some energy-intensive battery recycling procedures for electric vehicles. The energy obtained from sources that are not renewable may cause some of the environmental advantages of recycling to be outweighed by the energy needed to extract valuable components from utilized batteries. As well as, by carrying out a life cycle assessment, the environmental impact of the battery lifecycle from production and usage to recycling can be thoroughly examined, including the energy intensity of the process. Lifecycle assessment helps find ways to make the recycling process more energy-efficient and cost-effective. In addition, customer engagement and understanding are other vital points where insufficient consumer knowledge about the value of recycling and the accessibility of appropriate disposal methods may result in low customer engagement. The success of a recycling policy depends on educating customers about proper battery disposal. Policies aimed at recycling the batteries used in electric cars must succeed in raising consumer knowledge and awareness and encouraging their involvement. Increasing consumer awareness of the value of adequately disposing and recycling EV batteries can significantly impact recycling rates and help create a more sustainable end-of-life battery management strategy [22]. Yet, customers must understand how incorrect disposal of electric car batteries affects their health and the environment. If recycled improperly, the compounds found in used batteries have the potential to cause serious environmental harm. Encouraging environmentally responsible behavior among consumers involves educating them about the potential implications of improper disposal. Also, battery manufacturers may help by clearly identifying their products, emphasizing the need for recycling, and offering guidance on how to do so. Consumers may make educated decisions and comprehend their part in the recycling process with straightforward information. Lastly and most importantly, using social media channels to interact with customers and share information about recycling batteries is a good strategy. Programs on social media can increase awareness, provide information, and promote community engagement in sustainability initiatives.

Recommendations
Typically, a lithium-ion battery can last between 8 to 10 years or 100,000 to 200,000 miles, whichever comes first. However, several factors can affect the life of an EV battery. Some factors contributing to reducing the EV battery's lifespan include frequent fast charging, high-speed driving, and exposure to extreme temperatures. It is always recommended to regularly check the

battery's state of charge and identify issues before they become severe. It is also recommended that the battery be appropriately charged to only 80% of its capacity and that frequent fast charging is avoided. Despite all these preventive measures to increase the battery's lifespan, the life of the EV battery will end and proper recycling should be implemented [23]. Some of the recommendations to optimally recycle EV batteries with the objective of conserving resources, reducing environmental impact, improving energy efficiency, and minimizing waste include:
1. Develop efficient and environmentally friendly recycling techniques for EV batteries.
2. Ensure collaboration between industries, governments, and researchers and continuously introduce innovative methods to drive the development of more effective recycling processes.
3. Develop consistent and clear technology-neutral recycling policies and guidelines. These policies should clearly define the responsibilities at each stage of the battery life cycle, including the recycling stage, roles and duties, and accountability and obligations of manufacturers, consumers, and recycling facilities to dispose of the batteries at the end of their useful lives.
6. Reduce waste, conserve resources, and promote a greener and more sustainable future by giving used EV batteries a second life.
7. Educate customers about proper battery disposal. Increasing consumer awareness of the value of properly disposing of and recycling batteries helps create a more sustainable end-of-life battery management strategy.

Concluding Remarks
In summary, it emphasizes an urgent need for comprehensive and effective sustainability policies and regulations to address the challenges of recycling electric vehicle (EV) batteries. The complexities arising from chemical composition, the lack of infrastructure, and regulatory gaps pose significant barriers. Clear roles, technology-neutral guidelines, and active involvement of authorities are highlighted as essential components. Despite the importance of recycling regulations, challenges such as high costs and energy-intensive processes exist. Consumer education and engagement are crucial for the success of recycling initiatives. Collaborative efforts among authorities, manufacturers, and consumers are imperative to develop and implement sustainable and responsible recycling frameworks for EV batteries. This is crucial for mitigating environmental impact, fostering innovation, and ensuring responsible end-of-life management in the era of electric mobility.

References
[1] Kuhlman, T.; Farrington, J. What is Sustainability? Sustainability 2010, 2, 3436-3448. https://doi.org/10.3390/su2113436

[2] Anne P.M. Velenturf, Phil Purnell, Principles for a sustainable circular economy, Sustainable Production and Consumption, Volume 27, 2021, Pages 1437-1457. https://doi.org/10.1016/j.spc.2021.02.018

[3] Perera, F. Pollution from Fossil-Fuel Combustion is the Leading Environmental Threat to Global Pediatric Health and Equity: Solutions Exist. Int. J. Environ. Res. Public Health 2018, 15, 16. https://doi.org/10.3390/ijerph15010016

[4] Anenberg, S.; Miller, J.; Henze, D.; Minjares, R. A global snapshot of the air pollution-related health impacts of transportation sector emissions in 2010 and 2015.

[5] Pavlínek, P. Transition of the automotive industry towards electric vehicle production in the east European integrated periphery. Empirica 50, 35–73 (2023). https://doi.org/10.1007/s10663-022-09554-9

[6] https://www.imagine.art/, Text to image with AI Art Generator, accessed on 24-March-2024.

[7] Gao, Z., Xie, H., Yang, X., Zhang, L., Yu, H., Wang, W., Liu, Y., Xu, Y., Ma, B., Liu, X., & Chen, S. (2023). Electric vehicle lifecycle carbon emission reduction: A review. Carbon Neutralization, 2(5), 528-550. https://doi.org/10.1002/cnl2.81

[8] Peters, D. R., Schnell, J. L., Kinney, P. L., Naik, V., & Horton, D. E. (2020). Public Health and Climate Benefits and Trade-Offs of U.S. Vehicle Electrification. GeoHealth, 4(10). https://doi.org/10.1029/2020GH000275

[9] Jamil, M., Ahmad, F., & Jeon, Y. (2016). Renewable energy technologies adopted by the UAE: Prospects and challenges – A comprehensive overview. Renewable and Sustainable Energy Reviews, 55, 1181-1194. https://doi.org/10.1016/j.rser.2015.05.087

[10] Bustinza, O. F., Vendrell-Herrero, F., & Chiappetta Jabbour, C. J. (2024). Integration of product-service innovation into green supply chain management: Emerging opportunities and paradoxes. Technovation, 130, 102923. https://doi.org/10.1016/j.technovation.2023.102923

[11] https://www.statista.com/outlook/mmo/electric-vehicles/worldwide#unit-sales, online statistics portal that provides access to a vast array of data and insights across various industries and topics, accessed on 24-March-2024.

[12] Wellbrock, W., Ludin, D., Röhrle, L., & Gerstlberger, W. (2020). Sustainability in the automotive industry, importance of and impact on automobile interior – insights from an empirical survey. International Journal of Corporate Social Responsibility, 5(1), 1-11. https://doi.org/10.1186/s40991-020-00057-z

[13] https://www.iea.org/data-and-statistics/charts/overall-supply-and-demand-of-lithium-for-batteries-by-sector-2016-2022, The International Energy Agency (IEA) is a prominent organization focused on energy policy and analysis, accessed on 24-March-2024.

[14] Costa, C., Barbosa, J., Gonçalves, R., Castro, H., Campo, F. D., & Lanceros-Méndez, S. (2021). Recycling and environmental issues of lithium-ion batteries: Advances, challenges and opportunities. Energy Storage Materials, 37, 433-465. https://doi.org/10.1016/j.ensm.2021.02.032

[15] Kang, Z., Huang, Z., Peng, Q., Shi, Z., Xiao, H., Yin, R., Fu, G., & Zhao, J. (2023). Recycling technologies, policies, prospects, and challenges for spent batteries. IScience, 26(11), 108072. https://doi.org/10.1016/j.isci.2023.108072

[16] Neumann, J., Petranikova, M., Meeus, M., Gamarra, J. D., Younesi, R., Winter, M., & Nowak, S. (2022). Recycling of Lithium-Ion Batteries—Current State of the Art, Circular Economy, and Next Generation Recycling. Advanced Energy Materials, 12(17), 2102917. https://doi.org/10.1002/aenm.202102917

[17] Slattery, M., Dunn, J., & Kendall, A. (2021). Transportation of electric vehicle lithium-ion batteries at end-of-life: A literature review. Resources, Conservation and Recycling, 174, 105755. https://doi.org/10.1016/j.resconrec.2021.105755

[18] Albertsen, L., Richter, J. L., Peck, P., Dalhammar, C., & Plepys, A. (2021). Circular business models for electric vehicle lithium-ion batteries: An analysis of current practices of vehicle manufacturers and policies in the EU. Resources, Conservation and Recycling, 172, 105658. https://doi.org/10.1016/j.resconrec.2021.105658

[19] Skeete, J., Wells, P., Dong, X., Heidrich, O., & Harper, G. (2020). Beyond the EVent horizon: Battery waste, recycling, and sustainability in the United Kingdom electric vehicle transition. Energy Research & Social Science, 69, 101581. https://doi.org/10.1016/j.erss.2020.101581

[20] Xie, Y., Yu, H., & Changdong, L. (2016). Research on systems engineering of recycling ev battery. https://doi.org/10.2991/ameii-16.2016.291

[21] Loganathan, M., Anandarajah, G., Tan, C., Msagati, T., Das, B., & Hazarika, M. (2022). Review and selection of recycling technology for lithium-ion batteries made for ev application - a life cycle perspective. Iop Conference Series Earth and Environmental Science, 1100(1), 012011. https://doi.org/10.1088/1755-1315/1100/1/012011

[22] Wang, S. (2022). Multi-angle analysis of electric vehicles battery recycling and utilization. Iop Conference Series Earth and Environmental Science, 1011(1), 012027. https://doi.org/10.1088/1755-1315/1011/1/012027

[23] Luo, C., Zhang, Z., Qiao, D., Lai, X., Li, Y., & Wang, S. (2022). life prediction under charging process of lithium-ion batteries based on automl. Energies, 15(13), 4594. https://doi.org/10.3390/en15134594

Renewable energy sources, sustainability aspects and climate alteration: A comprehensive review

M. Amin Mir[1,a*], M. Waqar Ashraf[1,b], Kim Andrews[1,c]

[1]Department of Mathematics & Natural Sciences, Prince Mohammad Bin Fahd University, Al-Khobar, Saudi Arabia

[a]mmir@pmu.edu.sa, [b]mashraf@pmu.edu.sa, [c]kandrews@pmu.edu.sa

Keywords: Sustainability, Renewable Energy, Climate, Future, Environment

Abstract. The increasing global demand for energy is transforming our world into a closely connected community, yet the Earth remains unchanged in its capacity. As the world population seeks more energy to fuel social, economic, and developmental needs, along with health and well-being, the call for sustainable solutions intensifies. Unfortunately, escalating energy consumption contributes to rising greenhouse gas emissions and environmental harm. Embracing renewable energies becomes crucial for combating climate change, but such a shift must be sustainable to fulfill the energy requirements of future generations. A comprehensive strategy combining energy management and renewable sources is required to address these issues. An overview of current energy consumption trends, energy management techniques, and renewable energy sources is provided in this article. The results show that an integrated strategy that includes renewable energy sources and energy management techniques can dramatically reduce energy consumption and greenhouse gas emissions while also providing economic benefits. The article's conclusion highlights how important it is to implement an integrated strategy for energy management and renewable energy sources in order to achieve efficient and sustainable energy use.

Introduction

Global energy consumption has surged due to increased urbanization, industrialization, and population growth, causing adverse environmental effects such as climate change, air pollution, and resource depletion [1]. An integrated strategy including energy management techniques and renewable energy sources is necessary to address these issues [2]. An extensive analysis of energy use, energy management techniques, and renewable energy sources is given in this article. Analyzing recent studies, case studies, and assessing the efficacy of different energy management and renewable energy technologies are all part of the study technique [3]. Energy is essential for economic expansion, but the present fossil fuel dependency is unsustainable, requiring a move to renewable alternatives in order to lessen environmental effects [4]. Energy consumption and greenhouse gas emissions can be considerably decreased while providing economic advantages by putting energy management concepts into practice and switching to renewable energy sources. Strategies include using energy-efficient technologies, adopting practices, and employing management systems for monitoring and controlling energy usage [5]. Renewable sources like solar, wind, hydro, and geothermal provide sustainable alternatives, with global adoption increasing due to government incentives and declining technology costs [6]. Energy consumption, particularly in sectors like transportation, residential, commercial, and industrial, contributes significantly to greenhouse gas emissions. The International Energy Agency (IEA) reported a 2.3% global energy consumption increase in 2019, with the transportation sector leading at 32%, followed by residential (23%), commercial (12%), and industrial (37%) sectors. Fossil fuels, constituting coal, oil, and natural gas, accounted for 84% of global energy consumption in 2019 [7].

Sources of Renewable Energy

Renewable energy emerges as a pivotal solution to address these challenges [8]. Notably, in 2012, these sources contributed 22% to global energy generation, indicating a significant shift. Reliable energy supply is indispensable for heating, lighting, industry, and transportation, playing a crucial role in global economies. The substitution of fossil fuels with renewables substantially reduces greenhouse gas emissions. However, challenges like intermittent generation due to seasonal variations exist, necessitating intricate design and optimization methods. Fortunately, advancements in computer hardware and software empower researchers to overcome these challenges, fostering progress in the renewable and sustainable energy field [9].

Technology and Renewable Energy
Renewable energy sources derive from continual natural energy flows in our environment: bioenergy, solar, geothermal, hydropower, wind, and ocean energy.

Hydro energy. One important energy source that is obtained from the flow of water from higher to lower elevations is hydropower, which is mainly used to turn turbines and produce electricity. There are many different types of hydropower projects, such as in-stream projects, run-of-river projects, and dam projects with reservoirs. Hydropower is a technologically advanced resource that uses a variable resource over time. Hydropower reservoirs are used for navigation, irrigation, drinking water production, flood and drought control, and other uses. The main source of energy for hydropower is gravity plus the height at which the water descends onto the turbine. The potential energy relies on the mass of the water, the gravity factor ($g = 9.81$ ms^{-2}), and the head, defined as the difference between the dam and tail water levels. Turbines are engineered to accommodate an optimal water flow. Hydropower exhibits minimal particulate pollution, rapid upgradability, and the capability to store energy for extended periods.

Energy of Biomass. Bioenergy, a renewable energy source sourced from biological materials, plays a vital role in various applications. It serves as a versatile energy provider for transportation through biodiesel, electricity generation, and heating for cooking. Electricity derived from bioenergy encompasses diverse sources like wood residues from forests, agricultural byproducts such as sugar cane waste, and animal husbandry residue like cow dung. A notable advantage lies in the fact that the fuel for biomass-based electricity often originates from by-products or waste, avoiding competition between land designated for food and that for fuel. While global biofuel production is currently modest, it exhibits a continuous upward trend. In the United States, annual biodiesel consumption reached 15 billion liters in 2006, with a growth rate of 30–50% annually, aiming for 30 billion liters by the end of 2012 [10].

Solar Power. "Direct" solar energy pertains to renewable energy technologies that directly harness the Sun's energy. Unlike some renewables like wind and ocean thermal, which utilize solar energy after its absorption on Earth, solar energy technologies directly capture sunlight. Photovoltaic (PV) systems convert solar irradiance into electricity, and concentrating solar power (CSP) generates thermal energy. Solar energy not only fulfills direct lighting needs but also has the potential to produce fuels for transportation and other purposes [11]. The World Energy Council (2013) notes that solar radiation falling on Earth exceeds 7,500 times the world's annual primary energy consumption of 450 EJ [9].

Geothermal power. Geothermal power is harnessed from the interior of the earth as a natural source of energy, rooted in the planet's internal structure and associated physical processes. Despite substantial heat existing in the Earth's crust, it is often not equally distributed, hardly concentrated, and frequently lies at depths challenging for mechanical exploitation. The geothermal gradient, averaging about 30 °C/km, varies across the Earth's interior, with some regions attainable by digging exhibiting gradients well above average [12]. Geothermal reservoirs, mined for heat through wells and other methods, include naturally hot and permeable hydrothermal reservoirs and enhanced geothermal systems (ESG), which are sufficiently hot but benefit from hydraulic stimulation. Extracted fluids, varying in temperature, can then be used for electricity generation and other applications requiring heat energy [9].

Wind energy. Wind has emerged as a leading global energy source among renewables due to its widespread presence, especially in areas with substantial energy density [13]. Harnessing kinetic energy from moving air, wind energy is vital for mitigating climate change by generating electricity through large turbines positioned onshore or offshore [14]. Large-scale production and implementation of onshore wind technology have previously occurred [9]. Wind turbines efficiently convert wind energy into electricity, marking a significant stride in renewable energy solutions.

Ocean energy (tide and wave). Surface waves form when wind moves over water, particularly in the ocean. The duration, speed, and distance of sustained wind directly impact wave height and the energy produced. The ocean possesses vast energy potential stored in waves, tides, currents, and temperature differences. In 2008, the first commercial sea energy devices debuted with installations like the UK's SeaGen and Portugal's Pelamis. Currently, ocean energy areas are derived through wind, tides, waves, and thermal disparities between deep and shallow sea water [15].

Sustainable development and Renewable energy
Because it promotes economic productivity and human growth, renewable energy is essential to sustainable development. Prospects for energy security, social and economic advancement, increased energy accessibility, reducing the effects of climate change, and improving the environment and human health are presented by these energy sources [16].

Energy security. The concept of energy security lacks a universally agreed-upon definition, leading to varied interpretations. Nevertheless, the underlying concern for energy security revolves around ensuring a continuous and reliable energy supply, a fundamental requirement for sustaining economic operations [17]. Given the intrinsic connection between economic growth and energy consumption, maintaining stable energy access is a significant challenge for both developed and developing nations, posing potential economic and functional challenges in case of prolonged disruptions [15]. Renewable energy sources, unlike fossil fuels, are globally distributed and less subject to market trading. Introducing renewables not only reduces dependence on energy imports but also diversifies the supply portfolio, lessening vulnerability to price volatility and fostering global energy security. Additionally, renewable energy integration enhances the reliability of energy services, particularly in areas with inadequate grid access, contributing to overall energy security through a well-managed and diversified energy portfolio [16].

Economic and Social development. Economic growth and rising energy consumption have historically been strongly correlated, making the energy sector essential to economic development. Per capita income and energy consumption are positively correlated globally, indicating that economic expansion is the main cause of the recent increase in energy consumption. This growth also generates employment, with a 2008 study indicating around 2.3 million jobs worldwide in renewable energy technologies, contributing to enhanced health, education, gender equality, and environmental safety [9].

Energy access. Sustainable Development Goal Seven focuses on ensuring universal access to clean, affordable, and available energy. The use of renewable energy sources, widely distributed globally, is key to achieving this goal. Addressing access discrepancies requires a local understanding, particularly in regions like sub-Saharan Africa and South Asia, where urban and rural electrification differences are evident [18]. Renewable energy-based distributed grids prove more competitive in rural areas, offering substantial opportunities for mini-grid systems to enhance electricity access.

Effect of Climate change on Environmental and health. The renewable energy source utilization in power generation plays a pivotal role in curbing the emission of greenhouse gases, mitigating climate change, and lessening effect on environment and health associated with fossil fuel-derived pollutants. GHG emissions per capita also saw a 22% reduction from 1990 to 2012, as shown in Figure 3 (EEA, 2016). As mentioned carbon dioxide emissions in the United States from 1990 to 2013, exemplifying a decrease in CO_2 levels due to a transition from fossil fuels to available energy sources [19].

Renewable Energy Source Challenges
In low-carbon economies, renewable energy sources have the potential to dominate the energy supply, requiring significant adjustments to all energy systems. The main problem of the first half of the twenty-first century is generally recognized to be the shift from non-sustainable to renewable energy [20]. Notably, a country's policies and instruments significantly impact the adoption of renewable energy, influencing costs and technological advancements. Technological innovations, in turn, influence costs, contributing to market failures and limited adoption. An effective renewable energy policy must consider these interconnected factors to foster sustainability.

The study proposes several policy recommendations to effectively alleviate changes in climate and its impacts.

- Encourage all fields and areas to invest in technologies of renewable energy and adopt policies promoting their use, fostering a collective effort in reducing carbon emissions.
- Advocate for lifestyle and behavioral changes to reduce individual carbon footprints, emphasizing the significant contribution of personal choices to climate change mitigation.
- Support research into revolution and mechineries that minimize land use, prevent accidents associated with renewable energy sources, and address resource competition, particularly in bioenergy where food production competes with energy generation.
- Strengthen international collaboration and assistance for developing countries, facilitating infrastructure expansion and technology upgrades to enable modern and sustainable energy services. This approach aims to mitigate climate change and its adverse effects on a global scale.

Conclusion
Energy is a fundamental requirement in our daily lives, crucial for human development, economic growth, and productivity. Shifting towards renewable energy is recognized as a significant step in mitigating climate change, but its sustainability is paramount for securing a future that meets energy needs for generations to come. The goal of this study is to ascertain the sustainability of renewable sources and their potential to mitigate the effects of climate change by examining the relationship between sustainable development and green energy. Qualitative research was conducted through a comprehensive review of relevant literature within the study's scope. While renewable energy sources exhibit no net emissions throughout their lifecycle, barriers such as cost, pricing structures, political environments, and market conditions hinder their full utilization in emerging, minimum-developed, and developed nations. To address these challenges, the study advocates for the creation of global opportunities through international interactions. Supporting developing nations in accessing green energy, enhancing energy productivity, investing in clean energy utilization, and encouraging research and energy infrastructure to lower the cost of renewable energy and remove inefficiencies, and contribute to climate change mitigation. Opportunities related to renewable energy sources are identified in the report, including enhanced energy security, social and economic development, and climate change mitigation. However, obstacles like communication gaps, market failures, raw material availability issues, and inefficient energy use pose a danger to renewable energy's sustainability and ability to mitigate climate change.

Future Recommendations
Policy formulation and technological improvement: Encourage policy discussions across sectors to enhance technologies in the renewable sector for sustained development.
- Advocate for more efficient energy use at individual and global levels. Implement global energy efficiency programs, offering tax exemptions for energy-efficient initiatives and product designs.
- Invest in research to address concerns and potential risks associated with renewable energy.
- Raise public knowledge and education about effect reduction, adaptation, and mitigation of climate change.

In addition to addressing the sustainability of renewable energy, putting these recommendations into practice would support the seventh and thirteenth goals of sustainable development, which call for preventing climate change and its effects as well as guaranteeing that everyone has access to affordable, dependable, and sustainable energy.

References

[1] Tomin, Nikita, et al. "Design and optimal energy management of community microgrids with flexible renewable energy sources." Renewable Energy 183 (2022): 903-921. https://doi.org/10.1016/j.renene.2021.11.024

[2] Kartal, Mustafa Tevfik, et al. "Do nuclear energy and renewable energy surge environmental quality in the United States? New insights from novel bootstrap Fourier Granger causality in quantiles approach." Progress in Nuclear Energy 155 (2023): 104509. https://doi.org/10.1016/j.pnucene.2022.104509

[3] Tiwari, Sunil, Joanna Rosak-Szyrocka. Internet of things as a sustainable energy management solution at tourism destinations in India." Energies 15.7 (2022): 2433. https://doi.org/10.3390/en15072433

[4] Natarajan, B., Obaidat, M.S., Sadoun, B., Manoharan, R. New Clustering-Based Semantic Service Selection and User Preferential Model. IEEE Systems Journal, (200).

[5] Sadiq, Muhammad, et al. "The influence of economic factors on the sustainable energy consumption: evidence from China." Economic research-Ekonomska istraživanja 36.1 (2023): 1751-1773. https://doi.org/10.1080/1331677X.2022.2093244

[6] M. Amin Mir. Adsorption of Heavy Metals by Chopped Human Hair: An Equilibrium and Kinetic Study. *Asian Journal of Chemistry.* 35(6). (2023). 1458 – 1462. https://doi.org/10.14233/ajchem.2023.27653

[7] Wang, Shuguang, Luang Sun, and Sajid Iqbal. "Green financing role on renewable energy dependence and energy transition in E7 economies." Renewable Energy 200 (2022): 1561-1572. https://doi.org/10.1016/j.renene.2022.10.067

[8] Tiwari, G. N., & Mishra, R. K. Advanced renewable energy sources. Royal Society of Chemistry. (2011). P007-P015. https://doi.org/10.1039/9781849736978-FP007

[9] Banos, R., Manzano-Agugliaro, F., Montoya, F., Gil, C., Alcayde, A., & Gómez, J. (2011). Optimization methods applied to renewable and sustainable energy: A review. Renewable and Sustainable Energy Reviews, 15, 1753–1766. https://doi.org/10.1016/j.rser.2010.12.008

[10] Ayoub, M., & Abdullah, A. Z. Critical review on the current scenario and significance of crude glycerol resulting from biodiesel industry towards more sustainable renewable energy

industry. Renewable and Sustainable Energy Reviews, 16, (2012). 2671–2686. https://doi.org/10.1016/j.rser.2012.01.054

[11] Edenhofer, O., Pichs-Madruga, R., Sokona, Y., Seyboth, K., Matschoss, P., Kadner, S., ... von Stechow, C. Renewable Energy Sources and Climate Change Mitigation. Cambridge: Cambridge University Press. (2011). https://doi.org/10.1017/CBO9781139151153

[12] Barbier, E. Geothermal energy technology and current status: An overview. Renewable and Sustainable Energy Reviews, 6, (2002). 3–65. https://doi.org/10.1016/S1364-0321(02)00002-3

[13] M. Amin Mir, Waqar Ashraf M. The challenges and potential strategies of Saudi Arabia's water Resources: A review in analytical way. *Environmental Nanotechnology, Monitoring and Management,* 20 (2023). https://doi.org/10.1016/j.enmm.2023.100855

[14] Asumadu-Sarkodie, S., & Owusu, P. A. The potential and economic viability of wind farms in Ghana Energy Sources, Part A: Recovery, Utilization, and Environmental Effects. (2016). https://doi.org/10.1080/15567036.2015.1122680

[15] Esteban, M., & Leary, D. Current developments and future prospects of offshore wind and ocean energy. Applied Energy, 90, 128–136. (2012). https://doi.org/10.1016/j.apenergy.2011.06.011

[16] Asumadu-Sarkodie, S., & Owusu, P. ACarbon dioxide emissions, GDP, energy use and population growth: A multivariate and causality analysis for Ghana, (2016). 1971–2013. Environmental Science and Pollution Research International. https://doi.org/10.1007/s11356-016-6511-x

[17] Kruyt, B., van Vuuren, D. P., de Vries, H., & Groenenberg, H. Indicators for energy security. Energy Policy, 37, (2009). 2166–2181. https://doi.org/10.1016/j.enpol.2009.02.006

[18] Brew-Hammond, A. Energy access in Africa: Challenges ahead. Energy Policy, 38, (2010). 2291–2301. https://doi.org/10.1016/j.enpol.2009.12.016

[19] United States Environmental Protection Agency. Carbon dioxide emissions.2, (2015) http://www3.epa.gov/climatechange/ghgemissions/ gases/co2.html

[20] Frederick Verbruggen, Adam R Aron, Michaël A Stevens, Christopher D Chambers. Theta burst stimulation dissociates attention and action updating in human inferior frontal cortex. Proc Natl Acad Sci U S A, (2010) 3;107(31):13966-71. https://doi.org/10.1073/pnas.1001957107

Potential uses of renewable energy in construction: Advantages and challenges

Essam ZANELDIN[1,a*], Waleed AHMED[2,b], Betiel PAULOS[3,c], Feruz GABIR[3,d], Makda ARAYA[3,e], El Bethel MULUYE[3,f] and Deborah DEBELE[3,g]

[1]Associate Professor, Department of Civil and Environmental Engineering, United Arab Emirates University, Al-Ain, United Arab Emirates

[2]Assistant Professor, Engineering Requirements Unit, United Arab Emirates University, Al-Ain, United Arab Emirates

[3]Student, College of Engineering, United Arab Emirates University, Al-Ain, United Arab Emirates

[a]essamz@uaeu.ac.ae, [b]w.ahmed@uaeu.ac.ae, [c]201950002@uaeu.ac.ae, [d]201950008@uaeu.ac.ae, [e]201950013@uaeu.ac.ae, [f]201950311@uaeu.ac.ae, [g]201950305@uaeu.ac.ae

Keywords: Construction Industry, Green Buildings, Renewable Energy, Sustainability

Abstract. The construction industry accounts for a high percentage of the total global energy consumption, placing it among the main sectors contributing to climate change, pollution, and energy-related problems. This fact has placed tremendous pressure on the construction industry to find solutions to this crucial problem and shift to more sustainable, energy-efficient, and cost-effective construction practices. In this study, the importance of using renewable energy in the construction sector, particularly building construction, is highlighted and a review of some emerging practices in using renewable energy in construction is presented. The paper also presents the various sources of renewable energy and their applications in construction along with their advantages and drawbacks. The paper highlights the importance of establishing standards and regulations related to the use of renewable energy in building projects.

Introduction

With the exponential growth of population in the world coupled with the increased needs of humans, the global consumption of energy is growing rapidly at an average annual rate of 2.2% [1, 2] and the construction industry accounts for around 40% of the total global energy consumption and 30% of global carbon dioxide emissions, placing it among the main sectors responsible for air pollution and environmental instability [1]. As a result of the rapid growth of population, more buildings will be needed and, if bult using the old traditional construction methods, the result will be more carbon emissions and environmental instability, such as the greenhouse effect and the extreme weather caused by energy. This has stimulated the importance of using green, low-carbon, sustainable, and other forms of renewable energy in construction. Serving the same purpose, the International Energy Agency has set a goal of net zero emissions by 2050, placing the construction industry under intense pressure to achieve this target [3]. To this end, a number of countries are announcing initiatives to achieve net zero emissions by 2050. However, the world is still behind in terms of the implementation of a clear and well-defined policy to achieve this objective, particularly in the construction industry. Europe and the USA, for example, have redefined regulations and policies related to the development of near-zero-energy buildings for the development of renewable energy [4, 5]. China is also committed to reaching peak carbon by 2030 and carbon neutrality by 2060 [6]. With the construction sector being a main player in this context, the application of renewable energy in construction to produce energy-efficient buildings by using natural materials has become a major driver to reduce the contribution of the building sector to climate change and energy use and promote sustainability [7].

It is well established in the literature that natural materials are good sources of renewable energy since they are green, environmentally friendly, and serve as an alternative to traditional energy sources [7, 8]. The use of renewable energy in construction projects becomes eminent as it promotes sustainability and reduces environmental impacts. This is clearly important considering the increasing need to reduce carbon emissions by mitigating the impact of construction activities on the environment. In building construction, renewable energy is an integration of sustainable sources of energy, such as water, wind, solar, plants, biomass, and geothermal in the building life cycle stages, including design, construction, and operation and maintenance to reduce the use of traditional sources of energy and, therefore, mitigate climate change and promote the environmental sustainability of buildings [7, 9, 10]. With the increasing awareness of the use of renewable energy in building construction, its application in modern buildings has also gained momentum. Architects, for example, have designed buildings with proper orientation to facilitate the use of natural sunlight for heating and ventilating [11, 12]. Other engineers involved in the design of buildings are also using sustainable sources in their designs such as solar panels to generate electricity, natural fibers and materials to replace traditional construction materials, and biomass boilers and heating systems to provide sustainable heating and hot water solutions contributing to energy efficiency and reducing carbon emissions. However, the application of renewable energy in buildings depends on the type of the source of energy and the characteristics of energy. As reported by Khan and Al-Ghamdi [9] and Wu and Skye [10], renewable energy sources such as solar, geothermal, wind, and biomass energy have the potential to satisfy the sustainable energy needs of buildings.

Potential Uses of Sustainable Renewable Energy Sources in Construction
The use of renewable energy sources has become a viable solution to the problem of air pollution and environmental instability resulting from construction projects. In this section, the most important renewable energy sources that can be utilized in the construction of buildings to promote sustainability and reduce environmental impact are presented. As shown in Fig. 1, renewable energy sources include solar energy, biomass energy, geothermal energy, wind energy, and hydro energy, in addition to the hybrid renewable energy systems.

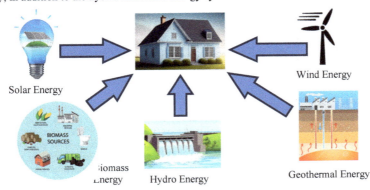

Fig. 1: Major Sources of Renewable Energy Used in Building Construction.

Solar Energy
It is well established that solar energy is the most commonly used source of renewable energy [7]. It is generated for building construction using a range of technologies such as solar power to generate electricity and solar thermal energy, including solar water heating. Solar energy technologies are either passive or active, depending on how they capture and distribute solar energy and convert it into solar power. Active techniques include the use of photovoltaic systems, concentrated solar power, and solar water heating to generate energy while passive techniques include the optimal building orientation towards the sun, selection of materials with favorable thermal mass or light dispersing properties, thermal biomass, and the design of spaces with natural air circulation [8]. Solar energy is known for its environmentally friendly attributes and unlimited supply, placing it among the leading and widely used sources of renewable energy in the world [13]. In the USA, for example, solar energy accounts for 31% of the total energy consumption [10].

Several researchers have addressed the use of solar energy to improve the environmental performance of buildings and facilitate climate circulation. Vassiliades et al. [14], for example, noticed that the use of solar energy in buildings helps reduce the negative impacts of building on the climate and the use of photovoltaic technology improves the energy utilization efficiency and reduces the consumption of energy. Another study by [15] showed that the use of solar systems in buildings can increase the renewable energy factor to 83% and reduce energy demand by 48%. It is, therefore, important to maximize the use of solar energy in buildings during the design phase as this will result in improved efficiency, reduce the operating costs, and enhance the functionality of buildings, in addition to its main benefit of being environmentally friendly as it does not produce greenhouse emissions. Despite all these advantages, the use of solar energy results in high maintenance costs and requires technological advancements.

Biomass Energy
Biomass is an organic material extracted from living organisms such as plants, animals, or microorganisms. Biomass is the oldest renewable energy source used by humans [16]. To lower emissions and reduce dependence on fossil fuels, biomass is usually utilized in the form of biomaterials in structural or non-structural elements of buildings. Biomass energy relies, in general, on natural resources such as wood, plant fibers, and organic waste materials, coming from human, plant, and animal wastes. Biomass includes other materials such as construction waste and animal excreta, which can also be used to generate electricity [9]. Several research efforts investigated the use of biomass energy in buildings. Rahman et al. [17] investigated the possibility of using biomass energy as the main source of power for a residential building. Allouhi et al. [18] and Wu and Skye [10] indicated that biomass can have different uses in buildings including utilizing it as gas, fuel, heat, and power generation. In addition, biomass is carbon–neutral efficient source of energy, and its efficiency can be further enhanced by compressing biomass wood into pellets under high pressure and temperature [19]. However, biomass combustion can lead to corrosion on heating surfaces due to boiler deposits [7]. While woodchip boilers improve local air quality and reduce local ground-level particulate matter concentrations, it may not be as environmentally friendly as compared to natural gas boilers. In addition to its use in buildings as a source of energy, biomass can also be used as a thermal insulation material or as a structural element in building construction, providing a sustainable and green alternative to some traditional building components, and improving energy supply chains by reducing dependence on imported fossil fuels [20].

Despite all the benefits resulting from the use of biomass in buildings, it is important to note that some biomass materials may be less durable and less resistant to some factors such as moisture, fire, and pests as compared to conventional building materials. This is in addition to the fact that biomass materials have limited availability and may vary in performance and quality [21]. To address these concerns, additional treatment and protection measures may be required.

Geothermal Energy
The stored internal heat of earth contributes to the generation of geothermal energy, which is a renewable source of energy that is not dependent on climate or time of day and can supply energy all day long, independently of external conditions [6]. Geothermal energy is mainly used for heat production and cooling and can work in combination with other energy systems, such as solar energy. Geothermal energy systems can improve energy efficiency while reducing energy costs and greenhouse gas emissions, as compared to traditional heating and cooling systems. A study by [22] confirmed that the use of geothermal energy significantly reduces energy demand, cost, and CO_2 emissions, as compared to conventional gas boilers, demonstrating its effective contribution to achieve the goal of net-zero-energy buildings. In addition to these advantages, geothermal systems require a relatively small land area, operate quietly without the noise generated by traditional HVAC systems, and provide design flexibility as they can be used with different architectural designs. Despite the mentioned advantages of geothermal energy systems, their installation cost is relatively high and the underground site conditions determine if a geothermal system is feasible or not [7]. It is, therefore, important to conduct a precise study on the subsurface conditions of the site to investigate the costs associated with installing a geothermal energy system for a building. This is in addition to the environmental impact of installing and operating geothermal systems, which are resulting from the noise associated with the installation of the systems and the treatment of geothermal fluids.

Wind Energy
The use of wind energy in buildings is considered one of the most widely used source of renewable energy sources [23]. The wind energy system consists of wind turbines, mechanical energy, heat pumps, and other required energy using wind vortex machines [7]. The most important advantage of using wind energy, as a source of renewable energy, is to reduce carbon emissions and consumption of energy. Statistics indicate that, as of 2017, the use of wind energy has reduced greenhouse gas emissions by at least 600 million tons [7]. According to [24], wind energy can provide around 15% of buildings' energy needs. To best utilize wind energy, architects design buildings in a way that they use natural ventilation for air circulation through natural wind power, which reduces the use of the air conditioning systems and, consequently, reduces energy consumption.

However, equipment needed to generate wind energy require high initial investment and maintenance cost, as compared with traditional energy sources. Another concern is the noise generated by the wind turbines, in addition to the fact that buildings layout limits the use of wind energy since the wind direction and speed between buildings may be affected by turbulence and blocking, which results in a reduced efficiency of generating the wind power [24]. In addition, the uncertain characteristics of wind make the power generated by wind irregular. It is, therefore, important to explore other new technologies to improve the efficiency of wind energy.

Hydro Energy
Hydropower is an energy generated by water. Hydropower is derived from the kinetic energy of falling or flowing water. It can be generated from streams, lakes and rivers or man-made structures such as dams, lagoons and reservoirs. It relies on the water cycle, which is driven by the sun, making it a clean and renewable source of energy. Hydropower is used to generate low-cast electricity, provide flood control, support irrigation, and produce clean drinking water. provides low-cost electricity and durability over time compared to other sources of energy. While

the initial construction cost of hydropwoer systems can be costly, the cost can be reduced by using existing structures such as bridges, tunnels, and dams. However, the amount of energy extracted from the water depends on the available water volume and the difference in height between the turbines and the elevated source, known as the hydraulic head.

Hybrid Renewable Energy Systems
Hybrid energy systems combine different energy technologies to reduce costs, reduce greenhouse gas emissions, and improve capability, value, energy efficiency, or environmental performance of buildings as compared with the use of independent renewable energy systems. Hybrid energy systems combine multiple sources of energy with traditional electricity to meet the energy demands of buildings. Examples of hybrid energy systems include combining solar energy with wind energy, solar energy with hydrogen energy, wind energy with hydrogen energy, geothermal energy with hydrogen energy, etc. Hydrogen hybrid energy, for example, can be used for heating houses, supplying hot water, cooking, and meeting electricity needs. The solar-hydrogen hybrid system, for example, is considered the most efficient system of generating renewable energy. As the benefits of hybrid energy systems become more widely recognized, they are becoming increasingly popular in residential commercial and buildings to reduce energy costs and to promote sustainable construction.

Summary and Concluding Remarks
The use of renewable energy in buildings directly addresses the increasing demand of energy worldwide and mitigates the crucial concern of global warming. This paper presents the various types of renewable energy and their application in construction, including energy sources such as solar, biomass, geothermal, wind, and hydro energy, in addition to the hybrid sources of renewable energy. It is quite obvious that each of these sources offer sustainable and environmentally friendly advantages that enhance building energy efficiency and reduce operational costs. However, it is highly recommended to formulate standards and regulations related to renewable energy that can be followed by construction practitioners. The implementation of such standards and regulations will encourage the development of renewable energy in the construction industry and promote sustainability and innovation. It is also important to note that the application of hybrid renewable energy systems in buildings can provide designers and constructors with alternative and cost-effective renewable energy solutions. This study provides construction practitioners with information about the various renewable energy systems that can be used in the design and construction of buildings along with their benefits, challenges, and disadvantages.

References
[1] M. Nazari-Heris, A.T. Esfehankalateh, P. Ifaei, Hybrid energy systems for buildings: a techno-economic-enviro systematic review, energies. 16(12) (2022). https://doi.org/10.3390/en16124725

[2] R.A. Salam, K.P. Amber, N.I. Ratyal, M. Alam, N. Akram, C.Q.G. Munoz, F.P.G. Marquez, An overview on energy and development of energy integration in major South Asian Countries: the building sector, Energies. 13 (2000). https://doi.org/10.3390/en13215776

[3] S. Zhang, P. Oclon, J.J. Klemes, P. Michorczyk, K. Pielichowska, K. Pielichowski, Renewable energy systems for building heating, cooling and electricity production with thermal energy storage, Renew Sustain Energy Rev. 165 (2022) 112560. https://doi.org/10.1016/j.rser.

[4] B. Liu, D. Rodriguez, Renewable energy systems optimization by a new multi-objective optimization technique: a residential building, J Build Eng. 35 (2021) 102094. https://doi.org/10.1016/j.jobe.

[5] M. Yang, L. Chen, J. Wang, G. Msigwa, A.I. Osman, S. Fawzy, D.W. Rooney, P-S Yap, Circular economy strategies for combating climate change and other environmental issues, Environ Chem Lett. (2022). https://doi.org/10.1007/s10311-022-01499-6

[6] A.I. Osman, L. Chen, M. Yang, G. Msigwa, M. Farghali, S. Fawzy, D.W. Rooney, P-S Yap, Cost, environmental impact, and resilience of renewable energy under a changing climate: a review, Environ Chem Lett 21 (2023) 741–764. https://doi.org/10.1007/s10311-022-01532-8

[7] L. Chen, Y. Hu, R. Wang, X. Li, Z. Chen, J. Hua, A. I. Osman, M. Farghali, L. Huang, J. Li, L. Dong, D.W. Rooney, P-S Yap, Green building practices to integrate renewable energy in the construction sector: a review, Environmental Chemistry Letters. (2023). https://doi.org/10.1007/s10311-023-01675-2

[8] S. Dey, A. Sreenivasulu, G.T.N. Veerendra, K.V. Rao, P.S. Babu, Renewable energy present status and future potentials in India: overview, Innov Green Dev. (2022). https://doi.org/10.1016/j.igd.2022

[9] S.A. Khan, S.G. Al-Ghamdi, Renewable and integrated renewable energy systems for buildings and their environmental and socio-economic sustainability assessment, In: (ed) Springer. (2021) 127–144. https://doi.org/10.1007/978-3-030-67529-5_6

[10] W. Wu, H.M. Skye, Residential net-zero energy buildings: review and perspective, Renew Sustain Energy Rev. 142 (2021). https://doi.org/10.1016/j.rser.2021.110859

[11] Q. Gong, F. Kou, X. Sun, Y. Zou, J. Mo, X. Wang, Towards zero energy buildings: a novel passive solar house integrated with flat gravity-assisted heat pipes, Appl Energy. 306 (2022). https://doi.org/10.1016/j.apenergy.2021.117981

[12] C. Ionescu, T. Baracu, G.E. Vlad, H. Necula, A. Badea, The historical evolution of the energy efficient buildings, Renew Sustain Energy Rev. 49 (2015) 243–253. https://doi.org/10.1016/j. rser.2015.04.062

[13] S.R. Aldhshan, K.N. Maulud, W.S. Jaafar, O.A. Karim, B. Pradhan, Energy consumption and spatial assessment of renewable energy penetration and building energy efficiency in Malaysia: a review, Sustainability. 13 (2021). https://d oi.org/10.3390/su13169244

[14] C. Vassiliades, A. Savvides, A. Buonomano, Building integration of active solar energy systems for facades renovation in the urban fabric: effects on the thermal comfort in outdoor public spaces in Naples and Thessaloniki, Renew Energy. 190 (2022) 30–47. https://doi.org/10.1016/j.renene.2022.03.094

[15] M. Bilardo, M. Ferrara, E. Fabrizio, Performance assessment and optimization of a solar cooling system to satisfy renewable energy ratio (RER) requirements in multi-family buildings, Renew Energy. 155 (2020) 990–1008. https://doi.org/10.1016/j.renene.2020.03.044

[16] C. Yang, H. Kwon, B. Bang, S. Jeong, U.D. Lee, Role of biomass as low-carbon energy source in the era of net zero emissions, Fuel. 328 (2022). https://doi.org/10.1016/j.fuel.2022.125206

[17] H. Rahman, M.R. Sharif, R. Ahmed, T. Nijam, M.A. Shoeb, Designing of biomass-based power plant for residential building energy system, International conference on electrical engineering and information communication technology (ICEEICT), 2015, pp. 1–6. https://doi.org/10.1109/ICEEICT.2015.73073 66

[18] A. Allouhi, S. Rehman, M. Krarti, Role of energy efficiency measures and hybrid PV/biomass power generation in designing 100% electric rural houses: a case study in Morocco, Energy Build. 236 (2021). https://doi.org/10.1016/j.enbuild.2021.110770

[19] H. Hartmann, V. Lenz, Biomass energy heat provision in modern small-scale systems. In: Kaltschmitt M (ed) Springer New York, New York, NY, 2019, pp. 533–586. https://doi.org/10.1007/978-1-4939-7813-7_248

[20] A. Behzadi, E. Thorin, C. Duwig, S. Sadrizadeh, Supply-demand side management of a building energy system driven by solar and biomass in Stockholm: a smart integration with minimal cost and emission. Energy Convers Manag. 292 (2023). https://doi.org/10.1016/j.enconman.2023.117420

[21] M. Hiloidhari, M.A. Sharno, D.C. Baruah, A.N. Bezbaruah, Green and sustainable biomass supply chain for environmental, social and economic benefits, Biomass Bioenerg. 175 (2023). https://doi.org/10.1016/j.biombioe.2023.106893

[22] D. D'Agostino, F. Minichiello, F. Petito, C. Renno, A. Valentino, Retrofit strategies to obtain a NZEB using low enthalpy geothermal energy systems. Energy. 239 (2022). https://doi.org/10.1016/j.energy.2021.122307

[23] Z. Zhang, X. Liu, D. Zhao, S. Post, J. Chen, Overview of the development and application of wind energy in New Zealand, Energy Built Environ. 4 (2023) 725–742. https://doi.org/10.1016/j.enbenv.2022.06.009

[24] K.C. Kwok, G. Hu, Wind energy system for buildings in an urban environment. J Wind Eng Ind Aerodyn. 234 (2023). https://doi.org/10.1016/j.jweia.2023.105349

Adaptive cooling framework for Photovoltaic systems: A seasonal investigation under the terrestrial conditions of Sharjah, UAE

Mena Maurice FARAG[1,a*], Abdul-Kadir HAMID[1,b*], Mousa HUSSEIN[2,c]

[1]Department of Electrical Engineering, College of Engineering, University of Sharjah, Sharjah, UAE

[2]Department of Electrical and Communication Engineering, College of Engineering, United Arab Emirates University, Al Ain, UAE

[a]u20105427@sharjah.ac.ae, [b]akhamid@sharjah.ac.ae, [c]mihussein@uaeu.ac.ae

Keywords: Adaptive Cooling, Photovoltaic Systems, Solar Energy, Temperature Regulation, Photovoltaic-Thermal Applications

Abstract. The increment of PV operating temperature has a significant impact on the overall efficiency, longevity, and degradation of PV systems. In this notion, researchers have sought to investigate different cooling methodologies to minimize the impact of abrupt operating temperatures. However, most investigations discuss the impact of the working base fluid in small periods without proposing methods for temperature regulation based on defined thresholds. In this study, an adaptive cooling framework is proposed through thermal and electrical modeling to examine the cooling effect on a 2.88 kW grid-connected PV system installed in Sharjah, UAE. An operating temperature threshold of 55°C is considered based on the annual average operating temperature, to facilitate adaptive cooling. The framework is modeled based on heat transfer thermodynamic laws and implemented on MATLAB using experimentally driven measurements collected from the above-mentioned system for December, March, June, and September. As a result, the proposed framework has presented notable merits in terms of electrical and thermal characteristics across the four different seasons. The highest heat extraction was observed in September, where a reduction of 25.36% was observed in PV operating temperature, showing the effectiveness of temperature regulation in harsh weather conditions. As a result, the electrical characteristics have improved significantly leading to an 8.79%, 6.39%, and 6.58% enhancement in maximum power output, maximum voltage, and electrical efficiency, respectively.

Introduction

The continuous depletion of conventional energy resources has led to the introduction of clean and renewable energy resources (RES) to tackle the escalating consequences of climate change [1–3]. Solar energy has been popular amongst other RES due to its reliability, abundance, and zero-net fuel dependence, hence leading the way for the development of photovoltaic (PV) systems for clean electricity generation [4].

PV modules generate electrical energy based on two environmental factors, solar irradiance, and ambient temperature [5]. Typically, there is a direct proportionality for power with respect to solar irradiance [6], while there is an indirect proportionality with respect to the ambient temperature [7]. The PV cell performance is heavily dependent on environmental conditions, specifically operating temperature [8]. The increment of the PV module operating temperature dramatically reduces the overall efficiency of the PV panel. In this notion, cooling of PV modules is essential to sustain the PV system's longevity and performance. The extraction of dissipated heat from PV modules is critical to enhancing the operation of the PV cell [9]. The employment of water as a working base fluid has been a popular and viable cooling technique amongst other available techniques due to its feasibility on different system levels [10].

Several studies in the literature have reported the enhancement in PV module performance through the significant reduction in PV module operating temperature. A study reported by [11], investigated the impact of temperature of the PV cell on its performance through a range of 25°C – 60°C while maintaining a constant solar irradiance exposure. A laboratory-based experiment also conducted in [12], investigated the impact of operating temperature from a range of 25°C – 55°C while maintaining a 1000 W/m² solar irradiance. A simulation study reported by [13], combined artificial neural networks (ANN) and capabilities of reduced module operating temperature. The study demonstrated that with a reduction of 10°C in module operating temperature, the number of PV modules can be reduced by 12 % while supplying the same electrical capacity. An experimental study reported in [14], investigated the impact of water as a working base fluid for front surface heat extraction on a large-scale PV system. The study reported an increase of 9.76% in output power generation and a maximum efficiency of up to 13.47% while generating a thermal difference of 2.3°C. A study investigated under both laboratory and experimental conditions is reported in [15]. The validation of the effectiveness of water as a working base fluid was observed as a reduction of 24 K in PV module operating temperature was observed. As a result, the output power was enhanced by 10% as compared to the uncooled PV module. The employment of water as the cooling technique was experimentally investigated during July as reported in [16]. The study reported that a 10% increase in PV output power was observed due to the reduction of 20% in PV surface temperature, thus improving the PV module conversion efficiency by 14%. Additionally, the optimization of the water cooling process during the harsh climatic conditions in the UAE is reported [17]. The experimental investigation proposes an adaptive technique based on a defined 55°C temperature threshold to facilitate automated front surface cooling. As a result, a decrement of 17.3°C in PV module operating temperature was observed with an extracted heat of 15.41°C. In this notion, the output power and output voltage increased by 39.21% and 16.31%, respectively achieving a 12.5% electrical conversion efficiency. Another study reported the employment of automated timers for the automatic pumping of water for the cooling process as reported in [18]. Under summer conditions of the UAE, the experimental study is conducted to reveal the performance of a timed cooling operation on PV module performance. As a result, the average output power increased by 1.6 %, while decreasing the operating temperature by 6% on average. Additionally, the investigation of a small-scale controlled water spraying system was investigated by [19]. The self-cleaning methodology observed an improved electrical efficiency of 2.53% and an overall system efficiency of 83.3%.

The scientific literature explored the significant impact of deploying water as a working base fluid to enhance the performance of PV systems, especially in harsh and arid weather conditions. However, most investigations report the operation of water as a working base fluid for a specific day of the year, while the performance may vary throughout the entire year. The concept of investigating water as a working base fluid for different seasons of the year as well under defined temperature thresholds needs to be highlighted, to optimize the cooling procedure.

In this notion, this study presents a modeling approach to characterize the thermal properties of water as a working base fluid, illustrating the cooling effect and its impact on the PV module's electrical and thermal properties. The developed model is based on a 55°C temperature threshold, which operates as a gradient for facilitating an adaptive cooling effect, to optimize the cooling process when operating temperature is critical. Moreover, the study considers a 2.88 kW grid-connected PV system as a practical case study installed in Sharjah, United Arab Emirates, to model the impact of adaptive front surface cooling across four different seasons based on experimental measurements retrieved from the system, presenting variable weather conditions.

Dependence of electrical performance on thermal parameters

Environmental parameters have a prominent impact on the electrical conversion efficiency of PV systems, and their irregularity leads to deviation and instability in their overall performance. The United Arab Emirates is blessed with large exposure to solar irradiance of an average of 2285 kWh/m² annually, presenting the promise of PV system deployment [20]. However, 20% of sunlight is converted and generated into electricity, while the remainder is dissipated as waste heat leading to degradation in PV system performance and lifespan. PV cell technologies are defined with a power temperature coefficient (β), that demonstrates a degradation rate in %/°C with every 1°C above the standard testing conditions (STC) i.e., 25°C.

The dramatic impact of temperature can be observed during the summer periods, when the PV module operating temperature may reach up to 70°C. Such abrupt operating temperature leads to a significant reduction in output power generation. A comparison between summer and winter temperature profiles for July and December, respectively as illustrated in Fig. 1.

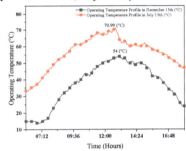

Fig. 1. PV module operating temperature profile comparison between summer and winter seasons for July 15th and December 15th.

In this notion, critical electrical parameters have been defined based on theoretical and mathematical interpretations as reported in [21], demonstrating the impact of temperature on maximum power output, maximum voltage output, and the electrical conversion efficiency for PV modules.

The maximum power output (P_{mpp}) at a single interval is demonstrated by Eq. 1 as follows:

$$P_{mpp} = P_{STC} \times \frac{G}{G_{STC}} \times \left[1 + \beta_{STC}\left(T_{PV} - T_{STC}\right)\right] \tag{1}$$

where P_{STC} is the power prescribed by the manufacturer at standard testing conditions (STC), G is the solar irradiance measured at actual conditions, G_{STC} is the solar irradiance at STC conditions, which coincides to 1000 W/m², β_{STC} represents the power temperature coefficient that typically represents a negative decrement in power by %/1°C as the operating PV temperature surpasses the prescribed STC conditions, i.e., 25°C. This metric is usually mentioned by the PV manufacturer and is dependent on the T_{PV} representing the measured operating PV module temperature and T_{STC} is the operating PV module temperature at STC conditions, which coincides to 25°C. The power temperature coefficient typically varies between various PV technologies as previously discussed in Table 1.

Similarly, the maximum voltage output (V_{mpp}) at a single interval is demonstrated by Eq. 2 as follows:

$$V_{mpp} = V_{STC} \times \frac{G}{G_{STC}} \times \left[1 + \gamma_{STC}\left(T_{PV} - T_{STC}\right)\right] \tag{2}$$

where V_{STC} is the voltage prescribed by the manufacturer at STC conditions, which coincides with the open circuit voltage (V_{oc}), γ_{STC} represents the V_{oc} temperature coefficient that typically represents a negative decrement in voltage by %/1°C as the operating PV temperature surpasses the prescribed STC conditions, i.e., 25°C.

Furthermore, the electrical conversion efficiency (η) at a single interval is demonstrated by Eq. 3 as follows

$$\eta = \eta_{STC} \times \left[1 + \beta_{STC}\left(T_{PV} - T_{STC}\right)\right] \tag{3}$$

where η is the electrical efficiency prescribed by the manufacturer at STC conditions.

Research Methodology
Experimental State-of-the-art

A 2.88 kW grid-connected PV system is established at the University of Sharjah main campus on the rooftop of the W-12 building (Lat. 25.34° N; Long. 55.42° E) as illustrated in Fig. 2. The system operates through a real-time data acquisition system, for instantaneous access of system data [22]. The system was previously discussed for its operation both in small and large periods, showing superior data recording capability and complete infrastructure [23]. In addition, the system operates as a complete grid-connected PV system, injecting AC electrical energy to the 3-phase local utility grid. Moreover, the on-grid PV system records electrical and environmental parameters at 5-minute intervals.

Fig. 2. Demonstration of (a) Established on-grid PV system (b) Power and Communication Loop

Adaptive Cooling Framework model

Based on the previously discussed state-of-the-art, a model is proposed and simulated through MATLAB, to depict the characteristics of water as a working base fluid based on the annual measured data retrieved from the real-time data acquisition hub. Water is selected as a working fluid due to its exceptional heat transfer characteristics and feasible implementation. The constant thermophysical properties are considered to show a uniform effect across all months, except with the change in inlet water temperature as per the given month. The months selected for data variability are December, March, June, and September demonstrating the variability in PV module operating temperature, validating the impact of the proposed temperature regulation technique.

The thermal modeling of water as a working base fluid is defined based on a set of thermodynamic laws, to depict the behavior of water. The heat absorption energy (Q) is defined by Eq. 4 as follows:

$$Q = m \times c \times \Delta T = m \times c \times \left(T_{PV} - T_{in}\right) \tag{4}$$

Where m is the mass of the water in kg, c is the specific heat of the water in J/kg.k, and ΔT is the temperature difference between the operating temperature and the water temperature in the Inlet tank assumed by its respective weather conditions.

The mass of water (m) is computed based on Eq. 5 as follows:

$$m = \rho \times A \times d \tag{5}$$

Where ρ is the density of water in kg/m³, A is the surface area of the PV system in m², and d is the thickness of the water film in millimeters (mm).

Therefore, the temperature reduction ($T_{reduction}$) is computed through Newton's law of cooling as depicted in Eq. 6 as follows:

$$T_{reduction} = \frac{Q}{h \times A \times t} \tag{6}$$

Where h represents the convective heat transfer coefficient in W/m².K, and t represents the time in seconds over which cooling occurs. The value considered is 250 W/m².K since it is typical for harsh weather conditions such as the UAE, where it typically ranges from 50 – 300 W/m².K.

In this notion, the thermal model proposed will simulate the cooling effect under a temperature threshold of 55°C, which is selected based on the annual average temperature. The parameters of the thermal model are selected and presented in Table 1, which is based on the experimental conditions and location of Sharjah, UAE.

Table 1. Selected thermal parameters for the proposed adaptive cooling framework

Description	Value
Specific Heat [J/kg.K]	4186
Density [kg/m³]	1000
Convective heat transfer coefficient [W/m²/k]	250
Surface Area of PV system [m²]	18.5
Thickness of water film [mm]	1
Time over which cooling occurs [s]	19.55
Average water temperature for cooling per month [°C]	December: 24.5 March: 32.5 June: 34.5 September: 34.5

Results and Discussion

The experimental data measured from the on-grid PV system is considered across an entire day from 6 AM to 6 PM during four distinct months. In principle, the thermodynamic laws discussed are implemented using MATLAB to assess the performance of the PV system across four distinct months both electrically and thermally, through the MATLAB script development.

The PV module operating temperature is the primary figure of merit to be assessed based on the proposed framework. Based on the computations discussed above, longer periods of cooling are required during June and September as compared to December and March. Thus, an average PV operating temperature reduction of 20.43% and 25.35% is observed during June and September, respectively. The reduction in operating temperature is illustrated in Fig 3 and a summary of the temperature reduction is presented in Table 2, presenting the notable heat extraction during all four seasons using the above-mentioned thermodynamic laws.

Moreover, the relative decrease in operating temperature has led to an increment in the maximum power point. The maximum power output as computed in Eq. 1, presents an average increment of 6.3% and 8.8% in June and September, respectively, as demonstrated in Fig. 4.

The relative enhancement is also observed across other parameters such as the maximum operating voltage and electrical conversion efficiency as computed by Eqs. (2)-(3), presenting the adaptability of temperature regulation across different seasons for maintaining PV system lifetime and longevity. Thus, the proposed adaptive cooling framework proves viability when implemented practically on large-scale PV systems. The performance assessment of the proposed adaptive cooling framework is presented in Table 3, providing improved electrical performance across all seasons of the year, and maintaining high electrical efficiency at longer periods.

Table 2. Summary of operating temperature assessment based on proposed adaptive cooling framework for large-scale on-grid PV system

Description	Period of Adaptive Cooling [Hours]	Operating Temperature [°C]		
		Uncooled Average	Cooled Average	Average Decrement
December	10:40 AM – 2:20 PM	41.05	32.78	20.13%
March	11:05 AM– 2:05 PM	45.24	40.98	9.41%
June	9:10 AM – 4:20 PM	54.24	43.16	20.43%
September	8:55 AM – 3:50 PM	55.32	41.29	25.36%

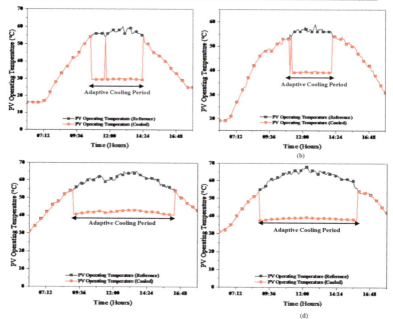

Fig. 3. Comparative assessment of operating temperature reduction during adaptive cooling process for large-scale on-grid PV system during selected days from (a) December (b) March (c) June (d) September

Table 3. Summary of electrical characteristics based on proposed adaptive cooling framework for large-scale on-grid PV system

Description	Maximum Power Point [W]			Maximum Operating Voltage [V]			Electrical Conversion Efficiency [%]		
	Uncooled Average	Cooled Average	Average Increment	Uncooled Average	Cooled Average	Average Increment	Uncooled Average	Cooled Average	Average Increment
December	1411.73	1497.88	6.10%	168.12	175.67	4.49%	15.42%	15.96%	3.50%
March	1656.48	1700.73	2.67%	197.39	201.26	1.96%	15.21%	15.41%	1.31%
June	1553.70	1651.51	6.30%	186.71	195.29	4.59%	14.51%	15.26%	5.17%
September	1661.28	1807.34	8.79%	200.35	213.16	6.39%	14.44%	15.39%	6.58%

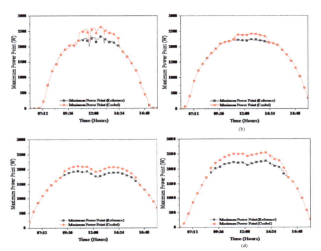

Fig. 4. Comparative assessment of M$_{PP}$ during adaptive cooling process for large-scale on-grid PV system during selected days from (a) December (b) March (c) June (d) September

Summary

This paper proposed a modeling framework for facilitating adaptive cooling for photovoltaic systems, for enhancing system efficiency and longevity. The proposed framework was implemented using MATLAB and experimentally driven environmental conditions, to simulate the thermal and electrical characteristics of a 2.88 kW grid-connected PV system in Sharjah, UAE. A temperature regulating threshold of 55°C was considered, to simulate the system characteristics for December, March, June, and September, providing a wide range of variability in PV module operating temperatures. The highest operating temperature reduction was observed in September with 25.36%, showing the effectiveness of temperature regulation in harsh weather conditions. As a result, the electrical characteristics have improved significantly leading to an 8.79%, 6.39%, and 6.58% enhancement in maximum power output, maximum voltage, and electrical efficiency, respectively. As a future work, the temperature threshold response time will be validated under experimental conditions and incorporation of advanced hardware. Additionally, the utilization of machine learning techniques to optimize the necessary thermal parameters will be investigated for improved energy efficiency and usage.

References

[1] N. Novas, R.M. Garcia, J.M. Camacho, A. Alcayde, Advances in Solar Energy towards Efficient and Sustainable Energy, Sustainability. 13 (2021) 6295. https://doi.org/10.3390/su13116295

[2] S. Paraschiv, L.S. Paraschiv, Trends of carbon dioxide (CO2) emissions from fossil fuels combustion (coal, gas and oil) in the EU member states from 1960 to 2018, Energy Reports. 6 (2020) 237–242. https://doi.org/10.1016/j.egyr.2020.11.116

[3] M.M. Farag, R.A. Alhamad, A.B. Nassif, Metaheuristic Algorithms in Optimal Power Flow Analysis: A Qualitative Systematic Review, Int. J. Artif. Intell. Tools. 32 (2023). https://doi.org/10.1142/S021821302350032X

[4] E. Kabir, P. Kumar, S. Kumar, A.A. Adelodun, K.-H. Kim, Solar energy: Potential and

future prospects, Renew. Sustain. Energy Rev. 82 (2018) 894–900. https://doi.org/10.1016/j.rser.2017.09.094

[5] M.M. Farag, N. Patel, A.-K. Hamid, A.A. Adam, R.C. Bansal, M. Bettayeb, A. Mehiri, An Optimized Fractional Nonlinear Synergic Controller for Maximum Power Point Tracking of Photovoltaic Array Under Abrupt Irradiance Change, IEEE J. Photovoltaics. 13 (2023) 305–314. https://doi.org/10.1109/JPHOTOV.2023.3236808

[6] M.M. Farag, F. Faraz Ahmad, A.K. Hamid, C. Ghenai, M. AlMallahi, M. Elgendi, Impact of Colored Filters on PV Modules Performance: An Experimental Investigation on Electrical and Spectral Characteristics, in: 50th Int. Conf. Comput. Ind. Eng., 2023: pp. 1692–1704.

[7] T. Salameh, A.K. Hamid, M.M. Farag, E.M. Abo-Zahhad, Energy and exergy assessment for a University of Sharjah's PV grid-connected system based on experimental for harsh terrestrial conditions, Energy Reports. 9 (2023) 345–353. https://doi.org/10.1016/j.egyr.2022.12.117

[8] N.K. Almarzooqi, F.F. Ahmad, A.K. Hamid, C. Ghenai, M.M. Farag, T. Salameh, Experimental investigation of the effect of optical filters on the performance of the solar photovoltaic system, Energy Reports. 9 (2023) 336–344. https://doi.org/10.1016/j.egyr.2022.12.119

[9] T. Salameh, A.K. Hamid, M.M. Farag, E.M. Abo-Zahhad, Experimental and numerical simulation of a 2.88 kW PV grid-connected system under the terrestrial conditions of Sharjah city, Energy Reports. 9 (2023) 320–327. https://doi.org/10.1016/j.egyr.2022.12.115

[10] H.M.S. Bahaidarah, A.A.B. Baloch, P. Gandhidasan, Uniform cooling of photovoltaic panels: A review, Renew. Sustain. Energy Rev. 57 (2016) 1520–1544. https://doi.org/10.1016/j.rser.2015.12.064

[11] S. Chander, A. Purohit, A. Sharma, Arvind, S.P. Nehra, M.S. Dhaka, A study on photovoltaic parameters of mono-crystalline silicon solar cell with cell temperature, Energy Reports. 1 (2015) 104–109. https://doi.org/10.1016/j.egyr.2015.03.004

[12] M. Al-Maghalseh, Experimental study to investigate the effect of dust, wind speed and temperature on the PV module performance, Jordan J. Mech. Ind. Eng. 12 (2018) 123–129.

[13] H. Attia, K. Hossin, Integrated Renewable PV System through Artificial Neural Network Based MPPT and Water Cooling Treatment, in: 2019 Int. Conf. Electr. Comput. Technol. Appl., IEEE, 2019: pp. 1–5. https://doi.org/10.1109/ICECTA48151.2019.8959581

[14] M.M. Farag, F.F. Ahmad, A.K. Hamid, C. Ghenai, M. Bettayeb, M. Alchadirchy, Performance Assessment of a Hybrid PV/T system during Winter Season under Sharjah Climate, in: 2021 Int. Conf. Electr. Comput. Commun. Mechatronics Eng., IEEE, 2021: pp. 1–5. https://doi.org/10.1109/ICECCME52200.2021.9590896

[15] K. Sornek, W. Goryl, R. Figaj, G. Dąbrowska, J. Brezdeń, Development and Tests of the Water Cooling System Dedicated to Photovoltaic Panels, Energies. 15 (2022) 5884. https://doi.org/10.3390/en15165884

[16] M. Al-Odat, Experimental Study of Temperature Influence on the Performance of PV/T Cell under Jordan Climate Conditions, J. Ecol. Eng. 23 (2022) 80–88. https://doi.org/10.12911/22998993/152283

[17] M.M. Farag, A.K. Hamid, Performance assessment of rooftop PV/T systems based on adaptive and smart cooling facility scheme - a case in hot climatic conditions of Sharjah, UAE, in: 3rd Int. Conf. Distrib. Sens. Intell. Syst. (ICDSIS 2022), Institution of Engineering and Technology, 2022: pp. 198–207. https://doi.org/10.1049/icp.2022.2448

[18] H. Attia, K. Hossin, M. Al Hazza, Experimental investigation of photovoltaic systems for

performance improvement using water cooling, Clean Energy. 7 (2023) 721–733. https://doi.org/10.1093/ce/zkad034

[19] K. Mostakim, M.R. Akbar, M.A. Islam, M.K. Islam, Integrated photovoltaic-thermal system utilizing front surface water cooling technique: An experimental performance response, Heliyon. 10 (2024) e25300. https://doi.org/10.1016/j.heliyon.2024.e25300

[20] M.M. Farag, R.C. Bansal, Solar energy development in the GCC region – a review on recent progress and opportunities, Int. J. Model. Simul. 43 (2023) 579–599. https://doi.org/10.1080/02286203.2022.2105785

[21] D.L. Evans, L.W. Florschuetz, Terrestrial concentrating photovoltaic power system studies, Sol. Energy. 20 (1978) 37–43. https://doi.org/10.1016/0038-092X(78)90139-1

[22] M.M. Farag, F.F. Ahmad, A.K. Hamid, C. Ghenai, M. Bettayeb, Real-Time Monitoring and Performance Harvesting for Grid-Connected PV System - A Case in Sharjah, in: 2021 14th Int. Conf. Dev. ESystems Eng., IEEE, 2021: pp. 241–245. https://doi.org/10.1109/DeSE54285.2021.9719385

[23] M.M. Farag, A.K. Hamid, Experimental Investigation on the Annual Performance of an Actively Monitored 2.88 kW Grid-Connected PV System in Sharjah, UAE, in: 2023 Adv. Sci. Eng. Technol. Int. Conf., IEEE, 2023: pp. 1–6. https://doi.org/10.1109/ASET56582.2023.10180880

A novel approach to address reliability concerns of wind turbines

Sorena ARTIN[1,a*]

[1]Prince Mohammad Bin Fahd University, Al Khobar, Eastern Province, Saudi Arabia

[a]sartin@pmu.edu.sa

Keywords: Wind Turbines, Reliability Analysis, Random Variables, Renewable Energy

Abstract. Designing and manufacturing a system in the current industrial world cannot be accomplished without addressing safety related issues. For this purpose, system reliability is a powerful tool to ensure that failure probability of the system is below an accepted level while the system is operational. A commonly used approach to deal with these considerations is to define a performance function for the system in order to investigate its reliability. In this case, renewable energy systems (RESs) are not different. When a wind turbine, as a RES, is designed, its reliability cannot be ignored or underestimated. Therefore, stable and efficient models are needed to make sure that the turbine remains operational and is able to safely generate electricity power. In this paper, a new approach is proposed to set up a reliability analysis model for the wind turbines. The introduced model takes two important factors, i.e. the wind speed and the wind angle, and their probability distributions into account. These two factors are indeed considered as random variables to design a new system performance function and set up the new model in order to investigate wind turbine's reliability.

Introduction

Climate change has recently attracted attentions around the world. This has made many researchers and organizations to focus on renewable energy systems (RESs) and their related issues. However, fossil fuels still play important roles, either in financial markets and/or international relationships, as RESs, such as wind turbines and solar panels, are not yet efficient enough to replace fossil fuels in a foreseeable future [1, 2].

There are always challenges when designing every single RES, regarding their economic feasibility, safety checks, engineering design, reliability features, etc. These challenges need to appropriately be addressed as otherwise they could bring catastrophic consequences, like shutting down entire renewable energy station [3].

Renewable energy resources often have random behavior. For instance, there needs to be irradiance and wind available and this availability depends on randomness of the resources. Data mining and machine learning techniques are often employed to predict this availability in order to investigate the system's behavior in the future [4, 5].

Moreover, the Internet of Things (IoT) has been used to oversee solar power stations [4]. As these panels often need a vast area to install large-scale systems, it is crucial to find the right place and also analyze functionality of the panels in real-time tasks. To do this, there are some algorithms which employ the IoT techniques to improve these systems' reliability. Also, a widely applied technique is to estimate their parameters in order to formulate deterministic optimization models of solar panels. A method has been introduced that uses stochastic fractal search algorithms to estimate current-voltage parameters for modelling purposes [6].

Also, various data mining methods are used to predict solar cell's energy production. However, a new finding proposes a hybrid method incorporating machine learning algorithms and statistical methods is going to perform better in terms of PV energy predictions [4, 5, 7].

Other RESs have their own associated challenges and risks. However, the main challenges to generate electricity power by wind turbines are again related to randomness of the main factor; i.e.

the wind. It has been reported that wind turbines have great potentials to generate green energy and help environment, as they produce low emission compared with traditional fossil fuels. On the other hand, there are new barriers to research and industry in order to design these systems due to randomness and intermittent characteristics associated with these turbines. As a result of these challenges, uncertainty level is high that needs to be addressed by introducing efficient reliability analysis models and methods [8].

Further, many researches work on offshore wind power stations as another way to produce renewable energy. It has been found that many untouched offshore locations are rich in terms of resources of available wind. Among available structures for offshore wind turbine stations, fixed structures, such as monopile, jacket and tripod support are often preferred. These structure however do have uncertainty problems. It has been reported that uncertainty of these systems can be considered by using semi-probabilistic nature of the existing frameworks [9].

There are also researches that have found offshore wind turbines would bring more reliability related consideration into energy production process. It has been reported that the Gaussian process regression is of the most stable and efficient methods to handle these problems [10]. Although it is believed that offshore wind resources have great potential for energy production, these stations often require huge maintenance costs which means it could be very challenging to deal with these systems and therefore there are projects have been abandoned due to the financial problems [11].

Wind Turbines' Reliability Considerations
Moving towards renewable energy has motivated researchers to work on mathematical models of RESs. It is reported that hybrid renewable energy systems (HRESs) could be more efficient than single renewable source systems. However, more challenging problems are expected when working with a HRES. This often happens due to a large number of factors that need to be considered to set up HRESs' mathematical models.

Reliability related issues are of the highest importance when it comes to developing new systems. In this case, reliability concerns of RESs are mainly originating from random nature of their required natural resources [12, 13]. A system's reliability is often studied by employing a reliability function which can be defined by the system's failure rate λ as below [14]:

$$R(t) = e^{-\lambda t} \qquad (1)$$

Reliability analysis problems are another tool to investigate a system's reliability. For this purpose, a reliability analysis problem, as an optimization problem, needs to be solved in order to find a reliability index. Analytical and simulation approaches are widely used to perform a reliability analysis problem, either by using a mathematical model to evaluate system reliability or by applying simulation techniques to approximate the reliability index [15].

A wind energy conversion system (WECS) can also be used to formulate a wind turbine's reliability analysis problem [12]. The wind speed is a major factor in this system and so its probability distribution needs to be determined. It has been reported that the Weibull distribution is an appropriate choice for this purpose, which is often shown as a matrix representing time. α_w and β_w, as a scale parameter and a shape parameter respectively, are applied in the Weibull density and distribution functions.

$$f_v(v) = \frac{\beta_w}{\alpha_w^{\beta_w}} v^{\beta_w - 1} e^{-\frac{v}{\alpha_w^{\beta_w}}} \qquad (2)$$

$$F_v(v) = 1 - e^{-\frac{v}{\alpha_w^{\beta_w}}} \qquad (3)$$

As random variables play significant roles in reliability analysis problems, it would be crucial to find these variables for a wind turbine reliability model. It is reported that wind speed and wind angle are two random variables of a reliability analysis problem of wind turbines [16]. To work out these variables, two boundaries of the wind speed must be considered as cut-in V_{ci} and cut-off speeds V_{co}. If the wind speed was below V_{ci} or above V_{co}, then wind turbine would stop working (either due to the lack of enough energy or because of safety concerns) [12, 13, 17]. The power generated by the turbine can be worked out as:

$$P_w(v) = \begin{cases} 0 & if\ v < V_{ci} \\ \dfrac{v - V_{ci}}{V_R - V_{ci}} P_R & if\ V_{ci} \leq v \leq V_R \\ P_R & if\ V_R \leq v \leq V_{co} \\ 0 & if\ V_{co} < v \end{cases} \quad (4)$$

The coefficient $\dfrac{v-V_{ci}}{V_R-V_{ci}}$ could be replaced by a quadratic polynomial and multiplied into the rated capacity [18].

Wind Turbines' Performance Function
A system performance function, which is also known as a limit-state function, is often employed to set up a reliability analysis problem and a reliability-based design optimization (RBDO) problem [19]. Once random variables of a wind turbine were determined, they would be used to define the system performance function. A probabilistic constraint as $P[G(t) \leq 0] \leq \Phi(-\beta)$ is then formulated in an RBDO problem to take reliability concerns into account.

This performenc function is then used to setting up a reliability analysis problem for the wind turbine. A first-order direct reliability analysis problem is defined as below [20]:

$$\begin{aligned} &\min \|(u_1, u_2, \dots, u_n)\| \\ &s.t.\ G_U(u_1, u_2, \dots, u_n) = 0 \end{aligned} \quad (5)$$

where $(u_1, u_2, \dots, u_n)$ represents standard normalized design variable and G_U is standard normalized system performance function. It has been reported that the conjugate gradient direction-based (CGDB) method is the most stable and efficient method to solve this problem and find a reliability index β [21].

Probability distribution and also their relevant density function are the basis of all reliability analysis problems. In this case, for power generated by the WECS, the following density function is available, in which F_v shows the probability distribution function of the WECS [12]:

$$f_{P_w}(P_w) = \begin{cases} F_1 = 1 - F_3 & if\ P_w = 0 \\ F_2 & if\ V_{ci} \leq v \leq V_R \\ F_3 & if\ P_w = P_R \end{cases} \quad (6)$$

where

$$F_2 = \frac{V_R - V_{ci}}{P_R} \frac{\beta_w}{\alpha_w^{\beta_w}} (V_{ci} + (V_R - V_{ci}) \frac{P_w}{P_R})^{\beta_w - 1} e^{-(\frac{V_{ci} + (V_R - V_{ci})\frac{P_w}{P_R}}{\alpha_w})^{\beta_w}}$$

and

$$F_3 = F_v(V_F) - F_v(V_{ci})$$

It is found that the random variable v in the above function can follow two probability distribution functions as the Weibull distribution [12] or the Normal distribution [22]. In the next section, it will be shown how to set up a reliability analysis problem of wind turbines based on the random variables and probability distributions discussed in this paper.

Introducing Reliability Analysis Problems for Wind Turbines
To investigate reliability concerns of a wind turbine, a performance function should first be defined. It is elaborated here how to formulate such a system performance function based on two previously discussed random variables and then a new approach is proposed to set up new reliability analysis problems for wind turbines.

Random Variables
The first task is to determine random variables. The wind speed and the wind angle are considered in this paper as two random variables of a wind turbine using which a performance function can be formulated. Once the performance function was available, then a reliability analysis problem can be set up.

It is required to find (or identify) probability distribution function(s) of each random variable. It is already known that the wind turbine, as the first random variable, may follow two probability distributions, either the Gaussian distribution (also known as the Normal distribution) or the Weibull distribution. Data collected in different geographical locations prove that these two distributions are appropriate choices to explain behavior of the wind speed.

The second random variable would be the wind angle. However, no valid data is yet available for this variable to determine its probability distribution function. Therefore, it remains a big challenge to determine the wind angle's distribution in order to set up a wind turbine's reliability analysis problem.

System Safety and Failure
The cut-in speed V_{ci} and the cut-off speed V_{co} should also be considered when setting up the reliability analysis problem. Considering V_{ci} and V_{co} as two thresholds and given the wind speed v must be between these thresholds, the system safety and system failure could be defined for a wind turbine.

Based on the Eq. 4, a turbine is supposed to work properly when $V_{ci} \leq v \leq V_{co}$. Therefore, this condition is assumed as system's safety which means the turbine is able to generate electricity power only if the wind speed is between its boundaries. However, when the wind speed is out of the above-mentioned range, i.e. $V_{co} < v$ or $v < V_{ci}$, then the turbine stops working and no power could be generated, and so it would be considered as system failure.

However, it must be mentioned that total failure of the system, which means the turbine needs maintenance or replacement, is different from the system failure defined in this Section.

For simplicity reasons, let's assume x is a random variable representing the wind speed. Also, d_1 and d_2 are the thresholds of random variable x, which represent the cut-in and cut-off speeds, respectively. Then, the discussed safety and failure conditions can be used to formulate the wind turbine's performance function(s) as below:

$$G_1(x) = x - d_1 \tag{7}$$

$$G_2(x) = d_2 - x \tag{8}$$

Therefore, $G_1(x) > 0$ and $G_2(x) > 0$ show the system safety which means electricity power can be generated, and so $G_1(x) < 0$ and $G_2(x) < 0$ indicate system failure meaning no power can be generated.

So, the failure surface of a reliability analysis problem (such as Eq. 5) or the probabilistic constraint of an RBDO problem for wind turbines can be set up as $G_1(x) = 0$ and $G_2(x) = 0$.

Future Works and Conclusion
Reliability has always been one of the most important considerations when designing any new system, and in this case, the RESs in general and wind turbines in particular are not different. In this paper, new approaches are proposed to deal with reliability issues of wind turbines.

The main focus here is to identify random variables of the system, i.e. a wind turbine, and then employ them to define a performance function for the turbine. This performance function can then be used to design a direct reliability analysis problem, or an inverse reliability analysis problem, or even an RBDO problem.

Based on the discussions on reliability issues in the previous Section, it can be concluded that an important step in the future works to setting up the required problem(s) is to figure out probability distribution of the wind angle, as a random variable of the system. The other random variable is the wind speed for which at least two probability distributions are available. Once probability distribution and statistical data of both random variables were available, then they can be used to set up a system performance function.

The wind speed characteristics, such as the cut-in and cut-off speeds, and the power function can also be used to define a performance function for a wind turbine. In all these cases, the wind turbine's performance function should be set up based on the safety conditions to ensure electricity power can be generated.

Once a system performance function was formulated using any of the above approaches, a reliability analysis problem, either direct or inverse, and/or an RBDO problem can be modeled to address reliability issues of a wind turbine.

References

[1] B. Morawski; Global Wind, Solar Production Hit Highest Benchmarks Ever in 2021, But Coal Also Kept Pace; World Economy Forum (2022).

[2] S. Costello; Neighbourhood Dispute Erupts Over Man's Backyard Wind Turbine; Nine News Australia (2022).

[3] J. Hill; Swedish Wind Turbine Collapses Days After Wind Farm Inauguration; Renew Economy (2022).

[4] S. Shakya; A self-monitoring and analysing system for solar power station using IoT and data mining algorithms. Journal of Soft Computing Paradigm, 3(2) (2021), 96-109. https://doi.org/10.36548/jscp.2021.2.004

[5] C. Vennila, A. Titus, T. Sri Sudha, U. Sreenivasulu, R. Pandu Ranga Reddy, K. Jamal, A. Belay; Forecasting Solar Energy Production Using Machine Learning; International Journal of Photoenergy, 7797488 (2022). https://doi.org/10.1155/2022/7797488

[6] H. Rezk, T. Sudhakar Babu, M. Al-Dhaifallah, H. Zeidan; A Robust Parameter Estimation Approach Based on Stochastic Fractal Search Optimization Algorithm Applied to Solar PV Parameters; Energy Reports, 7 (2021), 620-640. https://doi.org/10.1016/j.egyr.2021.01.024

[7] S. Artin; Reliability Enhancement of Solar Panels based on the Photocurrent Equality; 2023 Asia-Pacific Conference on Applied Mathematics and Statistics; Nanjing, China, June 2023.

[8] J. Deng, H. Li, J. Hu, Z. Liu; New Wind Speed Scenario Generation Method Based on Spatiotemporal Dependency Structure; Renewable Energy, 163 (2021), 1951-1962. https://doi.org/10.1016/j.renene.2020.10.132

[9] S. Okpokparoro, S. Sriramula; Uncertainty Modelling in Reliability Analysis of Floating Wind Turbine Support Structure; Renewable Energy, 165 (2021), 88-108. https://doi.org/10.1016/j.renene.2020.10.068

[10] D. Wilkie, C. Galasso; Gaussian Process Regression for Fatigue Reliability Analysis of Offshore Wind Turbines; Structural Safety, 88 (2021), 102020. https://doi.org/10.1016/j.strusafe.2020.102020

[11] C. Clark, B. DuPont; Reliability-Based Design Optimization in Offshore Renewable Energy Systems; Renewable and Sustainable Energy Reviews, 97 (2018), 390-400. https://doi.org/10.1016/j.rser.2018.08.030

[12] G. Tina, S. Gagliano, S. Raiti; Hybrid Solar/Wind Power System Probabilistic Modelling for Long-Term Performance Assessment; Solar Energy, 80 (2006), 578-588. https://doi.org/10.1016/j.solener.2005.03.013

[13] S. Eryilmaz, I. Bulanik, Y. Devrim; Reliability Based Modelling of Hybrid Solar/Wind Power System for Long Term Performance Assessment; Reliability Engineering and System Safety, 209 (2021), 107478. https://doi.org/10.1016/j.ress.2021.107478

[14] G. Ezzati, A. Rasouli; Evaluating System Reliability Using Linear-Exponential Distribution Function; International Journal of Advanced Statistics and Probability, 3(1) (2015), 15-24. https://doi.org/10.14419/ijasp.v3i1.3927

[15] M. Zarmai, C. Oduoza; Impact of Intermetallic Compound Thickness on Thermo-Mechanical Reliability of Solder Joints in Solar Cell Assembly; Microelectronics Reliability, 116 (2021), 114008. https://doi.org/10.1016/j.microrel.2020.114008

[16] S. Al Sanad, L. Wang, J. Parol, A. Kolios; Reliability-Based Design Optimization Framework for Wind Turbine Towers; Renewable Energy, 167 (2021), 942-953. https://doi.org/10.1016/j.renene.2020.12.022

[17] A. Askarzadeh, L. Coelho; A Novel Framework for Optimization of a Grid Independent Hybrid Renewable Energy System: A Case Study of Iran; Solar Energy, 112 (2015), 383-396. https://doi.org/10.1016/j.solener.2014.12.013

[18] B. Zhang, M. Wang, W. Su; Reliability Analysis of Power Systems Integrated with High-Penetration of Power Converters; IEEE Transactions on Power Systems, 36(3) (2021), 1998-2009. https://doi.org/10.1109/TPWRS.2020.3032579

[19] G. Ezzati; A Reliability-Based Design Optimization Model for Electricity Power Networks; Dynamics of Continuous, Discrete and Impulsive Systems, Series B: Applications & Algorithms, 22 (2015), 339-357.

[20] G. Ezzati, M. Mammadov, S. Kulkarni; Solving Reliability Analysis Problems in the Polar Space: International Journal of Applied Mathematical Research, 3(4) (2015), 353-365. https://doi.org/10.14419/ijamr.v3i4.3302

[21] S. Artin, S. Salimzadeh; A Conjugate Gradient Direction-Based Method to Evaluate Reliability Analysis Problems; IAENG International Journal of Applied Mathematics, 52(3) (2022), 659-666.

[22] G. Ochoa, J. Alvarez, M. Chamorro; Data Set on Wind Speed, Wind Direction and Wind Probability Distributions in Puerto Bolivar – Colombia; Data in Brief, 27 (2019). https://doi.org/10.1016/j.dib.2019.104753

Solar radiation forecasting using attention-based temporal convolutional network

Damilola OLAWOYIN-YUSSUF[1,a], Mohamad MOHANDES[1,b*], Bo LIU[1,c], and Shafiqur REHMAN[1,d]

[1]Department of Electrical Engineering, King Fahd University of Petroleum and Minerals, Academic Belt Road, Dhahran, 31261, Eastern Region, Saudi Arabia

[a]g202110610@kfupm.edu.sa, [b]mohandes@kfupm.edu.sa, [c]boliu@kfupm.edu.sa, [d]srehman@kfupm.edu.sa

Keywords: Global Solar Radiation, Temporal Convolutional Network (TCN), LSTM

Abstract. Solar energy, an inexhaustible and pristine power source, harbors the capability to mitigate the emissions of greenhouse gases and the dependency on fossil fuels, thereby playing a pivotal role in the conservation of our ecosystem. Nevertheless, the process of harnessing solar energy from sunlight is subject to the capricious characteristics of weather conditions, which include variables such as the density of cloud cover, levels of atmospheric moisture, and fluctuations in temperature. Hence, the task of prognosticating solar radiation holds significant importance for the strategic planning and efficient management of solar power systems. The current machine-learning methods for predicting global solar radiation make use of recurrent networks. One major downside of recurrent-based models is that they are exposed to vanishing gradients and stagnant performance over longer available input sequences. The model showcased is an attention-fueled Temporal Convolutional Network (TCN) intertwined with Convolutional Neural Network (CNN). The suggested method merges the advantages of the feature extraction proficiencies of a TCN and the aggregation capabilities of a CNN. The method has been tested for up to 24 hours of future time sequence prediction and it has been noted that its performance is unmatched.

Introduction
Owing to the unpredictable characteristics of weather conditions, the production of solar energy harnessed from sunlight cannot be pre-established with absolute certainty. Many methods for predicting solar radiation have been employed because it is critical in many industries, including solar energy production, agriculture, and weather forecasting[1]. Accurate predictions of solar radiation help these industries to plan and make informed decisions. Historically, the prediction of solar radiation was reliant on physical models that considered a multitude of factors. These included the position of the sun, the angle of incidence, the extent of cloud cover, and the prevailing atmospheric conditions [1]. These models require significant expertise and resources to develop, and the quality and availability of input data can limit their accuracy. Furthermore, they may be unable to capture the complex relationships between input variables and output predictions [1]. In recent years, there has been a growing interest in using machine learning (ML) algorithms to improve solar radiation prediction [2]. ML algorithms can model complex relationships between input and output by learning from the provided data. An ML technique that has an effective handle on getting better performance with big data is deep learning because of its ability to incorporate large datasets and interpret complicated correlations between variables in a versatile, trainable way [3]. Deep learning techniques like multi-layered perceptron [4], convolutional neural networks [5], and recurrent neural network [6, 7] have been used to explore predicting global solar radiation. While these methods have no doubt produced encouraging results, they do so with some challenges. Firstly, some of the methods used in predicting solar radiation fail to provide an

estimation over multiple time steps. Multiple-time step prediction provides the advantage of being used in early warning applications and predictive planning. Secondly, although other models based on recurrent networks like Recurrent Neural Network (RNN), Long Short-Term Memory (LSTM), and gated recurrent units (GRUs) have the capability of storing significant information and prediction over multiple time steps, they do suffer from degradation when making predictions over multiple time steps.

In response to these challenges, a multi-time step prediction architecture is proposed using a two-stage approach. The first stage extracts feature from the input dataset by using an attention mechanism with a temporal convolutional network (TCN) backbone. The first stage acts as an encoder as it outputs a fixed-length representation of the input. The second stage acts as a decoder using the fixed length representation to predict the desired output. It does so by leveraging a CNN and a dense layer to forecast solar radiation for multiple time steps.

Related Work
The application of ML techniques has brought about significant advancements in the study of solar radiation [8]. ML, a smaller category under artificial intelligence, revolves around the principle of instructing algorithms to identify patterns and formulate predictions predicated on data. These algorithms, with their ability to learn and adapt, offer a more efficient and precise approach to data analysis compared to traditional methods. [9] proposed a novel data preprocessing approach that aims to reduce forecasting errors, which are often associated with traditional prediction methods such as Markov chains or k-Nearest Neighbors (KNN). They engineered an enhanced multi-layer perceptron (MLP) model, incorporating three neurons within the concealed layer. This model demonstrated the capacity to yield predictions that were on par, if not surpassing, those generated by techniques such as Bayesian inference, Markov chains, and the KNN algorithm. Xing et al. introduced an innovative hybrid stack autoencoder LSTM (SAELSTM) architecture, specifically de-signed for predicting daily global solar radiation (GSR) [10]. This architecture harnesses the power of deep learning and incorporates a feature selection technique grounded on Manta Ray Foraging Optimization (MRFO). The utilization of this architecture in the context of GSR forecasting is further elaborated in the work of Ghimire et al. [11]. The deep learning hybrid SAELSTM model outperformed other models and persistence methods in simulations in terms of accuracy. The model generates intervals for high-quality solar energy predictions with a high likelihood of coverage and minimal interval errors. The study found that deep learning models, such as Bidirectional LSTM [12], perform better than traditional ML for forecasting daily GSR models. In a study conducted by Alizamir et al., wavelet transformation was utilized to break down different meteorological parameters to predict daily solar radiation [7]. The decomposed signals were then used as input into an LSTM recurrent network. While this approach improved network performance, it also increased the number of input parameters needed, thereby increasing the complexity of the optimization process. In another study, [13] employed CNN and an amalgamation of CNN and LSTM to predict monthly radiation at multiple steps. The study inferred that CNN outperformed other models such as MLP, LSTM, GRU, and CNN-LSTM. However, it's important to note that the receptive fields of CNN do not consider the sequence progression of time series data, which could limit its effectiveness in certain applications. Ghimire et al. predicted solar radiation by selecting features using a random forest recursive feature elimination [5]. The convolutional neural network extracted features which were then fed as input to a multilayer perceptron to generate a predicted output. However, using a multilayer perceptron for prediction limited the model's capability of predicting global solar radiation over multiple time steps. This highlights the need for models that can effectively handle time series data and make accurate predictions over multiple time steps.

Methodology

Symbol Definitions and Issue Formulation:

Consider an exogenous series, denoted as $X = (X_1, X_2, \ldots, X_n)$ and $X \in \mathbb{R}^{T \times n}$ where n represents the number of features and T signifies the time steps. The $i - th$ exogenous series, expressed in terms of time steps, can be represented as $X_i = (X_i^1, X_i^2, \ldots, X_i^T)$ or $X_i \in \mathbb{R}^T$. The objective of a time series prediction network is to train a function that, given a specific set of previous time series features $X = (X_1, X_2, \ldots, X_n)$ and their corresponding outputs within that time steps $\hat{Y} = \hat{y}^{T+1}, \hat{y}^{T+2}, \ldots, \hat{y}^{T+k}$ where $Y \in \mathbb{R}^k$. This can be mathematically expressed as:

$$\hat{y}^{T+1}, \hat{y}^{T+2}, \ldots, \hat{y}^{T+k} = F(X^1, X^2, \ldots, X^T, Y) \tag{1}$$

In this equation, the function $F(.)$ is the function whose parameters are learnable. This means that the function can adapt and improve its performance based on the data it is trained on, thereby enhancing the accuracy of the predictions it makes.

Model:

The design of the proposed model as outlined in Figure 1 takes a series of driving input sequences as its input. These sequences are then passed through an LSTM block, which acts like a translator, converting the input sequences into a form that the model can understand better; this

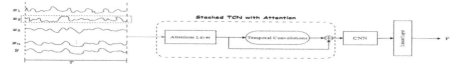

Figure 1 Graphical illustration of the proposed model architecture. It consists of an input layer (the blue line leading to the attention block signifies an LSTM layer used for embedding input), a stacked attention based TCN, a CNN block for merging the TCN output, and finally a linear output layer.

is known as embedding. The decision to use recurrent layers for extracting embeddings was inspired by the work of Gugulothu et al., where GRUs were utilized to generate embeddings for decoding multiple sequences in a multivariate time series network [14].

The output gate of an LSTM is expressed as:

$$o_n = F_{emb}(.) = \sigma(W_o * X_n + U_o * h_{n-1} + b_o) \tag{2}$$

X_n: the input vector at time n
h_n: the hidden state vector at time n
W: input-to-hidden weight matrix
b: hidden layer bias vector
σ: sigmoid activation functions.

From equation 2, the output o_n gives the temporal input embedding. The temporal input can be rewritten as:

$$x_{emb,i} = F_{emb}(X_i^1, X_i^2, \ldots, X_i^T) \tag{3}$$

The embedded input is then fed into a feature extraction network. This network is made up of an attention block stacked on top of a temporal convolution network.

Attention Block:

The Attention Block [15], works by first obtaining an attention weight vector from the provided input. This input could be represented as $x_{emb,i} = F_{emb}(X_{emb,i}^1, X_{emb,i}^2, \ldots, X_{emb,i}^T)$. The attention block helps the model focus on the most important parts of the input. It's postulated that the embeddings possess the same dimensional attributes as the input, albeit this is typically not the scenario. The attention weight vectors are calculated using the following equations:

$$u_i = W_u^T x_{emb,i} + b_u \tag{4}$$

Where $W_u \in \mathbb{R}^{T \times 1}$, and $b_u \in \mathbb{R}$ are parameters to be learned. These attention weight vectors are then normalized using a SoftMax function to ensure they all sum to unity. The normalization SoftMax functions can be expressed as:

$$\mu_i^t = \frac{\exp(u_i^t)}{\sum_{t=1}^{T} \exp(u_i^t)} \quad (5)$$

where $t \in [1, T]$. These normalized softmax values represent the distribution of the input that should be paid attention to. The attention output, defined by the function $F_{att}(\cdot)$, can be calculated by multiplying the normalized SoftMax by the input:

$$x_{att,i} = \mu_i^t \cdot x_{emb,i} \quad (6)$$

Extracting Features with TCN:
In sequence modelling, recurrent networks such as RNN and its variants have been traditionally used until Bai et al. introduced the concept of using generic convolutional networks for sequence modelling tasks. This new approach, called Temporal Convolutional Networks (TCN), outperformed the LSTM [16]. The structure of the TCN is a simple modification of the conventional CNN. The TCN uses causal convolutions, which ensure that the model's prediction at a given time does not depend on future values of the input and makes them much faster to train compared to recurrent models as they do not have recurrent connections. A further enhancement of the causal convolution, known as dilated causal convolution, allows convolution over a wider window by skipping some input values. The receptive field of the dilated causal convolution is much wider than that of the causal convolution, making it more efficient. Figure 2 shows the dilated causal convolutions for different levels of dilation. Given the input $x \in \mathbb{R}^T$ and a filter $f : \{0, \ldots, \alpha, \ldots, m-1\}$ of size m, the dilation convolution operator on α within the sequence can be defined as:

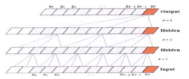

Figure 2 TCN Architecture

$$F_{TCN}(\alpha) = \sum_{j=0}^{m-1} f(j) \cdot x^{(\alpha - d \cdot j)} \quad (7)$$

Where d is the dilation factor, and $\alpha - d \cdot j$ explains the orientation of the past.
The Temporal Convolutional Network (TCN) block, as depicted in figure 1, is a composite of several components. These encompass a dilated causal convolution, weight standardization, a Rectified Linear Unit (ReLU) activation function, and dropout strata incorporated to augment the resilience of the network. The extent of the TCNs could potentially cause the vanishing gradient problem. This is a difficulty encountered during the training of artificial neural networks with gradient-based learning methods and backpropagation. To mitigate this, a skip or residual connection has been incorporated. In a residual block, as described by He et al. [17], there's a pathway that leads us through a series of transformations, denoted as F_o. The results of these transformations are then seamlessly integrated with the block's original input, x. The output after a residual connection, O_{res}, is given by the equation:

$$O_{res} = \sigma(x + F_o(x)) \quad (8)$$

Here, σ represents the activation function. This function introduces non-linearity into the output of a neuron. This non-linearity helps the network learn from the error so that the model can classify inputs that are not linearly separable.

Aggregating extracted features with CNN:
The output from the stack of attention TCN block is of the form $\mathbb{R}^{N \times T \times L}$ where N is the batch size, T is the number of the input sequence, and L is the number of channels in each TCN layer. There exists a requirement to transform the output of this Temporal Convolutional Network (TCN)

into a vector of dimension $\mathbb{R}^{N\times 1\times K}$, where K signifies the number of anticipated output sequences. To achieve an output of this dimension two different approaches can be followed: flattening the last two dimensions and using a linear layer, or a convolutional layer. In this study, the path of using a convolutional layer was followed because of the added advantage of reducing the number of parameters needed for computation. If a linear layer was used, the number of parameters required would be of dimension $T \times L \times K$ while when a 1-dimensional convolutional layer is used the number of weights required would be $(L \times m) + (L \times K) = L \times (m + K)$, where m is the size of kernel used in the convolutional layer. The $L \times K$ accounts for the number of weights needed in the linear layer for the output to be in the right dimension after the convolutional layer. The convolution operation over an input X can be expressed mathematically as:

$$O_{conv} = \sigma_{lr}(W_{conv}X + B_{conv}) \qquad (9)$$

where σ_{lr} is the leaky rectified linear unit (leakyReLU) activation function, W_{conv} is the convolutional weight, and B_{conv} is the convolutional bias. The dense layer utilized after the CNN is an affine amalgamation of the output after the convolution layer and a certain bias, devoid of an activation layer.

Experimental data description, training settings, and evaluation metrics
The empirical dataset employed in this investigation, amassed in the northern region of Saudi Arabia, comprises 75133 data entries encapsulating variables such as solar radiation, atmospheric temperature, relative humidity, velocity and direction of wind, and precipitation, extending over a period from 2012 to 2021. However, it's important to note that there are gaps in the dataset between

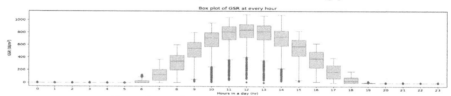

Figure 3 Box plots of GSR at every hour of the day.

October 3rd, 2019, and January 1st, 2020, as well as between March 29th, 2020, and June 4th, 2020. Given the sequential nature of the dataset, the data was not shuffled when loaded and was divided into training, validation, and testing portions. The split was done in a 70:15:15 ratio, ensuring a substantial amount of data for each phase of the model development and evaluation process.

In this study, we considered the lags of Global Solar Radiation (GSR) as input to the model, to envisage the next one hour, six hours, 12 hours, and 24 hours respectively. The lags used are 24, 48, and 72 hours respectively, or in mathematical notation as $T \in \{24, 48, 72\}$. To understand the effect of the time of day on global solar radiation, a box plot showing the distribution of GSR at every hour of the day is presented in Figure 3. The time of day significantly affects global solar radiation, with some outliers observed. A comparison of the maximum values of the variables in our dataset reveals the need for input scaling. Various scaling methods like the standard scaler, robust scaler, and min-max scaler have been used in different studies [18]. The best scaling method for a dataset depends on the variable distribution in that dataset, which can be visualized using box plots. The robust scaler, which uses the interquartile range and median, is used for this problem. The robust scaler calculates the scaled value of an item V in a series of inputs as:

$$V_{scaled} = \frac{V_{original} - \tilde{\eta}}{IQR} \qquad (10)$$

where the scaled value is V_{scaled}, the original value is $V_{original}$, $\tilde{\eta}$ represents the input median, and IQR is the input's interquartile range.

The forecast was made for the upcoming 1, 6, 12, and 24 hours, focusing on the variable k in equation 1, where k can be any of the values 1, 6, 12, or 24. This was done to test the theory of

Figure 4 Prediction results for 24 future time steps.

Temporal Convolutional Network (TCN) structure. The model's hidden size was determined to be 64 through a grid search, and this size was also used for the RNN and LSTM models. The model structure consists of 1 CNN layer and 6 TCN blocks each having a dropout rate of 0.2. An Adam optimizer with weight decay of 1×10^{-6} was used with a learning rate scheduler having a patience of 2, and a threshold of 0.01. All models underwent training under homogeneous conditions, commencing with a learning rate of 1×10^{-3} and persisting for 40 epochs within a Google Colab environment, utilizing the Torch library of Python-3. A v100 GPU with a 15GB RAM infrastructure was used for the entire experiment. To evaluate the models, three widely used metrics for time series prediction were employed: R-squared, Mean Absolute Error (MAE), and Root Mean Squared Error (RMSE). Lower values of RMSE and MAE and a higher R-squared indicate a better model.

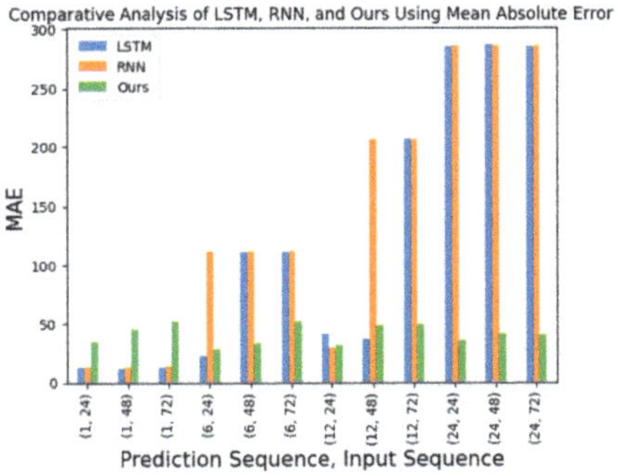

Results

The outcomes of applying the proposed model to predict global solar radiation over several time steps are detailed in Table 1 and Figure 5. It's plausible that employing exclusively recurrent architectures, such as LSTM and RNN, might culminate in optimal outcomes for a solitary temporal increment. However, it's also arguable that the model introduced here can confidently compete at that single time step. When predictions are made over multiple time steps like that shown in Figures 4, and regardless of the amount of historical time sequence data used as input, the performance of the recurrent models (i.e., LSTM and RNN) declines significantly. The extent to which they are less accurate compared to the proposed model is quite substantial. There were some unexpected performances when the RNN was predicting 12 future time steps, and the LSTM was predicting 6-time steps. Even under these conditions, the proposed model achieved performance measures close to the best possible outcomes. The consistent performance of this model, either close to the best or the best, attests to its balance and robustness.

Figure 5 MAE Comparison

Table 1. Juxtaposition of Different Methods with 1 and 6 Prediction Steps.

Mthds	Pred. Steps	1			6			12			24		
	Inp. Seq.	24	48	72	24	48	72	24	48	72	24	48	72
	Metric												
LSTM	RMSE	35.765	39.639	38.988	**55.651**	158.812	159.395	84.145	89.088	259.090	323.678	323.688	323.576
	MAE	**11.719**	**11.400**	**11.699**	**21.768**	110.315	110.644	40.773	**36.344**	206.024	284.396	286.481	284.912
	R-Sq.	0.988	0.985	0.986	**0.971**	0.735	0.733	0.932	0.926	0.345	0.032	0.031	0.032
RNN	RMSE	30.437	**32.067**	**33.161**	159.285	159.385	159.293	**67.259**	258.887	258.753	323.627	323.554	323.531
	MAE	12.195	12.0160	12.911	110.162	110.404	110.693	**27.966**	204.729	205.334	284.612	284.856	284.684
	R-Sq.	**0.991**	**0.990**	**0.990**	0.733	0.733	0.733	**0.957**	0.345	0.347	0.032	0.032	0.0319
Ours	RMSE	71.96	95.539	107.244	72.619	**72.770**	**87.462**	75.711	**87.858**	**81.709**	**79.112**	**81.334**	**91.994**
	MAE	33.81	44.961	51.358	27.406	**32.155**	**51.358**	30.793	**47.847**	**48.853**	**35.018**	**40.539**	**39.420**
	R-Sq.	0.952	0.915	0.893	0.951	**0.951**	**0.928**	0.946	**0.928**	**0.938**	**0.941**	**0.938**	**0.921**

Conclusion

This work suggested the application of an attention-fueled temporal convolutional network in conjunction with a convolutional neural network to predict global solar radiation (GSR). This approach is particularly effective when the accessible historical sequence of GSR spans durations of 24, 48, and 72 hours. We then compared the proposed model with other ML models used for GSR, including RNN and LSTM. The models were evaluated using RMSE, MAE, and R2 metrics. The empirical findings demonstrated that the suggested model exhibited superior performance compared to other models in most of the instances. Potential avenues for subsequent research could encompass broadening the temporal scope of the models to forecast one week or one month into the future and juxtaposing their performance with other established methodologies. Another direction is to investigate further feature generation and manipulation and to include other meteorological variables as input features for the models.

References

[1] Makade, R.G., Chakrabarti, S., Jamil, B.: Development of global solar radiation models: A comprehensive review and statistical analysis for indian regions. Journal of Cleaner Production 293, 126208 (2021) https://doi.org/10.1016/j.jclepro.2021.126208

[2] Solano, E.S., Dehghanian, P., Affonso, C.M.: Solar radiation forecasting using machine learning and ensemble feature selection. Energies 15(19), 7049 (2022). https://doi.org/10.3390/en15197049

[3] Ngiam, K.Y., Khor, W.: Big data and machine learning algorithms for health-care delivery. The Lancet Oncology 20(5), 262–273 (2019). https://doi.org/10.1016/S1470-2045(19)30149-4

[4] Rehman, S., Mohandes, M.: Artificial neural network estimation of global solar radiation using air temperature and relative humidity. Energy policy 36(2), 571–576 (2008). https://doi.org/10.1016/j.enpol.2007.09.033

[5] Ghimire, S., Nguyen-Huy, T., Prasad, R., Deo, R.C., Casillas-Perez, D., SalcedoSanz, S., Bhandari, B.: Hybrid convolutional neural network-multilayer perceptron model for solar radiation prediction. Cognitive Computation 15(2), 645–671 (2023). https://doi.org/10.1007/s12559-022-10070-y

[6] Pang, Z., Niu, F., O'Neill, Z.: Solar radiation prediction using recurrent neural network and artificial neural network: A case study with comparisons. Renewable Energy 156, 279–289 (2020). https://doi.org/10.1016/j.renene.2020.04.042

[7] Alizamir, M., Shiri, J., Fard, A.F., Kim, S., Gorgij, A.D., Heddam, S., Singh, V.P.: Improving the accuracy of daily solar radiation prediction by climatic data using an efficient hybrid deep learning model: Long short-term memory (lstm) network coupled with wavelet transform. Engineering Applications of Artificial Intelligence 123, 106199 (2023). https://doi.org/10.1016/j.engappai.2023.106199

[8] Ali-Ou-Salah, H., Oukarfi, B., Bahani, K., Moujabbir, M.: A new hybrid model for hourly solar radiation forecasting using daily classification technique and machine learning algorithms. Mathematical Problems in Engineering 2021, 1–12 (2021). https://doi.org/10.1155/2021/6692626

[9] Paoli, C., Voyant, C., Muselli, M., Nivet, M.-L.: Forecasting of preprocessed daily solar radiation time series using neural networks. Solar energy 84(12), 2146–2160 (2010). https://doi.org/10.1016/j.solener.2010.08.011

[10] Xing, X., Li, Z., Xu, T., Shu, L., Hu, B., Xu, X.: Sae+ lstm: A new framework for emotion recognition from multi-channel eeg. Frontiers in neurorobotics 13, 37 (2019). https://doi.org/10.3389/fnbot.2019.00037

[11] Ghimire, S., Deo, R.C., Wang, H., Al-Musaylh, M.S., Casillas-P´erez, D., SalcedoSanz, S.: Stacked lstm sequence-to-sequence autoencoder with feature selection for daily solar radiation prediction: a review and new modeling results. Energies 15(3), 1061 (2022). https://doi.org/10.3390/en15031061

[12] Liu, G., Guo, J.: Bidirectional lstm with attention mechanism and convolutional layer for text classification. Neurocomputing 337, 325–338 (2019). https://doi.org/10.1016/j.neucom.2019.01.078

[13] Azizi, N., Yaghoubirad, M., Farajollahi, M., Ahmadi, A.: Deep learning based long-term global solar irradiance and temperature forecasting using time series with multi-step multivariate output. Renewable Energy 206, 135–147 (2023). https://doi.org/10.1016/j.renene.2023.01.102

[14] Gugulothu, N., Tv, V., Malhotra, P., Vig, L., Agarwal, P., Shroff, G.: Predicting remaining useful life using time series embeddings based on recurrent neural networks. arXiv preprint arXiv:1709.01073 (2017)

[15] Vaswani, A., Shazeer, N., Parmar, N., Uszkoreit, J., Jones, L., Gomez, A.N., Kaiser, L., Polosukhin, I.: Attention is all you need. Advances in neural information processing systems 30 (2017)

[16] Bai, S., Kolter, J.Z., Koltun, V.: An empirical evaluation of generic convolutional and recurrent networks for sequence modeling. arXiv preprint arXiv:1803.01271 (2018)

[17] He, K., Zhang, X., Ren, S., Sun, J.: Deep residual learning for image recognition. In: Proceedings of the IEEE Conference on Computer Vision and Pattern Recognition, pp. 770–778 (2016). https://doi.org/10.1109/CVPR.2016.90

[18] Balabaeva, K., Kovalchuk, S.: Comparison of temporal and non-temporal features effect on machine learning models quality and interpretability for chronic heart failure patients. Procedia Computer Science 156, 87–96 (2019). https://doi.org/10.1016/j.procs.2019.08.183

Application of artificial intelligence (AI) in wind energy system with a case study

Fay ALZAHRANI[1,a], Feroz SHAIK[1,b*], Nayeemuddin MOHAMMED [2,c], Nasser Abdullah Shinoon AL-NA'ABI[3,d]

[1]Department of Mechanical Engineering, Prince Mohammad Bin Fahd University, Al Khobar, Kingdom of Saudi Arabia

[2]Department of Civil Engineering, Prince Mohammad Bin Fahd University, Al Khobar, Kingdom of Saudi Arabia

[3]Department of Mechanical Engineering, University of Technology and Applied Sciences, Sultanate of Oman

202000892@pmu.edu.sa[a]; fshaik@pmu.edu.sa[b], mnayeemuddin@pmu.edu.sa[c]; nasser.alnaabi@utas.edu.om[d]

Keywords: Wind Energy, Artificial Intelligence, Wind Turbine, Windmill, Genetic Algorithm, Wind Speed, Current Output

Abstract. Renewable energy is the fastest growing source of clean energy worldwide. The employment of wind energy is expected to increase dramatically over the next few years. There is a good source of wind power on the highways due to the movement of vehicles. A small windmill could utilize the wind power generated by passing vehicles and produce electricity that can power the lights on the highway. This paper presents the application of artificial intelligence to predict the current output from a small windmill placed on the highway. The results show a good concurrence between the experimental and predicted values.

Introduction

The main energy source is from fossil fuels, which is extensively used to meet the demand. The usage of fossil fuels directly harms the clean environment and also leads to global warming. Fossil fuels are non-renewable and get depleting, which makes people to focus on renewable energy sources. All over the world harnessing of solar and wind as a sustainable source of energy gained popularity to curtail the heavy dependency on fossil fuels and also to counter the global warming. When wind energy is used to produce electricity, less pollution from conventional power plants will be released into the environment. The need of concentrating on renewable energy resources has increased, particularly in the wake of the Gulf of Mexico oil leak and the Japanese nuclear accident. The installed wind energy capacity reached 196,630 MW globally in 2010, with 37,642 MW added in that year, according to the World Energy Association's report on wind energy for 2010. Enhancing wind farm design and layout; boosting wind turbine accessibility, dependability, and efficiency; streamlining the upkeep, assembly, and installation of offshore and onshore turbines and their substructures; showcasing massive wind turbine prototypes and expansive, interconnected wind farms, etc. are the main research areas that should be prioritized in the wind energy industry [1]. The first wind-powered generator was invented by Charles F. Brush, an electrical pioneer from America, and it produced energy in his backyard. He built a windmill that was 40 tons in weight and stood 60 feet tall. The actual wind mill measured 56 feet in diameter. The wind mill had a total of 144 separate blades. 500 revolutions per minute was the turbine's peak rotational speed. Everything in his basement was wired up to 408 batteries. With this technology, he was able to power his entire house, including the lab. Up until 1909, his wind mill operated for 20 years [2].

The wind turbine's size is determined by its intended use. Typically, tiny turbines have a power output between 20 and 100 kW. The 20- to 500-watt "micro" turbines are smaller and have a wider range of uses, including the charging of sailboat and recreational vehicle batteries. Water pumping is one use for turbines ranging from one to ten kW. Grain mills and water pumps have been powered by wind for millennia. While mechanical windmills remain a cost-effective and practical choice for water pumping in wind-free regions, farmers and ranchers are discovering that wind-electric pumping offers greater versatility and doubles the volume of water pumped for the same initial outlay. Furthermore, mechanical windmills have to be positioned straight above the well, which could not maximize the wind resources that are available. Electric cables can be used to link wind-electric pumping systems to the pump motor, which can be installed where the best wind resource is available. Depending on how much power you wish to create, household turbines can range in size from 400 watts to 100 kW (100 kW for extremely big loads). An average household consumes around 10,000 kWh (kilowatt-hours) of power year, or 830 kWh each month. To significantly meet this requirement, a wind turbine with a rating of between 5 and 15 kW would be needed, depending on the typical wind speed in the region. If the average yearly wind speed in the area is 14 miles per hour (6.26 meters per second), a 1.5 kW wind turbine can supply the energy needed for a house that uses 300 kWh per month. Automatic overspeed-governing mechanisms are included in most turbines, which prevent the rotor from spinning uncontrollably in extremely strong winds [3].

Harrous and Ahshan [4], [5] developed a hybrid solar/wind system for his home. The hybrid system consists of Bergey XL-1, a 1000-watt wind generator mounted on a tower 104 feet tall along with 300 watts of solar, which is a stand-alone system with batteries. The battery bank is a 220-amp system made up of eight 6-volt batteries wired as a 24-volt system. The system runs incandescent lights and a well pump at the barn, as well as water through heaters. The cost of the complete system was around $ 10,000 including equipment, trenching for wires, building permit, etc. For wind energy uses, there must be open space or accessible coastlines for wind energy plants. Saudi Arabia is a large nation with extensive coastlines and open spaces. In the majority of these locations, the wind speed is sufficiently high to make using wind energy cost-effective. Saudi Arabian authorities will invest billions in this potential field of electricity since they understand the value of renewable energy, particularly wind energy. Despite its vast oil reserves, Saudi Arabia is very interested in actively participating in the development of new technologies for the exploitation and use of renewable energy sources. Despite Saudi Arabia's substantial wind resource potential, there are several obstacles to its development. These comprise the resource's erratic nature, its seasonal and diurnal fluctuations, its isolated geographic position, and the electrical grid infrastructure required to transfer wind energy to load regions. Significant technological obstacles must be overcome in order to fully utilize Saudi Arabia's wind potential. The energy balance between the needed load and the generated power, as well as the matching of the wind turbine and location with an appropriate economic position, remain a significant problem. By matching the locations and wind turbines, the researcher created an extensive computer program that does all the calculations and optimization needed to precisely build the Saudi Arabian wind energy system [6].

Eltamaly et al [7] built and examined the dynamic performance of a novel wind turbine producing system using a thyristor inverter. The system is basically based on shaft generators, which are highly reliable and produce high-quality power output and are frequently employed in big ships. It was looked into if this innovative method could provide low-distortion electric power at a steady frequency even when the natural wind's velocity fluctuated. Additionally, a dynamic model was created, and it was discovered to have good agreement with the system's experimental and simulated results. Zemamou et al [8] investigated the remarkable performance of savonius wind turbines and how they might be used as an alternative to normal wind turbines to extract valuable energy from air streams. Some benefits of employing this kind of machine include its straightforward design, high starting and full operation moment, ability to receive wind from any direction, minimal noise and angular velocity when operating, and reduced wear on moving components. There have been many suggested modifications for this gadget over time. Another benefit of employing such a machine is the range of possible rotor designs. The performance of a Savonius rotor is impacted by each configuration. The performance of a Savonius rotor is influenced by air flow, geometric, and operational factors. For the majority of settings, the quoted range for the highest averaged power coefficient is between 0.05 and 0.30. The usage of stators has also been shown to result in performance increases of up to 50% for the tip speed ratio of the highest averaged power coefficient.

Renewable energy technologies affect how household power demands are met. Since most of the energy produced by fossil fuels is used in buildings and their unchecked use is linked to environmental risks, global warming, and the possibility of their depletion, it will be advantageous to replace the conventional energy generation system with renewable energy sources [9]. Globally, there is a growing need to transition from fossil fuels to renewable energy sources. The main causes of this transformation are the lack of fossil fuels and their detrimental consequences on the environment, particularly the climate. As a result, interest in renewable energy sources such as solar, wind, and wave energy is growing around the world [10]. Converters for multiphase generators, back-to-back linked converters, passive generator-side converters, and converters without an intermediary dc-link for high-power wind energy conversion systems (WECS) are all included in the low and medium voltage category. The series/parallel connection of wind turbine ac/dc output terminals and high voltage ac/dc gearbox are taken into consideration while evaluating the onshore and offshore wind farm layouts [11].

Artificial Intelligence (AI) in Wind Energy Systems
The majority of wind farms are situated in isolated areas or several miles offshore, thus it is vital to monitor their mechanical parts for maintenance in order to keep them from breaking down mechanically and perhaps cutting themselves off from the electricity grid [12]. Machine learning algorithms, particularly artificial neural network ANNs, are commonly used to process gathered data. The ANN's structure is inspired by real neurons, with basic processing units coupled by weighted linkages. It contains three major layers: input, concealed, and output. Furthermore, the number of hidden layers may be increased to construct the deep neural network (DNN) architecture [13]. The Artificial Neural Network (ANN), Backpropagation Neural Network (BPNN), Adaptive Neuro-Fuzzy Inference System (ANFIS), and Genetic Algorithm (GA) are some of the most often used and proven AI approaches. The level of technology today and potentially uses tried-and-true methods to create AI-powered renewable energy systems, particularly solar energy systems. To ascertain the state and progress of AI approaches in the field of renewable energy systems (RES), particularly solar power systems, a number of peer-reviewed journal publications were analyzed [14].

The physical methods forecast wind energy using meteorological data, such as topography, atmospheric pressure, and ambient temperature; the hybrid methods combine the advantages of multiple single forecasting models to obtain the final prediction results through various weighting strategies; and the intelligent methods process and optimize the integration of external and internal big data to estimate future wind energy. The statistical approaches anticipate wind energy time series by an assessment of the probability distribution and random process of the samples. Since intelligent approaches and AI-based hybrid methods are more efficient at analyzing the complex connections present in huge data sets, they are essential for increasing energy efficiency, decreasing energy usage, and allowing real-time decision-making in the wind energy business. [15].

Case Study

A small wind mill was fabricated using wind turbine mounted inside the tube, generator and a battery pack. The fabricated tubular wind mill was flexible and can withstand turbulence. The turbine inside the tube rotates in the direction of wind turbulence. Standard generator system was used which can deliver a power output of 1 kW along with a maintenance free battery pack, inverter and charge controller. Figure 1 shows the schematic diagram of the wind mill. Experimental data was recorded by keeping the wind mill on road side platform based on vehicular movements for 7 days. Duration of data recording on each day varies from 30 to 180 min.

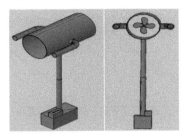

Fig 1. Schematic Diagram of Wind Mill

Artificial Neural Networks (ANN) use genetic algorithms that make use of adaptive heuristic search methods. A genetic algorithm is a better method for achieving the global optimum's convergence. Chromosome initialization is the first step in the genetic algorithm's operation, after which fitness is assessed using an objective function [16]. Chromosomes are genetically propagated by first selecting the most fit individuals and then using operators such as crossover and mutation. A multi-objective solution from the optimization toolkit and a genetic algorithm were used to optimize the process output variable models [17]. Ten neurons or nodes made up the hidden layer, the output current serving as the dependent output neuron, and time and wind speed serving as independent input variables were used to create the ANN network model [18]. Ten neurons made up the hidden layer of the neural network, which was trained until the mean squared error between the target and model output was as little as possible. The comparison between the goal values of the present output and the output values of the ANN network model is displayed in Figure 2 [19]. A high correlation coefficient value across training, validation, testing, and overall comparison shows that the model can accurately forecast the wind mill's current production value. The comparison output variables between the experimental investigations and the ANN-GA projected values are displayed in Table 1 [20], [21].

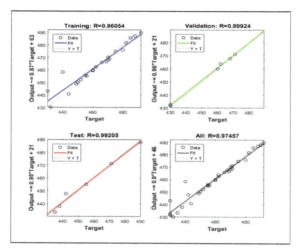

Fig 2. ANN network model output values and the target values of the current output

Table 1. Experimental data and ANN-GA (Genetic Algorithm) Predicted Data

Time	Wind Speed [m/s]	Current [Amp] (Experimental)	Current [Amp] (Predicted from ANN)
30	6.44	460	459.562
60	6.58	470	469.965
90	6.86	490	487.948
120	6.72	480	475.955
150	6.58	470	470.052
180	6.78	485	485.718
30	6.3	450	450.116
60	6.37	455	455.799
90	6.52	466	466.029
120	6.47	462	463.786
150	6.44	460	460.033
180	6.59	471	471.153
30	6.88	492	489.558
60	6.83	488	487.069
90	6.59	471	468.879
120	6.56	469	466.526

150	6.32	452	452.211
180	6.28	449	448.877
30	6.44	460	459.562
60	6.02	430	431.558
90	6.04	432	430.039
120	6.09	435	433.239
150	6.13	438	437.854
180	6.18	442	447.680
30	6.38	456	455.237
60	6.37	455	455.799
90	6.55	468	467.877
120	6.58	470	468.046
150	6.65	475	474.828
180	6.74	482	481.832
30	6.37	455	454.937
60	6.16	440	458.332
90	6.21	444	440.683
120	6.02	430	432.400
150	6.44	460	460.033
180	6.02	430	442.955
30	6.3	450	450.116
60	6.37	455	455.799
90	6.52	466	466.029
120	6.46	462	462.543
150	6.44	460	460.033
180	6.59	471	471.153

Results and Discussion
The simple design of rotor blades enhanced the wind velocity and thereby generate high currents. Artificial Intelligence (AI) methods, including machine learning and neural networks, have been used successfully to improve wind turbine performance and efficiency. The implementation of AI in wind energy systems has shown promising outcomes, as demonstrated by the case study done in this paper. The case study entailed gathering real-time data from a wind turbine and using AI models to forecast wind speed and direction, optimize turbine performance, and boost energy output. It was shown by the results that artificial intelligence (AI) may greatly raise the overall efficiency of wind energy systems, which will raise output and lower operating expenses. Additionally, by offering more precise forecasts and enhanced grid stability, AI can help with the better integration of wind energy into the system. In summary, this study shows how artificial

intelligence (AI) has the power to completely transform the wind energy industry, making it more dependable, effective, and sustainable.

Conclusions

A small fabricated windmill was used to measure the current out on a highway during vehicle movement. The experimental data for seven days at an interval of 30 min. The obtained data was used in Artificial Neural Network-Genetic Algorithm and training and testing. The ANN predicted values were found to be in good concurrence with the experimental data.

References

[1] Y. Eroğlu and S. U. Seçkiner, "Wind farm layout optimization using particle filtering approach," *Renewable Energy*, vol. 58, pp. 95–107, Oct. 2013. https://doi.org/10.1016/j.renene.2013.02.019

[2] National Renewable Energy Laboratory (NREL), Golden, CO., "Small Wind Electric Systems: An Ohio Consumer's Guide," DOE/GO-102005-2077, 15016004, Mar. 2005. https://doi.org/10.2172/15016004

[3] T. J. Wenning and J. K. Kissock, "Methodolgy for Preliminary Assessment of Regional Wind Energy Potential," in *ASME 2009 3rd International Conference on Energy Sustainability, Volume 2*, San Francisco, California, USA: ASMEDC, Jan. 2009, pp. 1031–1040. https://doi.org/10.1115/ES2009-90469

[4] A. Harrouz, I. Colak, and K. Kayisli, "Control of a small wind turbine system application," in *2016 IEEE International Conference on Renewable Energy Research and Applications (ICRERA)*, Birmingham, United Kingdom: IEEE, Nov. 2016, pp. 1128–1133. https://doi.org/10.1109/ICRERA.2016.7884509

[5] R. Ahshan, A. Al-Badi, N. Hosseinzadeh, and M. Shafiq, "Small Wind Turbine Systems for Application in Oman," in *2018 5th International Conference on Electric Power and Energy Conversion Systems (EPECS)*, Kitakyushu, Japan: IEEE, Apr. 2018, pp. 1–6. https://doi.org/10.1109/EPECS.2018.8443520

[6] P. Rogowski, M. Rogowska, T. Smaz, and F. Grapow, "Small Wind Turbine Off-Grid Power Generation Optimization," in *2021 IEEE 4th International Conference on Renewable Energy and Power Engineering (REPE)*, Beijing, China: IEEE, Oct. 2021, pp. 355–359. https://doi.org/10.1109/REPE52765.2021.9617019

[7] A. M. Eltamaly, "Design and implementation of wind energy system in Saudi Arabia," *Renewable Energy*, vol. 60, pp. 42–52, Dec. 2013. https://doi.org/10.1016/j.renene.2013.04.006

[8] M. Zemamou, M. Aggour, and A. Toumi, "Review of savonius wind turbine design and performance," *Energy Procedia*, vol. 141, pp. 383–388, Dec. 2017. https://doi.org/10.1016/j.egypro.2017.11.047

[9] S. Vahdatpour, S. Behzadfar, L. Siampour, E. Veisi, and M. Jahangiri, "Evaluation of Off-grid Hybrid Renewable Systems in the Four Climate Regions of Iran," *JREE*, vol. 4, no. 1, Feb. 2017. https://doi.org/10.30501/jree.2017.70107

[10] M. Abdelateef Mostafa, E. A. El-Hay, and M. M. ELkholy, "Recent Trends in Wind Energy Conversion System with Grid Integration Based on Soft Computing Methods: Comprehensive Review, Comparisons and Insights," *Arch Computat Methods Eng*, vol. 30, no. 3, pp. 1439–1478, Apr. 2023. https://doi.org/10.1007/s11831-022-09842-4

[11] V. Yaramasu, B. Wu, P. C. Sen, S. Kouro, and M. Narimani, "High-power wind energy conversion systems: State-of-the-art and emerging technologies," *Proc. IEEE*, vol. 103, no. 5, pp. 740–788, May 2015. https://doi.org/10.1109/JPROC.2014.2378692

[12] N. O. Farrar, M. H. Ali, and D. Dasgupta, "Artificial Intelligence and Machine Learning in Grid Connected Wind Turbine Control Systems: A Comprehensive Review," *Energies*, vol. 16, no. 3, p. 1530, Feb. 2023. https://doi.org/10.3390/en16031530

[13] F. Elyasichamazkoti and A. Khajehpoor, "Application of machine learning for wind energy from design to energy-Water nexus: A Survey," *Energy Nexus*, vol. 2, p. 100011, Dec. 2021. https://doi.org/10.1016/j.nexus.2021.100011

[14] J. T. Dellosa and E. C. Palconit, "Artificial Intelligence (AI) in Renewable Energy Systems: A Condensed Review of its Applications and Techniques," in *2021 IEEE International Conference on Environment and Electrical Engineering and 2021 IEEE Industrial and Commercial Power Systems Europe (EEEIC / I&CPS Europe)*, Bari, Italy: IEEE, Sep. 2021, pp. 1–6. https://doi.org/10.1109/EEEIC/ICPSEurope51590.2021.9584587

[15] E. Zhao, S. Sun, and S. Wang, "New developments in wind energy forecasting with artificial intelligence and big data: a scientometric insight," *Data Science and Management*, vol. 5, no. 2, pp. 84–95, Jun. 2022. https://doi.org/10.1016/j.dsm.2022.05.002

[16] N. Mohammed, P. Palaniandy, and F. Shaik, "Pollutants removal from saline water by solar photocatalysis: a review of experimental and theoretical approaches," *International Journal of Environmental Analytical Chemistry*, vol. 103, no. 16, pp. 4155–4175, Dec. 2023. https://doi.org/10.1080/03067319.2021.1924160

[17] N. Mohammed, A. Asiz, M. A. Khasawneh, H. Mewada, and T. Sultana, "Machine learning and RSM-CCD analysis of green concrete made from waste water plastic bottle caps: Towards performance and optimization," *Mechanics of Advanced Materials and Structures*, pp. 1–9, Aug. 2023. https://doi.org/10.1080/15376494.2023.2238220

[18] N. Mohammed, P. Palaniandy, F. Shaik, and H. Mewada, "Experimental and computational analysis for optimization of seawater biodegradability using photo catalysis," *IIUMEJ*, vol. 24, no. 2, pp. 11–33, Jul. 2023. https://doi.org/10.31436/IIUMEJ.v24i2.2650

[19] N. Mohammed, P. Palaniandy, F. Shaik, H. Mewada, and D. Balakrishnan, "Comparative studies of RSM Box-Behnken and ANN-Anfis fuzzy statistical analysis for seawater biodegradability using TiO2 photocatalyst," *Chemosphere*, vol. 314, p. 137665, Feb. 2023. https://doi.org/10.1016/j.chemosphere.2022.137665

[20] N. Mohammed, P. Palaniandy, F. Shaik, B. Deepanraj, and H. Mewada, "Statistical analysis by using soft computing methods for seawater biodegradability using ZnO photocatalyst," *Environmental Research*, vol. 227, p. 115696, Jun. 2023. https://doi.org/10.1016/j.envres.2023.115696

[21] N. Mohammed, P. Palaniandy, F. Shaik, and H. Mewada, "Statistical Modelling of Solar Photocatalytic Biodegradability of Seawater Using Combined Photocatalysts," *J. Inst. Eng. India Ser. E*, Sep. 2023. https://doi.org/10.1007/s40034-023-00274-8

Prediction of distillate output in photocatalytic solar still using artificial intelligence (AI)

Reyouf ALQAHTANI[1,a], Feroz SHAIK[1,b*], Nayeemuddin MOHAMMED[2,c], Mohammad Ali KHASAWNEH[2,d], Tasneem SULTANA[3,e]

[1] Department of Mechanical Engineering, Prince Mohammad Bin Fahd University, Al Khobar, Kingdom of Saudi Arabia

[2] Department of Civil Engineering, Prince Mohammad Bin Fahd University, Al Khobar, Kingdom of Saudi Arabia

[3] Artificial Intelligence and Machine Learning, Godutai Engineering College for Women, Sharnbasva University, Kalaburagi, India

202000155@pmu.edu.sa[a]; fshaik@pmu.edu.sa[b*], mnayeemuddin@pmu.edu.sa[c]; mkhasawneh@pmu.edu.sa[d], tasneemsultana841@gmail.com[e]

Keywords: Solar Still, Photocatalytic, Artificial Intelligence, Distillate, Desalination

Abstract. Solar desalination is widely employed technology to separate potable water from saline water. In this study a solar still with one slope was employed to desalinate the saline water. The bottom plate of the solar still was coated with titanium dioxide to improve its performance. The distillate output was collected at three depths of water level in the still for different time intervals. Artificial Intelligence-Levenberg Marquardt (AI-LM) method was employed to predict the distillate output. The predicted values for the response were found to be in good agreement (R^2 = 0.997) with the experimental data.

Introduction

Water is one of the most abundant resources on Earth since it is necessary for human activity and all ecosystems. As the population grows, so does the demand for drinkable water. Large amounts of fresh water are required in every corner of the world for agricultural, industrial, and household purposes [1]. Oceans are a large supply of water, but they are not suitable for human consumption due to their salt [2]. Several writers reported on research in the field of solar desalination, which is the most prominent and cost-effective method, requiring simple technology and maintenance [3]. A single slope solar still is one of the best solar desalination devices, and numerous efforts are being made to improve its productivity [4].

The light absorber's evaporation efficiency remained practically unchanged after thirty cycles of evaporation and condensation. The photothermal layer's excellent light-harvesting properties, its ability to withstand heat, and the substrate's abundance of open channels allow for the system's exceptional photothermal performance for long-term solar desalination [5]. As the contact period extended up to six hours, the salt content progressively reduced by more than 25%. There is no discernible shift in the water's pH has occurred. Salinity in seawater can be efficiently decreased by the hybrid titanium di oxide (TiO_2) [6]. The combined photocatalytic and photothermal system has a high solar energy utilization efficiency and may be used to clean wastewater and produce potable water in one unit [7]. The cost of generating water using solar-based desalination techniques remains greater than that of typical fossil-fuel-based desalination facilities, owing mostly to the high cost of solar collectors. This is one of the primary issues limiting commercialization rates. Residue removal may be expensive in terms of water use and have unfavorable consequences on the environment [8].

Due to the synergistic impact of Polydopamine (PDA) / TiO$_2$ nanoparticles, the Janus evaporator can breakdown over 95% of organic dyes (such as Congo red and trypan blue) and efficiently desalinate and purify other unusual water sources. Potential uses for this Janus structured hydrogel evaporator include desalination and wastewater treatment [9]. The most intriguing aspect is that the solar-driven photothermal effect may potentially be used by a number of other water purification technologies [10]. With the help of an efficient monolithic material platform and a straightforward, reusable, portable, and reasonably priced solar thermal water purification system, water filtration for a range of environmental conditions is revolutionized [11]. Under the sun (i.e., 1 kW m^{-2}), the TiO$_2$-CuO-Cufoam evaporator concurrently exhibits high solar evaporation efficiency of 86.6% and efficiency of 80.0% for the elimination of volatile organic compounds (VOCs) [12]. The photocatalytic hydrogen generation process is a straightforward and economical technique used to produce solar hydrogen by imitating artificial photosynthesis [13]. The traditional method of desalination using passive solar stills relies solely on sun radiation as a source of thermal energy [14]. Seawater desalination by membrane distillation is believed to be attainable by the direct joule heating of the water-hydrophobic membrane interface utilizing a porous thin-film carbon nanotube [15].

Artificial Intelligence-Levenberg Marquardt (AI-LM) in Solar Still Systems

An artificial neural network (ANN) was created to investigate the role of the photocatalyst in the desalination process and supply the quantity of distillate needed for solar photocatalytic modelling. Water is produced via desalination for a variety of uses, including home usage, industrial processing, and water delivery [16]. Desalination methods most commonly used include membrane processes Reverse Osmosis (RO) or thermal desalination [17]. Three data sets training, validation, and test were created using the Fujairah sea water reverse osmosis (SWRO) plant's one-year operating data (n = 200) in order to create the ANN model. Good agreement was produced between the simulated and observed data in the test data set by the trained ANN model (TDS: R^2 = 0.96; flow rate: R^2 = 0.75) [18].

Reverse/Back propagation radial basis function (RBF) neural networks and a multilayer perception (MLPs) trained with the Rprop approach are used to forecast pH values. When the created models were compared to the linear regression methodology, it was found that the MLP and RBF neural network forecasts outperformed those of the conventional methods [19]. An effective method for handling complex and stochastic systems that only accept time series data as input variables and don't fully comprehend physical or hydrogeological components is the artificial neural network (ANN) [20]. With the Nash-Sutcliffe efficiency coefficient of 0.964, the correlation coefficient of 0.983, and the root mean square error of 1.052 km^2, the integrated model showed that this approach effectively reproduced static integrated analysis (SIA) variations [21]. Moreover, the ANN algorithm's performance was assessed in October 1999, during a super storm over the Bay of Bengal [22].

Al-Ghamdi et al., assess the effectiveness of ANNs for short-term water demand forecasting in Jeddah, Saudi Arabia, using a variety of normalization techniques (min-max, z-score, decimal, median, and Median Absolute Deviation (MAD) [23]. The ANN approach has been used to the calibration of an industrial prototype microwave six-port equipment, yielding great accuracy across a broad dynamic range [24] and [25]. Freshwater shortage is a major global concern due to the dramatic increase in demand for freshwater for drinking and personal use that has occurred as a result of population growth. The usefulness of machine learning in forecasting the performance of solar stills has been updated to the point where it is now a central component of numerous studies. Regardless of the size of the dataset, multiple regression models can display a moderate level of prediction accuracy. As opposed to multiple regression models, ANN (Artificial Neural Network) models have an accuracy that is greater and are influenced by dataset sizes ranging from 100 to 400. The models varied in terms of prediction accuracy. Support vector machines, artificial neural networks (ANNs), back-propagation ANNs, and random forests based on Bayesian optimization demonstrated strong prediction ability for hemispheric, inched, tubular, double-slope, and single-slope SSs, respectively. Several models were tested to anticipate thermal efficiency; the highest accurate prediction was produced at six input neurons using an ANN in combination with the Imperialist Competition Algorithm, with an RMSE (root mean square error) of 1.3673 [26].

To anticipate the solar still performance characteristics, a back propagation artificial neural network model was created. For the purpose of predicting solar still performance, the applicability and efficacy of artificial neural networks (ANNs) considering a number of operational and meteorological factors have been assessed. The five input variables pertaining to weather conditions were air temperature (T_o), relative humidity (RH), wind speed (U), solar radiation (Rs), and ultraviolet index (UVI); the four variables pertaining to system operational conditions were brine temperature, feed water temperature, total dissolved solids of brine (TDSB), and brine temperature. The last variable was the number of days, or Julian day (J). Testing and validation processes utilizing statistical criteria indicate that the created artificial neural network (ANN) model may provide extremely high efficiency results, confirming its usability and usefulness in solar desalination prediction. These results bolster the hypothesis that the built artificial neural network (ANN) model correctly anticipated the performance parameters of the solar still.

The primary advantage of the artificial neural network (ANN) model for solar desalination performance prediction is its ease of use, since it can be implemented with ease using any spreadsheet or computer language. The ANN model produced the contribution ratio, which shows how each input variable affects the outputs. In the ANN model for MD and gth prediction, TF is the input parameter with the highest contribution ratio. However, when it comes to objective response rate (ORR) prediction, ultra violet index (UVI) has the biggest contribution ratio. The study also demonstrated the usefulness and effectiveness of artificial neural networks (ANNs), which can forecast solar still performance without the need for additional trials and may result in time, effort, and resource savings [27].

Artificial neural networks, or ANNs, are sophisticated mathematical representations of the nervous system of humans. Over the last three decades, there has been a noticeable surge in the use of artificial neural networks (ANNs) for problem classification, pattern recognition, regression, and forecasting. The input layer, which is the initial layer in a multilayer perception (MLP) architecture, provides input variables to the network. The layers that are located between the input and output levels are known as hidden layers, and the output layer is the last layer. One of the most popular FFNNs is the multilayer perceptron (MLP) neural network. In multilevel perception (MLP), the unidirectional connections between neurons are represented by weights, which are the actual numbers present in the interval [28].

A thorough analysis of recent advancements using Nano/micro materials in solar stills is given. The majority of current efforts have focused on enhancing solar evaporation, which is only one of the fundamental processes in a solar still. To increase the system's productivity and efficiency, a variety of materials were used, including paper-based film, synthetic aerogel, and natural biomaterials. When combined with optimum thermal design and heat localization at the air-water interface, suitable materials can potentially attain an efficiency of over 90% when exposed to 1 kw/m^2 of solar energy. However, even with the application of these materials, productivity is still low. This demonstrates how different solar evaporation systems are from sun stills [30].

Case Study
The still was filled with saline water and exposed to natural sunlight. Three different levels of water were considered and at each level the distillate output was measured for various time intervals. The bottom plate of the still was coated with photocatalyst (Titanium dioxide) to enhance the performance of the still. Figure 1 displays the schematic diagram for the still.

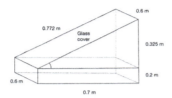

Fig 1. Schematic Diagram of Solar Still

The experimental data was utilized to test, train and predict the distillate output using ANN-LM algorithm. The Levenberg-Marquardt algorithm (LM) and damped least squares (DLS) approach were utilized in MATLAB R2023 by Math Works, Inc. to optimize the input variables [31]. A MATLAB tool called the Artificial Neural Network is used to computationally model data variables in order to investigate the links between input and output in systems or processes [32]. The size of the network is similar to the number of neurons in the brains of living beings that enable intelligent behavior. It consists of input nodes for the input independent variables of the experimental data and output nodes for the factors or dependent variables of the response [33]. S neurons, multilayer neural networks, and a multi-layer network with R input components were all used in the current optimization technique. Three subsets of the data were used, comprising of 70%, 15%, and 15% of the total [34]. Two subsets are utilized for testing and validation, and the first subset is used for training. The distillation target values and model output are displayed in Figure 2. The comparison of the experimental and expected values is displayed in Table 1.

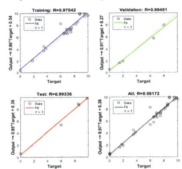

Fig 2. ANN network model output values and the target values of distillate still

Table 1. Experimental and ANN-LM Predicted Data

Depth, m	Time, hours	Solar Radiation W/m²	Amount of Distillate m³	Prediction
0.0333	8	500	0	0.00000402
0.0333	9	727	0.000015	0.00001605
0.0333	10	932	0.000035	0.00003705
0.0333	11	1045	0.00006	0.00005340
0.0333	12	1136	0.000065	0.00006545
0.0333	13	1114	0.00008	0.00007937

0.0333	14	1023	0.000095	0.00009445
0.0333	15	886	0.000085	0.00008508
0.0333	16	682	0.000075	0.00006860
0.0333	17	409	0.00007	0.00007149
0.0333	18	200	0.00006	0.00008233
0.025	8	500	0	0.00000509
0.025	9	727	0.00002	0.00001913
0.025	10	932	0.000038	0.00004068
0.025	11	1045	0.000065	0.00005742
0.025	12	1136	0.000075	0.00007279
0.025	13	1114	0.000085	0.00008831
0.025	14	1023	0.000098	0.00009792
0.025	15	886	0.000088	0.00009828
0.025	16	682	0.000076	0.00007577
0.025	17	409	0.00007	0.00006599
0.025	18	200	0.000065	0.00006684
0.0133	8	500	0	0.00000598
0.0133	9	727	0.000018	0.00002056
0.0133	10	932	0.000035	0.00004447
0.0133	11	1045	0.000068	0.00006771
0.0133	12	1136	0.000086	0.00008650
0.0133	13	1114	0.0001	0.00009622
0.0133	14	1023	0.000095	0.00009924
0.0133	15	886	0.000088	0.00009948
0.0133	16	682	0.000082	0.00008161
0.0133	17	409	0.000074	0.00007310
0.0133	18	200	0.000069	0.00007328

Results and Discussion
The use of Artificial Intelligence (AI) Levenberg Marquardt method in predicting the distillate output for a photocatalytic solar still has yielded promising results. The AI models, including machine learning algorithms and neural networks, demonstrated high accuracy in forecasting distillate output based on various input parameters such as time and solar radiation. These models were able to effectively capture the complex relationships between these factors and distillate output, leading to precise predictions. The application of AI in this context offers several advantages, including improved operational efficiency, better planning of distillation processes, and enhanced performance of photocatalytic solar stills.

Conclusion
Solar photocatalytic still was employed to desalinate the saline water. The output distillate was collected at three depths of water levels for various time intervals. The experimental data was utilized to train, test and predict the distillate output using artificial neural network-Levenberg Marquardt ANN-LM algorithm. The expected values and the experimental values were found to be in good agreement for the response with an average of $R^2 = 0.997$.

References

[1] N. Mohammed, P. Palaniandy, F. Shaik, and H. Mewada, "Statistical Modelling of Solar Photocatalytic Biodegradability of Seawater Using Combined Photocatalysts," *J. Inst. Eng. India Ser. E*, Sep. 2023. https://doi.org/10.1007/s40034-023-00274-8

[2] N. Mohammed, P. Palaniandy, F. Shaik, H. Mewada, and D. Balakrishnan, "Comparative studies of RSM Box-Behnken and ANN-Anfis fuzzy statistical analysis for seawater biodegradability using TiO2 photocatalyst," *Chemosphere*, vol. 314, p. 137665, Feb. 2023. https://doi.org/10.1016/j.chemosphere.2022.137665

[3] F. Shaik, S. S. R. Al Siyabi, N. Mohammed, and M. Eltayeb, "Assessment of ground and surface water quality and its contamination," *International Journal of Environmental Analytical Chemistry*, vol. 103, no. 6, pp. 1449–1467, May 2023. https://doi.org/10.1080/03067319.2021.1873975

[4] N. Mohammed, P. Palaniandy, F. Shaik, and H. Mewada, "EXPERIMENTAL AND COMPUTATIONAL ANALYSIS FOR OPTIMIZATION OF SEAWATER BIODEGRADABILITY USING PHOTO CATALYSIS," *IIUMEJ*, vol. 24, no. 2, pp. 11–33, Jul. 2023. https://doi.org/10.31436/iiumej.v24i2.2650

[5] M. Shafaee, E. K. Goharshadi, M. M. Ghafurian, M. Mohammadi, and H. Behnejad, "A highly efficient and sustainable photoabsorber in solar-driven seawater desalination and wastewater purification," *RSC Adv.*, vol. 13, no. 26, pp. 17935–17946, 2023. https://doi.org/10.1039/D3RA01938A

[6] R. Isha and N. H. Abd Majid, "A Potential Hybrid TiO$_2$ in Photocatalytic Seawater Desalination," *AMR*, vol. 1113, pp. 3–8, Jul. 2015. https://doi.org/10.4028/www.scientific.net/AMR.1113.3

[7] L. Shi *et al.*, "An Integrated Photocatalytic and Photothermal Process for Solar-Driven Efficient Purification of Complex Contaminated Water," *Energy Tech*, vol. 8, no. 9, p. 2000456, Sep. 2020. https://doi.org/10.1002/ente.202000456

[8] Y. Zhang, M. Sivakumar, S. Yang, K. Enever, and M. Ramezanianpour, "Application of solar energy in water treatment processes: A review," *Desalination*, vol. 428, pp. 116–145, Feb. 2018. https://doi.org/10.1016/j.desal.2017.11.020

[9] J. Wen *et al.*, "Architecting Janus hydrogel evaporator with polydopamine-TiO2 photocatalyst for high-efficient solar desalination and purification," *Separation and Purification Technology*, vol. 304, p. 122403, Jan. 2023. https://doi.org/10.1016/j.seppur.2022.122403

[10] Y. Lu, H. Zhang, D. Fan, Z. Chen, and X. Yang, "Coupling solar-driven photothermal effect into photocatalysis for sustainable water treatment," *Journal of Hazardous Materials*, vol. 423, p. 127128, Feb. 2022. https://doi.org/10.1016/j.jhazmat.2021.127128

[11] Y. Yang *et al.*, "Graphene-Based Standalone Solar Energy Converter for Water Desalination and Purification," *ACS Nano*, vol. 12, no. 1, pp. 829–835, Jan. 2018. https://doi.org/10.1021/acsnano.7b08196

[12] Y. Tian *et al.*, "High-performance water purification and desalination by solar-driven interfacial evaporation and photocatalytic VOC decomposition enabled by hierarchical TiO$_2$-CuO nanoarchitecture," *Intl J of Energy Research*, vol. 46, no. 2, pp. 1313–1326, Feb. 2022. https://doi.org/10.1002/er.7249

[13] M. S. Yesupatham *et al.*, "Photocatalytic seawater splitting for hydrogen fuel production: impact of seawater components and accelerating reagents on the overall performance," *Sustainable Energy Fuels*, vol. 7, no. 19, pp. 4727–4757, 2023. https://doi.org/10.1039/D3SE00810J

[14] R. Djellabi *et al.*, "Recent advances and challenges of emerging solar-driven steam and the contribution of photocatalytic effect," *Chemical Engineering Journal*, vol. 431, p. 134024, Mar. 2022. https://doi.org/10.1016/j.cej.2021.134024

[15] M. Fujiwara and M. Kikuchi, "Solar desalination of seawater using double-dye-modified PTFE membrane," *Water Research*, vol. 127, pp. 96–103, Dec. 2017. https://doi.org/10.1016/j.watres.2017.10.015

[16] N. Mohammed, A. Asiz, M. A. Khasawneh, H. Mewada, and T. Sultana, "Machine learning and RSM-CCD analysis of green concrete made from waste water plastic bottle caps: Towards performance and optimization," *Mechanics of Advanced Materials and Structures*, pp. 1–9, Aug. 2023. https://doi.org/10.1080/15376494.2023.2238220

[17] N. Mohammed, P. Palaniandy, and F. Shaik, "Pollutants removal from saline water by solar photocatalysis: a review of experimental and theoretical approaches," *International Journal of Environmental Analytical Chemistry*, vol. 103, no. 16, pp. 4155–4175, Dec. 2023. https://doi.org/10.1080/03067319.2021.1924160

[18] Y. G. Lee *et al.*, "Artificial neural network model for optimizing operation of a seawater reverse osmosis desalination plant," *Desalination*, vol. 247, no. 1–3, pp. 180–189, Oct. 2009. https://doi.org/10.1016/j.desal.2008.12.023

[19] H. Adel Zaqoot, A. Baloch, A. Khalique Ansari, and M. Ali Unar, "APPLICATION OF ARTIFICIAL NEURAL NETWORKS FOR PREDICTING pH IN SEAWATER ALONG GAZA BEACH," *Applied Artificial Intelligence*, vol. 24, no. 7, pp. 667–679, Aug. 2010. https://doi.org/10.1080/08839514.2010.499499

[20] S. Al Ajmi, M. Ali Syed, F. Shaik, M. Nayeemuddin, D. Balakrishnan, and V. R. Myneni, "Treatment of Industrial Saline Wastewater Using Eco-Friendly Adsorbents," *Journal of Chemistry*, vol. 2023, pp. 1–11, Apr. 2023. https://doi.org/10.1155/2023/7366941

[21] D. Li *et al.*, "Simulation of Seawater Intrusion Area Using Feedforward Neural Network in Longkou, China," *Water*, vol. 12, no. 8, p. 2107, Jul. 2020. https://doi.org/10.3390/w12082107

[22] B. Jena, D. Swain, and A. Tyagi, "Application of Artificial Neural Networks for Sea-Surface Wind-Speed Retrieval From IRS-P4 (MSMR) Brightness Temperature," *IEEE Geosci. Remote Sensing Lett.*, vol. 7, no. 3, pp. 567–571, Jul. 2010. https://doi.org/10.1109/LGRS.2010.2041632

[23] A.-B. Al-Ghamdi, S. Kamel, and M. Khayyat, "Evaluation of Artificial Neural Networks Performance Using Various Normalization Methods for Water Demand Forecasting," in *2021 National Computing Colleges Conference (NCCC)*, Taif, Saudi Arabia: IEEE, Mar. 2021, pp. 1–6. https://doi.org/10.1109/NCCC49330.2021.9428856

[24] Yi Liu, "Calibrating an industrial microwave six-port instrument using the artificial neural network technique," *IEEE Trans. Instrum. Meas.*, vol. 45, no. 2, pp. 651–656, Apr. 1996. https://doi.org/10.1109/19.492804

[25] N. N, A. Siva Kumaran K, A. A, A. V. S, and B. M. J, "Convolutional Neural Networks (CNN) based Marine Species Identification," in *2022 International Conference on Automation, Computing and Renewable Systems (ICACRS)*, Pudukkottai, India: IEEE, Dec. 2022, pp. 602–607. https://doi.org/10.1109/ICACRS55517.2022.10029109

[26] A. F. Mashaly, A. A. Alazba, A. M. Al-Awaadh, and M. A. Mattar, "Predictive model for assessing and optimizing solar still performance using artificial neural network under hyper arid environment," *Solar Energy*, vol. 118, pp. 41–58, Aug. 2015. https://doi.org/10.1016/j.solener.2015.05.013

[27] F. A. Essa, M. Abd Elaziz, and A. H. Elsheikh, "An enhanced productivity prediction model of active solar still using artificial neural network and Harris Hawks optimizer," *Applied Thermal Engineering*, vol. 170, p. 115020, Apr. 2020. https://doi.org/10.1016/j.applthermaleng.2020.115020

[28] A. O. Alsaiari, E. B. Moustafa, H. Alhumade, H. Abulkhair, and A. Elsheikh, "A coupled artificial neural network with artificial rabbits optimizer for predicting water productivity of different designs of solar stills," *Advances in Engineering Software*, vol. 175, p. 103315, Jan. 2023. https://doi.org/10.1016/j.advengsoft.2022.103315

[29] G. Peng *et al.*, "Micro/nanomaterials for improving solar still and solar evaporation -- A review." arXiv, Jun. 20, 2019. https://doi.org/10.48550/arXiv.1906.08461

[30] G. Peng *et al.*, "Potential and challenges of improving solar still by micro/nano-particles and porous materials - A review," *Journal of Cleaner Production*, vol. 311, p. 127432, Aug. 2021. https://doi.org/10.1016/j.jclepro.2021.127432

[31] N. Mohammed, P. Palaniandy, and F. Shaik, "Optimization of solar photocatalytic biodegradability of seawater using statistical modelling," *Journal of the Indian Chemical Society*, vol. 98, no. 12, p. 100240, Dec. 2021. https://doi.org/10.1016/j.jics.2021.100240

[32] N. Mohammed, P. Palaniandy, F. Shaik, B. Deepanraj, and H. Mewada, "Statistical analysis by using soft computing methods for seawater biodegradability using ZnO photocatalyst," *Environmental Research*, vol. 227, p. 115696, Jun. 2023. https://doi.org/10.1016/j.envres.2023.115696

[33] N. Mohammed, P. Palaniandy, and F. Shaik, "Solar photocatalytic biodegradability of saline water: Optimization using RSM and ANN," *AIP Conference Proceedings*, vol. 2463, no. 1, p. 020027, May 2022. https://doi.org/10.1063/5.0080297

[34] N. Mohammed, P. Palaniandy, and F. Shaik, "Advanced Oxidation Processes for the Treatment of Pollutants from Saline Water," *Removal of Pollutants from Saline Water: Treatment Technologies*, p. 95, 2021.

Power systems stability of high penetration of renewable energy generations

Ahmad HARB[1,a*], Hamza NAWAFLEH[2,b]

[1] German Jordanian University, Amman-Jordan

[2] University of Wisconsin Millwakuaee, USA

[a] ahmad.harb@gju.edu.jo, [b] Alnawf2@uwm.edu

Keywords: Power Systems Stability, Renewable Energy, Impacts of Renewable Energy on Power Systems

Abstract. In this Paper a comprehensive analysis of the Jordanian Power Grid (JPG)'s stability under various practical scenarios, including load disturbances and the integration of Renewable Energy Sources (RES). A key focus of the study is the impact of RES on the stability of the JPG, especially during unexpected disturbances. The findings reveal a notable trend: higher RES integration tends to decrease the grid's stability under certain conditions. Additionally, the report explores the effects of interconnecting the JPG with neighboring countries, such as Egypt. This connection is shown to potentially enhance the JPG's stability, both with and without the involvement of RES. The report delves into numerous cases, providing detailed discussions and insights. The conclusions drawn emphasize the critical importance of carefully managing the proportion of RES in the JPG to maintain its stability against various disturbances. This study offers valuable recommendations for future strategies to ensure the robustness and reliability of the Jordanian Power Grid in the face of evolving energy landscapes.

Introduction

The Hashemite Kingdom of Jordan is significantly reliant on foreign energy sources, with an import ratio of approximately 92% [1] The country's energy demand is on the rise, spurred by factors such as population growth and the influx of refugees from neighboring regions. Jordan faces substantial challenges in meeting its electricity needs due to limited local fossil fuel resources, inadequate conversion capacities, and the financial constraints of its energy sector. Renewable Energy Sources (RES) have emerged as a pivotal solution for Jordan, offering a means to secure electricity supply while protecting the environment [2–4] The nation's abundant solar and wind resources make this a particularly viable option. While RES contributes to reducing emissions and enhancing supply security, they also introduce greater uncertainty and variability in the transmission and distribution of power [5 - 7]. This dynamic has been complicated by the current economic impracticality of large-scale energy storage, which exacerbates the challenge of balancing generation with real-time demand. Moreover, while Distributed Generation curtails losses associated with electricity transport and transformation, it adds complexity to the system, necessitating advanced, research-based solutions [8 - 10]

Globally, there is a growing interest in transitioning to RES, and Jordan has made significant strides in this direction. Over the past decade, approximately 3500 MW of RES has been integrated into the Jordanian Power Grid (JPG), accounting for 30% of total generation as of 2022 [8 - 10] However, this increased reliance on RES presents operational challenges for the JPG. The main objective of this research is to investigate the effects of integrating large-scale renewable energy sources on the transient behavior and sustainability of the JPG. Utilizing the Power analysis program DIGSILENT, along with current system steady-state and dynamic models, this study will focus on the impacts of RES integration on JPG's voltage profile, sustainability, and overall stability. The analysis will shed light on the challenges and dynamics introduced by the increased share of renewable energy within the Jordanian power landscape.

Methodology
The methodology of this study centers on evaluating the stability of the Jordanian Power Grid (JPG) with a high Renewable Energy Sources (RES) share. To achieve this, the study employs DIGSILENT power software, a tool adept at assessing electrical systems, particularly in power transmission and distribution contexts. The JPG's operational model is meticulously simulated within DIGSILENT to investigate whether the system can maintain synchronism following significant transient disturbances. The key to this study is the simulation and analysis of various challenging scenarios, such as abrupt generation loss and frequency mismatches, within the JPG model. The robustness of the system's synchronism is rigorously tested under scenarios like sudden loss of generation. These tests are conducted across different grid load conditions—specifically peak and low load periods—under high RES integration scenarios. Additionally, the stability of the system's frequency is thoroughly examined. This aspect is crucial, given that system frequency significantly influences the performance of power generating units at all voltage levels, especially during extreme imbalances.

Results and Discussions
In this section, we explore two distinct scenarios simulated for the Jordanian Power Grid (JPG) in the year 2025, under conditions of high Renewable Energy Sources (RES) integration. These simulations are aimed at assessing the impact of increased RES on the stability of the JPG.

Case 1: Peak Load Scenario in 2025
Under peak load conditions with high integration of renewable projects, the network analysis reveals a robust system. A hypothetical disturbance, such as a complete loss of power generation from all wind farms due to a wind storm, was simulated. Despite this extreme scenario, the Jordanian electrical system maintained stable performance in terms of voltage, frequency, and generator power angles and speeds. No Under Frequency Load Shedding (UFLS) was necessary as the frequency remained above the critical threshold. The National Grid summary for this category in Year 2025 in peak load situation and high renewable projects integration is shown in Table 1.

Table 1: JPG summary for case 1

National Grid summary	
Grid Demand	4670MW
Total Installed Capacity	8200MW
Wind integration	774MW
Solar integration	1710MW
Spinning Reserve	820 MW
Grid Losses	96 MW
External In feed (Egypt)	7MW (to Egypt)

In this specific case, no major problems have been detected and the Jordanian electrical system shows stable behavior in voltage, frequency and generators power angle and speed, No UFLS detected since the frequency did not reach or drop below the first stage threshold of 49.1HZ, as shown in Figs1, 2, and 3.

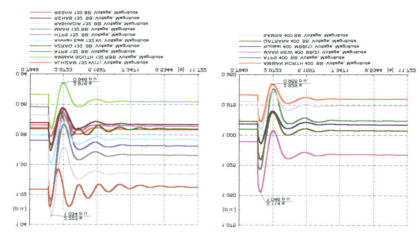

Figure 1: Voltage behavior for selected 132kV and 400kV busbars in the JPG [10]

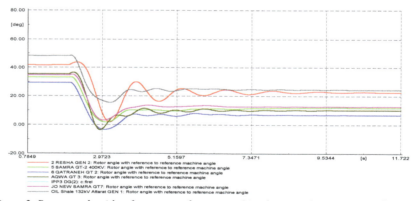

Figure 2: Power angle with reference to reference machine for several generators in the JPG

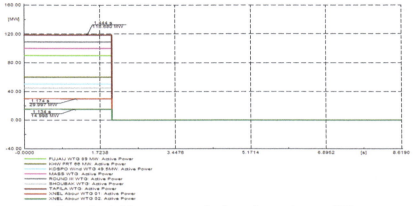

Figure 3: wind farms response after Loss of generation case [10]

Case 2: Low Load Scenario in 2025
In this scenario, the study considers a low load situation with high RES integration. The simulation predicts an increased reliance on RES, including wind and solar, alongside new primary energy resources like oil shale and nuclear power. When simulating a complete loss of power generation from all wind farms during a wind storm, it was observed that the system could handle the disturbance without major issues. The JPG summary for this category in Year 2025 low load situation and high renewable projects integration is shown in Table 2 below.

Table 2: JPG summary in 2025

National Grid summary	
Grid Demand	2100MW
Total Installed Capacity	8200MW
Wind integration	774MW
Solar integration	1710MW
Spinning Reserve	930MW
Grid Losses	91.4MW
External In feed (Egypt)	200MW (to Egypt)

In case the system forced to operate at 80% of renewable integration where only 500MW of conventional generation presented and the rest of power is covered by wind and PV, (the planning value of renewable is about 3000MW of approved projects until year 2025), the system will never converge or solve, the below errors as shown in Figure 4 will occur, where the inertia in the system is not sufficient to develop any response to the studied events [10]

Figure 4: Error massage in Simulation of 80% RES (Wind and PV integration).

In this case the assumption is to operate several conventional generation in the system in the minimum possible power dispatch condition in order to start the initial condition for simulation and to increase system inertia, the simulation is started with 57% of RES integration by making some of wind farms in the southern area in operation condition and switching off more PV projects in the grid, then implement wind and Solar power curtailment scheme, Results of case 4 (loss of several wind farms) are shown below, in this case the system is considered stable after loss of wind event.

Here below selected monitored elements and simulation results are shown. In this specific case, no major problems have been detected and the Jordanian electrical system shows stable behavior in voltage, frequency and generators power angle and speed, No UFLS detected even the frequency drop below the fourth stage threshold of 48.4HZ but the drop duration was less than 0.2 seconds. The simulation results of this case are presented, for low spring load of 2150MW in year 2025 with maximum RES integration. The most critical contingency in the low load condition appears to be the simultaneous trip of all wind farms, particularly in the southern green corridor. This leads to a significant power deficit of 612 MW. Nevertheless, the system's resilience is primarily due to the support from interconnected systems, namely Egypt and Syria. The simulation results show that peak power transfers, particularly from the Egyptian system, could exceed protective settings, posing a risk of separation from the Egyptian grid. This highlights the importance of careful management and coordination in interconnected grid operations, especially under high RES integration scenarios.

Conclusion

The study of the Jordanian Power Grid (JPG) for 2025 underlines its robustness amidst high Renewable Energy Sources (RES) integration. While the grid-maintained stability during peak loads even with extreme disruptions like total wind power loss, challenges arose in low load conditions at 80% RES integration due to insufficient system inertia. Strategic adjustments, including a balanced mix of renewable sources and curtailment strategies, were key to stabilizing the grid. The study highlights the grid's vulnerability to simultaneous wind farm disconnections, underscoring the importance of meticulous planning, balanced energy mix, and strong interconnectivity protocols in ensuring grid resilience in a renewable-dominant future.

References

[1] H. Alnawafah and A. Harb, "Modeling and Control for Hybrid Renewable Energy System in Smart Grid Scenario - A Case Study Part of Jordan Grid," in 2021 12th International Renewable Energy Congress (IREC), 2021, pp. 1–6. https://doi.org/10.1109/IREC52758.2021.9624739

[2] L. Xiong, X. Liu, D. Zhang, and Y. Liu, "Rapid Power Compensation Based Frequency Response Strategy for Low Inertia Power Systems," IEEE J Emerg Sel Top Power Electron, vol. 6777, no. c, pp. 1–1, 2020. https://doi.org/10.1109/jestpe.2020.3032063

[3] L. Xiong, X. Liu, D. Zhang, and Y. Liu, "Rapid Power Compensation-Based Frequency Response Strategy for Low-Inertia Power Systems," IEEE J Emerg Sel Top Power Electron, vol. 9, no. 4, pp. 4500–4513, Aug. 2021. https://doi.org/10.1109/JESTPE.2020.3032063

[4] M. Alkasrawi, E. Abdelsalam, H. Alnawafah, F. Almomani, M. Tawalbeh, and A. Mousa, "Integration of solar chimney power plant with photovoltaic for co-cooling, power production, and water desalination," Processes, vol. 9, no. 12, 2021. https://doi.org/10.3390/pr9122155

[5] B. Kroposki et al., "Achieving a 100% Renewable Grid: Operating Electric Power Systems with Extremely High Levels of Variable Renewable Energy," IEEE Power and Energy Magazine, vol. 15, no. 2, pp. 61–73, Mar. 2017. https://doi.org/10.1109/MPE.2016.2637122

[6] M. Velasco, P. Marti, J. Torres-Martinez, J. Miret, and M. Castilla, "On the optimal reactive power control for grid-connected photovoltaic distributed generation systems," in IECON 2015 - 41st Annual Conference of the IEEE Industrial Electronics Society, IEEE, Nov. 2015, pp. 003755–003760. https://doi.org/10.1109/IECON.2015.7392686

[7] E. Abdelsalam et al., "Performance analysis of hybrid solar chimney–power plant for power production and seawater desalination: A sustainable approach," Int J Energy Res, vol. 45, no. 12, pp. 17327–17341, 2021. https://doi.org/10.1002/er.6004

[8] Hamza Alnawafah, "MODELING AND CONTROL FOR HYBRID RENEWABLE ENERGY SYSTEM IN SMART GRID SCENARIO A CASE STUDY PART OF JORDAN GRID," German Jordanian University, Amman, 2020. Accessed: Sep. 08, 2023. [Online]. Available: http://hip.jopuls.org.jo/c/portal/layout?p_l_id=PUB.1009.1&p_p_id=search_WAR_fusion&p_p_action=1&p_p_state=normal&p_p_mode=view&p_p_col_id=column-1&p_p_col_pos=0&p_p_col_count=2&_search_WAR_fusion_action=navigate&_search_WAR_fusion_navigationData=search%7E%3D1%7E%21BIB%7E%212147483647%7E%217123071

[9] H. Alnawafah, R. Sarrias-Mena, A. Harb, L. M. Fernández-Ramírez, and F. Llorens-Iborra, "Evaluating the inertia of the Jordanian power grid," Computers and Electrical Engineering, vol. 109, Aug. 2023. https://doi.org/10.1016/j.compeleceng.2023.108748

[10] A. T. M. Abu Dyak, "The Effect of Connecting Large Scale Renewable Project on Power Quality And Voltage Control Of The Jordanian Electrical Grid," German Jordanian University, Amman, 2017.

Assessment of utilizing solar energy to enhance the performance of vertical aeroponic farm

Esam Jassim[1,a,*], Bashar Jasem[2,b], Faramarz Djavanroodi[1,c]

[1]College of Engineering, Prince Muhammad Bin Fahd University, 31952 Al-Khobar, Kingdom of Saudi Arabia

[2]Department of Computer Technology, Al-Hadba University College, Mosul, Iraq.

[a]ejassim@pmu.edu.sa, [b]adamalzuhairee@gmail.com, [c]fdjavanroodi@pmu.edu.sa

Abstract. The manuscript presents the significance of utilizing renewable energy in vertical aeroponic farming system. Components such as the tower garden, the solar panels, the water pump, and the control unit are integrated and designed to generate sufficient solar energy to power the system. The environmental conditions inside the chamber are optimized for plant growth, to maintain 18-24 °C temperature and relative humidity of 75-85 %. The paper also discusses the challenges, limitations, and future recommendations for improving the system. Finally, the paper provides some future recommendations for improving the design and performance of the solar powered vertical farm.

Introduction

Vertical farming (VF) is a process to produce food via utilizing vertical dimension for hydroponic growing of crops with indoor controlled-environment agriculture. Vertical gardens play a key role in tackling the ongoing challenges that modern cities encounters, due to a rapid growing in urban environment associated with the scarcity of water and the major growing in population, crises in food demand are emerged. Researchers predict about 25%-70% increasing in the crops needs by 2050 [1]. However, with the decrement in the availability of growing lands together with the worsen of climate condition, rises the demand of VF. The term "vertical farming" refers to the utilization of several vertical layers of crops planted within a warehouse to create an artificial growth environment that mimics horizontal farming.

Food costs remain augmented for several factors such as including population growing and food shortage. Although traditional Horizontal Farming HF are less expensive to maintain and build than greenhouses, such method still requires large landscape and fertilized soil. VF, on the other hand, is notorious for using a large amount of electricity, triggering the alarm of the essential of finding alternative source of energy. Renewable resources are currently drawing attention to sub the conventional electrical power. Hence, solar power became a popular choice for supplying energy to such devices since it is both practical and readily available. Solar lighting systems have a lot of promise, as demonstrated by a UK research that examined the potential cost reductions associated with their use. The analysis reveals 56% to 89% of possible savings [2]. The use of solar energy in vertical farming would thus benefit the entire process and make it even more dependable and sustainable.

Incorporating solar energy into vertical farming systems is a sustainable approach that enhances the overall efficiency and environmental impact of the agricultural process. Research indicates that integrating solar panels into vertical farming structures can provide a reliable and renewable energy source, powering LED lights, climate control systems, and other essential equipment. Such incorporation not only reduces the dependence on conventional energy grids but also mitigates the contribution of carbon by-product associated with the production of food. However, power intermittency and energy storage of solar panels are still of ongoing challenges, in particular integration into vertical farming. Hence, more studies are in need to optimize a sustainable and self-sufficient agricultural model.

Furthermore, studies emphasized that the incorporation of solar energy in vertical farming can lead to increased energy autonomy and cost-effectiveness over the long term. Solar panels strategically positioned on the vertical farming structures, harness sunlight for electricity and minimizing reliance on conventional power sources. This sustainable energy approach associates with the continuous demand for eco-friendly agricultural practices, addressing concerns related to climate change and conventional energy consumption in food production.

Innovations in solar technology, such as advancements in photovoltaic efficiency and energy storage solutions, are crucial in overcoming challenges associated with intermittent sunlight availability. Ongoing research explores the optimization of energy capture and storage mechanisms to ensure continuous and reliable power supply, regardless of weather conditions or time of day.

Moreover, incorporating solar energy in vertical farming is vital to the development of off-grid farming solutions, particularly in remote areas. This approach not only facilitates increased food production in regions with limited access to conventional power but also promotes sustainable agriculture practices that prioritize environmental stewardship.

As the field continues to evolve, interdisciplinary collaboration between agriculture, engineering, and renewable energy sectors becomes pivotal. This collaborative effort aims to refine and scale up solar-powered vertical farming systems, fostering a more feasible and sustainable future for agriculture.

Although one of the most flaws in VF systems is the high energy consumption, incorporation of renewable energy sources (RE) such as solar and wind could mitigate the capital and operation costs. Hence, utilizing renewable energy in VF systems have intensively addressed by recent literatures to assess the promotion of sustainability in urban agriculture.

System design
The prototype was initially built based on the standards of greenhouses, but it was customized to meet certain specifications. The final dimensions and design of the building are illustrated in Fig (1). Carbon steel, a material with high tensile strength and rigidity, was selected as the main structural material for the prototype. Insulation was selected to minimize the energy interaction between the system and the surroundings. Factors such as thermal resistance, ageing, moisture accumulation and thickness are prioritized during the selection process.

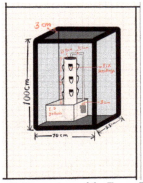

Fig. (1) Dimensions of the Tower Farm

LED lighting is the best choice for indoor farming for various reasons. It can reduce energy consumption for different crops as well as easy to control. The quality can be tailored to suit the photosynthetic pigments and photoreceptors of the crop by adjusting its intensity and spectrum[2]. This is important in vertical farming since plants can sense and respond to changes in light intensity and illuminance. LED lights can also influence the flavor, nutritional value and shelf life of crops.

Red/blue LED lights can enhance the growth and biomass of C. LED lights can also improve the growth, color, flavor, and phytonutrient content of leafy greens in controlled environments by adjusting the light spectrum [2].

Furthermore, LED lights are more energy-efficient than other forms of lighting, as they consume less energy for running lights and climate management [3]. LED lights also affect the heat load of the room. The heat load was lower when LED lights were off, and higher when they were on [4].

A recent experiment showed that LED grow lights accounted for 78% of the daily heat load on average, and also accelerated the processes of respiration and evaporation in plants. Moreover, an energy balance analysis revealed that LED lighting increased the evapotranspiration and heat loads by up to five times compared to a scenario without LED lighting. Therefore, it is important to choose the right type and intensity of LED grow lights for different plants and growth stages [4].

Solar resource assessment
Saudi Arabia, has an excellent solar power potential due to its long hours of light, few days with clouds (particularly, in eastern province), and elevated Direct Normal radiation DNI, which reaches a peak power density of 4800 kWh/m^2/year [5].

Such places are viable and highly positioned among the forefront of countries with the ability to produce electrical power using solar energy [6].

The DNI is a crucial factor to consider when evaluating the viability of a geographical area for the implementation of CSP technology. It is worth mentioning that CSP systems are generally considered viable in locations where the direct normal irradiance (DNI) surpasses 1800 kWh/m^2/year [7, 8].

FPS modeling:
This section examines the viability of the suggested FSP systems according to the conditions of east province in Saudi Arabia. Annual energy output (GWh), which refers to the sum of the hourly energy produced over the course of a year. The following equation is used to calculate annual energy output:

$$AE = \sum_{t=1}^{t=8760} Q_t$$

where t represents time (hour) and Q_t stand for the generated energy. The capacity factor, defined as a ratio of the total number of hours of electricity generated in an FSP plant relative to its nominal capacity over the course of one year, can be computed by the following equation:

$$Capacity\ Factor = Net\ Annual\ Energy \frac{\frac{kWh_{ac}}{yr}}{\frac{System\ Capacity\ (kWh_{ac})}{8760}}$$

$$CF = \frac{\frac{AE}{system\ capacity}}{8760\ (hr/year)}$$

The annual energy and capacity factor in eastern province in Saudi Arabia are determined to be 209 GWhr and 72.5% respectively.

Two 300-watt solar panels were utilized to provide the power demanded by the VF system. The size of each panel is 164 x 99.2 x 4 cms. The output electrical energy (E), in kWh, is determined from the following expression:

$$E = A \times \eta \times I \times PR$$

where: A= area of the solar panel;
η= solar panel efficiency (roughly 15%);
I = average annual solar radiation on titled panels (5.337 kWh/m² day for the optimal tilt angle 20.833°)
PR= performance ratio (average value = 0.75)

Results and discussion
A series of tests have been conducted on the system, including lighting, cooling, humidifier, dehumidifier and irrigation system, as illustrated in Fig. (2). The system run twice for each test: full day and half day. To create a suitable environment for lettuce growth, Styrofoam insulation was utilized to line a wooden frame that holds the nutrient solution as well as serves as a floating platform for the plants. Since the optimal conditions for lettuce are 75-85% humidity and 18-24 °C temperature, growth lights, an evaporative cooler, a humidifier, and a dehumidifier were installed to control the system thermal conditions. The growth lights have a power of 16W, while the evaporative cooler consumes 65W. A submersible pump utilized to circulate the nutrient solution requires 13W to operate at maximum flow rate of 700L/h and vertical head (H_{max}) of 1.0m. The humidifier has a wattage of 25W, and it can filter impurities from water and adjust the fog intensity. The dehumidifier, on the other hand, operates at consumption up to 25W and it can recover about 0.4kWh to dry.

The nutrient solution was pumped from the reservoir to the top of the garden tower, where it drips down over the exposed roots of the plants. A timer ensures that this process is repeated regularly to provide adequate oxygen, water, and nutrients to the plants.

The energy produced by each panel with specification elaborated earlier is 0.977 kWhr for a total of 1.95 kW hr. Equipment of the system needs 1.728 kWhr to operate at full load condition for 12 hrs. Hence, the system is durable and the technologies performed well from both the intended purpose and the eco-friendly environment.

Fig. (2) Lighting and humidification processes

Conclusion

This paper presents a design of a solar powered vertical farm that uses aeroponic tower garden technology to grow lettuce in a closed chamber. The main objectives of this design are to address the scarcity of power source in remote area, the reduction of environmental impact resulted by conventional agriculture, and the provision of fresh and healthy food for local consumption by incorporating of renewable energy. The paper describes the design process, the prototype fabrication, and the testing and analysis of the system. The design process involved selecting the appropriate materials and components for the aeroponic tower garden, the solar system, and the control system. The aeroponic tower garden consists of a vertical structure with multiple planting ports that spray nutrient-rich water to the roots of the plants. Design of the solar system consists of photovoltaic panels, batteries, charge controller, and inverter that provide electricity to power the pumps, AC, humidifier, dehumidifier, sensors, and lights.

The prototype was tested twice and produced the required environment for producing healthy and fresh lettuce. The paper discusses the results and analyses of the prototype performance, comparing it with other similar systems and highlighting its advantages and limitations.

References

[1] Yahya, O., Yahya, G., Al-Omair, A., Yahya, E., Jassim, E., Djavanroodi, F. (2023). "Aeroponic tower garden solar powered vertical farm". Volume 31, Pages 287 - 2982023 International Conference on Advanced Topics in Mechanics of Materials, Structures and

Construction, AToMech1 2023Al Khobar 12 March 2023. https://doi.org/10.21741/9781644902592-30

[2] Jahan, N., Moody, B., Kerr, J., Cynthia, D, & Lawson, C. "Grow lights for your vertical garden: A complete guide. Garden Tabs. Retrieved", February 16, 2022.

[3] SharathKumar, M., Heuvelink, E., & Marcelis, L. F. M. (2020). Vertical Farming: Moving from Genetic to Environmental Modification. Trends in Plant Science, 25(8), 724–727. https://doi.org/10.1016/j.tplants.2020.05.012

[4] Omar Yahya, Gharam Yahya, Ahmad Al-Omair, Emad Yahya, Esam Jassim, Faramarz Djavanroodi, Aeroponic tower garden solar powered vertical farm, Materials Research Proceedings, Vol. 31, pp 287-298, 2023. https://doi.org/10.21741/9781644902592-30

[5] Esam Jassim, "Improvement of Indoor Air Distribution by Utilizing Small Fans Powered by Solar Energy", International Journal of Computational Physics Series 1(2), 1-8, 2017. https://doi.org/10.29167/A1I2P1-8

[6] Kazem, Hussein A., et al. "A review of dust accumulation and cleaning methods for solar photovoltaic systems." *Journal of Cleaner Production* 276 (2020): 123187. https://doi.org/10.1016/j.jclepro.2020.123187

[7] Erica Zell, Sami Gasim, Stephen Wilcox, Suzan Katamoura, Thomas Stoffel, Husain Shibli, Jill Engel-Cox, Madi Al Subie, "Assessment of solar radiation resources in Saudi Arabia", Solar Energy, Volume 119, 2015, Pages 422-438. https://doi.org/10.1016/j.solener.2015.06.031

[8] M. Enjavi-Arsanjani, K. Hirbodi, M. Yaghoubi, "Solar Energy Potential and Performance Assessment of CSP Plants in Different Areas of Iran", Energy Procedia, Volume 69, 2015, Pages 2039-2048. https://doi.org/10.1016/j.egypro.2015.03.216

Energy efficiency and sustainability enhancement of electric discharge machines by incorporating nano-graphite

Asaad A. ABBAS[1,a], Raed R. SHWAISH[1,b*], Shukry H. AGHDEAB[1,c], and Waleed AHMED[2,d*]

[1]Production Engineering and Metallurgy Department University of Technology- Iraq, Baghdad, Iraq

[2]ERU, COE, UAE University, Al Ain, UAE

[a] asaad.a.abbas@uotechnology.edu.iq, [b] raed.r.shwaish@uotechnology.edu.iq, [c] Shukry.H.Aghdeab@uotechnology.edu.iq, [d] w.ahmed@uaeu.ac.ae

Keywords: EDM, Nano-Powder Mixed Electro-Discharge Machining, Electrode Wear Rate, Full Factorial

Abstract. The research paper endeavored to significantly improve the energy efficiency and sustainability of the electrode wear rate (EWR) in electric discharge machining (EDM) through a particular exploration of various parameters. These parameters included nano-powder concentration, electric current, and pulse-on time. By systematically analyzing and manipulating these variables, the study aimed to optimize the EDM process, thereby reducing energy consumption and enhancing the overall sustainability of the machining operation. We employed the Powder Mixed Electro-Discharge Machining (PMEDM) technique, a refined iteration of electric discharge machining. This innovative method blended graphite nano-powder with transformer oil to serve as the dielectric medium. The primary objective of this investigation is to examine the outcomes concerning crucial process parameters such as graphite powder concentration, electric current, and pulse-on time when machining high-speed steel, focusing mainly on electrode wear rate (EWR). Incorporating nano graphite (Gr) mixed powders into dielectric liquids has demonstrated a discernible enhancement in EWR across diverse operational conditions. The study found that the electrode wear rate (EWR) varied based on nano-powder concentration, electric current, and pulse-on time. Maximum EWR was observed at a concentration of 10 g/L, a current of 30 A, and a pulse-on time of 70 μs, while minimum EWR occurred at 0 g/L, 10 A, and 60 μs. The Full Factorial design, executed with MINITAB 17 software, validated these findings. The optimal EWR recorded was 1.154 mm3/min. The coefficient of determination (R-sq) for surface roughness prediction also stood at 91.15%.

Introduction

An essential player in non-traditional machining methodologies, electrical discharge machining (EDM) stands out as a proven, logical, and economical approach to the precision machining newly developed high-strength alloys. Its success lies in delivering tailored machining solutions with exceptional dimensional accuracy while reducing production costs. This technology holds promise in enhancing energy efficiency and sustainability in manufacturing processes due to its precise material removal capabilities and potential for optimizing resource utilization. Despite their inherent hardness and brittleness, conductive materials can undergo efficient and effective thermal energy treatment. This method mitigates wear and minimizes the expenses typically incurred in conventional machining processes. Ceramics, titanium, and steel are just a few materials amenable to this treatment [1]. Leveraging thermal energy in material treatment enhances the machining process and potentially reduces energy consumption and environmental impact, contributing to sustainability in manufacturing practices. Emerging materials with challenging machining characteristics continually evolve within electric discharge machining (EDM). These encompass ceramics, tool steel, superalloys, carbides, stainless steel, heat-resistant steel, and more. Widely applied across diverse sectors, including aerospace, nuclear, and die and mold-making industries, these materials pose intricate machining demands and underscore the necessity for advanced EDM techniques and technologies. By addressing the machining complexities of these materials, EDM plays a pivotal role in facilitating innovation and progress across various industrial domains. Furthermore, electric discharge machining (EDM) extends its influence into novel domains, encompassing sports equipment, medical and surgical instruments, optical devices, dental appliances, and jewelry manufacturing. Moreover, it permeates the research and development sector of the automotive industry [2-6]. This broadening scope reflects EDM's adaptability and versatility in addressing diverse machining needs across various industries, driving innovation and advancements in different sectors beyond traditional manufacturing applications. The swift solidification during the erosion process induces substantial internal thermal stress within the upper layers of the workpiece surface, thereby influencing the component properties. The cessation of discharge during erosion fundamentally characterizes a thermal erosion process [7]. This phenomenon underscores the intricate interplay between thermal dynamics and material behavior, highlighting the importance of understanding and managing thermal effects in erosion processes for effective material treatment and component performance enhancement.

Erosion arises from heat generation. Abrupt temperature surges in the machining process profoundly impact the physical attributes of the machined surface layer, leading to residual stress. This pivotal factor significantly influences machined surfaces quality and functional attributes, underscoring its importance in surface engineering and manufacturing practices. Understanding and effectively managing these thermal effects are paramount for achieving desired surface characteristics and optimal component performance [8-10]. A recent advancement in electric discharge machining technology is powder-mixed electric discharge machining (PMEDM). The finely graded powder material is meticulously mixed with the dielectric liquid to enhance its breakdown characteristics. This additional powder component improves the dielectric breakdown characteristics [11]. The study explored the effects of blending an Al and Cr powder mixture with kerosene. Findings indicated that this combination decreased the thermal insulating properties of the kerosene while widening the spark gap. Consequently, this stabilized the machining process and significantly boosted the material removal rate (MRR). Tzeng and Lee delved into the influence of different powder characteristics on the machining of SKD11 materials, which are equivalent to AISI H13 tool steel [12]. Despite promising results, the adoption of powder mix EDM in the industry has been gradual. A key contributing factor is the lack of clarity surrounding various aspects of this emerging technology, including the editing process. To address this gap, researchers employed response surface methodology (RSM) to examine the effects of process variables such as peak powder concentration, tool diameter, and current on the material removal rate (MRR) for EN8 steel [13], [14]. Researchers investigated the influence of fine metal powder grains such as aluminum (Al) and copper (Cu) when introduced into dielectric fluids during the EDM process of AISI D3 and EN31 steels using Taguchi design experiments. Numerous endeavors have aimed to simulate the EDM process and evaluate its effectiveness [15].

This study aimed to analyze the impact of EDM parameters on the induced electrode wear rate when employing copper electrodes, with or without the incorporation of Nano-graphite powder mixed in transformer oil dielectric, for A 240 stainless steel 304. The experimental design matrices for the materials were generated using the Full Factorial Design (FFD) approach. The electrode wear rate was analyzed using ANOVA models, with separate models developed for two sets of trials. The first set utilized transformer oil as the dielectric, while the second set employed nano-graphite particles mixed with the dielectric fluid (PMEDM) at concentrations of 5 and 10 g/l. This approach aimed to enhance machining efficiency, mitigate instability in arcing effects, and analyze the electrode wear rate resulting from process modifications.

Experimental work
As depicted in Figure 1, an experimental investigation was conducted using a CHEMER EDM machine (CM323C). The workpiece employed in the experiment was stainless steel 304, ASTM A 240, measuring (40 x 30 x 1.7 mm) in dimensions.

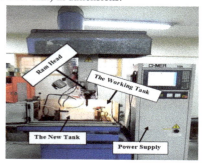

Fig. 1. Model of the EDM machine (CM 323C).

Chemical composition percentages of the 304 stainless steel workpiece materials per the ASTM E415 standard are summarized in Table 1 (tested at the Central Organization for Standardization and Quality Control Center).

TABLE 1. Stainless steel 304 Chemical composition of samples.

Material	C	Si	Mn	P	S	V	Cr	Mo	Ni	Fe
Weight (%)	0.06	0.6	1.1	0	0	0.14	19	0.2	9	Balance

Table 2 outlines numerous mechanical and physical characteristics of the workpiece. The electrode comprises 99.74% pure copper and possesses a diameter of 5 mm. Additionally, the dielectric solution, a form of transformer oil, incorporates a nano-powder mix of graphite .

TABLE 2. The mechanical and physical characteristics of the samples

Ultimate Strength	621 MPa
Density	8030 kg/m³
Hardness	1667 N/ mm²
Ductility	60 mm/mm
Melting point	2552-2642°F

The experiment occurred in a newly designed tank, which housed a blend of graphite nano-powder and transformer oil. A small pump was integrated into the reservoir to put off powder accumulation at the bottom or the formation of insulating surface deposits. This measure ensured the efficient separation of the internally mixed nano-powder. The viscosity of the transformer oil used in the practical experiments was measured at 28.01 Pa.s at a room temperature of 23°C. A sample of graphite nano-powder was obtained, with particle sizes measuring 80 nanometers. Before and after machining, the weight of the sample and electrode was thoroughly determined using an electronic weighing scale at a precision of 0.0001g. Following NPMEDM machining, the surface roughness of each workpiece was evaluated using a portable surface roughness tester, boasting an accuracy of 0.05 μm. The operational mechanism of nano-powder-mixed EDM is depicted in the following figure.

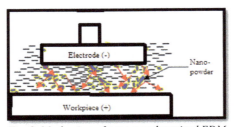

Fig. 2. Mechanism of nano-powder-mixed EDM.

The machining parameters encompassed varying concentrations of nano-powder (graphite), current, pulse-on time, and pulse-off time. Besides, nano-powder graphite concentrations ranged from 0, 5, to 10 g/l, while 10, 20, and 30 amperes discharge currents were employed. Furthermore, pulse-on times were selected at 50, 60, and 70 μs, with a pulse-off time set at 55 μs. Following the manufacturer guidelines, the remaining electric impulse parameters were maintained constant, including an open gap voltage of 140V and a negative tool electrode polarity. The goal is to explore the impact of nano-powder-mixed dielectric on the electrode wear rate (EWR) by varying input machining parameters. Table 3 shows the three parameters used in this study, each with three levels. Furthermore, Table 4 displays the input machining settings that haven't changed.

TABLE 3. *Machining Parameters and their corresponding stages.*

Parameters	Stages		
	1	2	3
Concentration (Nano-Graphite), gram/Liter	0	5	10
Current, Ampre	10	20	30
Pulse-on time, micro-Second	50	60	70

TABLE 4. *unchanged input parameters.*

Workpiece polarity	Positive
Electrode polarity	Negative
Dielectric fluid	Transformer oil with Mixture nano-powder of Gr
High voltage	140 V 1.2 A
Pulse-off time	25 μsec
Working time	5.0 sec
Jumping time	2.0 mm
Depth of cut	1 mm
Gap code	10
Servo feed	75 %

Results and discussion

The impact of process variables like powder concentration, discharge current, and pulse-on time on the response variable, namely the electrode wear rate (EWR), was investigated. Analysis of variance (ANOVA) was conducted using MINITAB 17 software to analyze the experimental data and ascertain the significance of these factors on EWR. The conclusions of the ANOVA analysis for EWR are presented in Table 5. ANOVA was employed to assess the importance of the model. Model terms are considered statistically significant if their "P-Value" is below 0.05, indicating a 95 percent confidence interval [16]. Furthermore, Table 6 presents the model summary, including the coefficients associated with the terms (Coef.), the standard error of each coefficient (SE Coeff), the t-statistic, and the p-value of each term. These values aid in assessing whether to accept or reject an invalid assumption. All of the terms' p-values are below the predetermined threshold, indicating their significance within the model. An R-squared value of 91.15% suggests that the predictors or factors in the model elucidate 91.15% of the total variance in the response. Additionally, the adjusted R-squared value of 88.50% factors in the number of predictors in the model and defines the significance of the association.

TABLE 5. *Analysis of Variance for EWR*

Source	Model	Linear	Nano Powder	Current	Ton	Error	Total
DF	6	6	2	2	2	20	26
Adj SS	15.741	15.741	7.727	3.033	4.981	1.528	17.269
Adj MS	2.62355	2.62355	3.86353	1.51673	2.49039	0.0764	
F-Value	34.34	34.34	50.57	19.85	32.6		
P-Value	0	0	0	0	0		

TABLE 6. *Model summary*

Term	Coeff	SE Coeff	T-Value	P-Value
Constant	2.7952	0.0532	52.55	0.000
Nano Graphite				
0	-0.6612	0.0752	-8.79	0.000
5	0.0121	0.0752	0.16	0.873
Current				
10	-0.3691	0.0752	-4.91	0.000
20	0.0731	0.0752	-0.97	0.343
Ton				
50	-0.5153	0.0752	-6.85	0.000
60	-0.0209	0.0752	-0.28	0.784
S	R-sq	R-sq (adj)	R-sq (pred)	
0.276406	91.15%	88.50%	83.87%	

The significant effects plot of the electrode wear rate (EWR) revealed that an increase in nano-powder concentration correlates with higher EWR. Similarly, elevating both currents (10, 20, and 30 A) and pulse-on times (50, 60, and 70 µs) and maintaining a pulse-off time of 55 µs contributed to this trend. This phenomenon can be attributed to the heightened spark energy between the workpiece and electrode due to increased temperature on the workpiece, resulting in an escalation of the EWR. Moreover, the longest pulse-off time (Toff) corresponds to the highest electrode wear rate (EWR) due to its role in prolonging the re-solidification period within the dielectric medium. This extension results in larger surface grains, leading to a higher EWR. Nano-powder concentration, current, and pulse-on time (Ton) are also observed to influence EWR, as depicted in Figure 3.

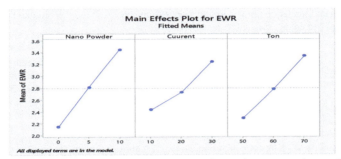

Fig. 3. Impact of Nano-graphite concentration, plus, and current on time on EWR

The nano-powder concentration correlates with the enhanced electrode wear rate (EWR) of the workpiece, along with increasing currents (10, 20, and 30 A) and pulse-on times (50, 60, and 70 µs). The impact of nano-graphite concentration, pulse-on time, and current on EWR is depicted in Figure 4.

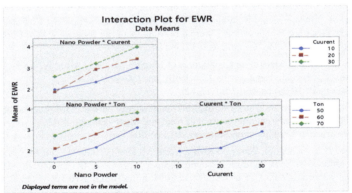

Fig. 4. The effect of Nano-graphite concentration, current, and pulse on time (T_{on}) on EWR.

Conclusions

In addition to exploring the impact of electrical process variables on the electrode wear rate (EWR) for high-speed steel with pure copper electrodes, this investigation highlights critical factors influencing machining efficiency and sustainability. The findings emphasize the significance of nano-powder concentration in transformer oil as a crucial determinant of electrode wear, with higher concentrations correlating with increased wear rates. Notably, the study reveals that the highest EWR occurs at ten g/l concentration, 70 µs at pulse-on time, and 30 Amp, current, emphasizing the sophisticated relationship of process parameters. Moreover, employing a Full Factorial design executed with MINITAB 17 software enhances our understanding of the machining process, facilitating the identification of optimal conditions for minimizing electrode wear. The finding of an optimal EWR value of 1.154 mm^3/min points out the potential for enhancing energy efficiency and sustainability in machining operations. The high coefficient of determination (R-sq) of 91.15% indicates the robustness of the predictive model, offering valuable understanding into future machining industries aimed at reducing wear rates while optimizing resource utilization. As industries strive for eco-friendly manufacturing practices, these findings provide a crucial foundation for developing more sustainable machining techniques that align with environmental and energy efficiency goals.

References

[1] Y. H. Guu, H. Hocheng C. Y. Chou and C. S. Deng, "Effect of electrical discharge machining on surface characteristics and machining damage of AISI D2 tool steel", Materials Science and Engineering, A 358, pp. 37-43, 2003. https://doi.org/10.1016/S0921-5093(03)00272-7

[2] Sundaram M. M. and K. Rajurkar, "Electrical and Electrochemical Processes, in Intelligent Energy Field Manufacturing", CRC Press. p. 173-212, 201, 2011. https://doi.org/10.1201/EBK1420071016-c6

[3] Klocke F., Zeis M., Klink A. and Veselova D., "Technological and Economical Comparison of Roughing Strategies via Milling, EDM and ECM for Titanium- and Niclelbased Blisks", Proceedings of the 1st CIRP Global Web Conference on Interdisciplinary Research in Production Engineering, Vol. 2, pp. 98-101, 2012. https://doi.org/10.1016/j.procir.2012.05.048

[4] Gu L., Le. L., Zhao W., and Rajurkar K. P., "Electrical Discharge Machining of Ti6Al4V with Bundled Electrode", International Journal of Machine Tools and Manufacturing, Vol. 53, pp. 100- 106, 2012. https://doi.org/10.1016/j.ijmachtools.2011.10.002

[5] Jahan M., Malshe A. and Rajurkar K., "Experimental Investigation and Characterization of Nano-scale Dry Electro-machining", Journal of Manufacturing Processes, Vol. 14 pp. 443-451, 2012. https://doi.org/10.1016/j.jmapro.2012.08.004

[6] Klocke F., "EDM Machining Capabilities of Magnesium (Mg) Alloy WE43 for Medical Applications", Procedia Engineering, Vol. 19(0): pp. 190-195, 2011. https://doi.org/10.1016/j.proeng.2011.11.100

[7] H. Singh, "Experimental study of distribution of energy during EDM process for utilization in thermal models", International Journal of Heat and Mass Transfer, 2012. https://doi.org/10.1016/j.ijheatmasstransfer.2012.05.004

[8] Ahmed N. Al-Khazraji, Samir A. Al-Rabii and Ali H. Al-Jelehawy," Fe Analysis of Residual Stresses Induced by Spot Welding of Stainless-Steel Type AISI 316" Engineering and Technology Journal, Vol. 32, Part (A), No. 2, 2014. https://doi.org/10.30684/etj.32.2A.7

[9] Ahmed N. Al-Khazraji, Farag M. Mohammed and Ali Riyadh A. Al-Taie," Residual Stress Effect on Fatigue Behavior of 2024- Aluminum Alloy" Engineering and Technology Journal, Vol. 29, No. 3, 2011. https://doi.org/10.30684/etj.29.3.13

[10] Ahmed N. Al-Khazraji, Samir A. Al-Rabii and Ali H. Fahem," Formation of Compressive Residual Stresses by Shot Peening for SpotWelded Stainless Steel Plates" Engineering and Technology Journal, Vol. 31, Part (A), No. 11, 2013. https://doi.org/10.30684/etj.31.11A14

[11] Wang, C. H., Lin, Y. C., Yan, B. H. and Huang, F. Y., "Effect of characteristics of added powder on electric discharge machining", J. Jpn. Inst. Light Met., Vol. 42 (12), pp. 2597–2604, 2001. https://doi.org/10.2320/matertrans.42.2597

[12] Tzeng, Y. F., Lee, C. Y., "Effects of powder characteristics on electro-discharge machining efficiency", Int. J. Adv. Manuf. Technol., Vol. 17, pp. 586–592, 2001. https://doi.org/10.1007/s001700170142

[13] C. Huang, J. Wang and X. Li, "Powder Mixed near Dry Electrical Discharge Machining", Advanced Materials Research, Vol. 500, 2012. https://doi.org/10.4028/www.scientific.net/AMR.500.253

[14] N. S. Khundrakpam, H. Singh, S. Kumar, and G. S. Brar," Investigation and Modeling of Silicon Powder Mixed EDM using Response Surface Method", International Journal of Current Engineering and Technology, Vol.4, No.2 (April 2014).

[15] B. Reddy, G.N. Kumar, and K. Chandrashekar, "Experimental Investigation on Process Performance of Powder Mixed Electric Discharge Machining of AISI D3 steel and EN-31 steel", International Journal of Current Engineering and Technology, Vol. 4, No. 3, June 2014.

[16] S. Assarzadeh and M. Ghoreishi, "A dual response surface-desirability approach to process modeling and optimization of Al2O3 powder-mixed electrical discharge machining (PMEDM) parameters," The International Journal of Advanced Manufacturing Technology, Vol. 64, No. 9-12, pp. 1459-1477, 2013. https://doi.org/10.1007/s00170-012-4115-2

Exploring sustainable micro milling: Investigating size effects on surface roughness for renewable energy potential

Ahmet HASCELIK[1,a*], Kubilay ASLANTAS[2,b], Waleed AHMED[3,c*]

[1] Department of Mechanical and Metal Technology, İscehisar Vocational School of Higher Education, Afyon Kocatepe University, Afyonkarahisar, Turkey

[2] Department of Mechanical Engineering, Faculty of Technology, Afyon Kocatepe University, Afyonkarahisar, Turkey

[3] ERU, College of Engineering, UAE University, Al Ain, UAE

[a]ahascelik@aku.edu.tr, [b]aslantas@aku.edu.tr, [c]w.ahmed@uaeu.ac.ae

Keywords: Sustainable Micro-Milling, Size Effect, Surface Roughness, Minimum Chip Thickness

Abstract. Micromilling is a helpful process empowering the fabrication of small-scale components characterized by complex geometries, heightened precision, and superior surface integrity. Widely embraced across aerospace, biomedical, and electronics sectors, it characterizes efficiency and sustainability in modern manufacturing paradigms. However, fundamental instabilities emerge during the cutting phase, particularly when the size effect diminishes below a critical threshold termed the minimum chip thickness, a parameter linked to the cutting-edge radius and feed rate dynamics. The micro-milling process enables the production of small-scale parts with complex geometries, high precision, and optimal surface quality. It stands as a preferred production method not only in the aerospace, biomedical, and electronics industries but also in the renewable energy sector, where its ability to create intricate components with precise dimensions and superior surface quality is crucial for optimizing efficiency and sustainability in energy harvesting and storage technologies. Instabilities are observed in the cutting process when the size effect falls below a critical value known as the minimum chip thickness. This critical value is related to the cutting-edge radius and feed rate. This study investigates the effect of size on surface roughness in micro-milling of Al6061-T6 workpiece. The results show that surface roughness is high at feed rates below the minimum chip thickness due to the ploughing mechanism. The shear mechanism is active at feed rates above the minimum chip thickness, but the ploughing effect is still observed at the 100μm edge of the cutting channel. The study revealed that surface roughness and height differences were high at feed rates significantly below or above the minimum chip thickness. However, surface quality was optimal at feed rates near the minimum chip thickness. Nevertheless, the study highlights an optimal peak in surface quality achieved at feed rates close to the minimum chip thickness, explaining a relationship between operational efficiency and sustainability in micro-milling endeavours.

Introduction

The demand for precision and small-scale parts in various industries, including electronics, medicine, biomedicine, aerospace, automotive, and telecommunications, is on the rise [1]. Micro-mechanical machining methods have been developed to meet these demands and produce parts with high precision and small dimensions [2]. Among these methods, micro-milling stands out as a way to produce precise and complex parts at a micro scale [3]. The micro-milling method is commonly utilised in various applications, including precision integrated circuits, microelectronic and medical devices, biomedical implants, and optical components like micro mirrors, micro lenses, sensors, microchips, micro propellers, and blades [1]. The micro-milling process differs fundamentally from conventional milling due to the size effect [4]. When the feed rate per tooth (fz) falls below a critical value, which depends on the cutting tool and workpiece pair, instabilities occur in the cutting process and surface roughness increases. The optimal ratio between the cutting tool's edge radius (Re) and the feed rate per tooth determines the critical value. The ratio defining the minimum chip thickness is called h_{min} [5]. In micromachining, the uncut chip thickness (h) is kept below h_{min}, and the negative rake angle effect caused by the cutting-edge radius results in ploughing during the cutting process [6, 7]. Experimental studies have been conducted to determine the minimum chip thickness. According to the literature, the minimum chip thickness varies between 20% and 30% of the cutting-edge radius [8]. In titanium alloys, this ratio is 30% [9], while in aluminium alloys, it can go up to 35-40% [10, 11]. Wu et al. (2020) discovered that the minimum chip thickness is 17% of the edge radius [12]. It is a well-established fact that cutting forces display unstable behaviour at feed rates below the minimum chip thickness [13, 14]. This is attributed to the ploughing mechanism that becomes active at feed rates below the minimum chip thickness during the cutting process. Figure 1a provides a schematic representation of full slot machining in the micro milling process, while Figure 1b illustrates the variation of the chip thickness based on the position of the cutting edge and the feed rate per tooth. As is commonly understood, the thickness of the chip during milling varies depending on the position of the tool. Initially, when the tool first makes contact with the workpiece, the chip thickness is almost zero. As the angular position of the cutting edge (φ) increases, the value of h also increases. When φ=90°, h=fz. If h<h_{min}, ploughing occurs, which results in a decrease in the quality of the machined surface and an increase in the amplitude of the cutting forces [15]. This study aimed to establish a relationship between the angular position of the tool (represented by the symbol φ) and the ploughing length (wp) by examining the change in surface roughness on the machined slot surface (Fig. 1).

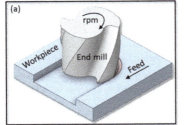
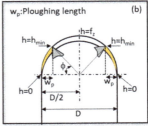

Figure 1. Relationship between uncut and minimum chip thickness.

The use of the ploughing mechanism instead of the shear mechanism during the cutting process has a negative impact on the surface quality of the workpiece. The high feed rate results in increased cutting forces, which in turn leads to a deterioration of the surface quality and an increase in surface roughness. To achieve the best surface quality during the cutting process, it is essential to use a feed rate that is large enough to prevent ploughing, yet small enough to keep cutting forces low. This study investigates the effect of size on surface roughness in the micro-milling process. Surface roughness measurements were taken from the cutting grooves of the micro-milled workpiece at different feed rates. The minimum chip thickness was determined, and the ploughing effect was identified.

Material and Method
This study used Al6061-T6 alloy, commonly used in the production of micro equipment in the defence, aerospace, and electronics sectors. Aluminium alloys are widely used due to their formability, light weight, high strength, and corrosion resistance. The T6 heat treatment improved

the mechanical and physical properties of the alloy. Compared to other aluminum alloys, Al6061-T6 stands out due to its high toughness, superior corrosion resistance, low density, high thermal conductivity, and low cost [17]. Table 1 shows the alloy's chemical and basic mechanical properties, where 10 mm x 10 mm x 15 mm specimens were used in cutting tests.

Table 1. Chemical composition and mechanical properties of Al6061-T6 material [17].

Elements	Chemical Composition							Mechanical Properties		
	Al	Mg	Fe	Cu	Zn	Si	Mn	UTS (MPa)	YS (MPa)	%Elongation
Alloying elements(%	Rest	1.14	0.35	0.3	0.25	0.67	0.12	307	275	20

The cutting tests employed a 975 µm diameter, 2-flute tungsten carbide cutting tool with AlTiSiN coating. The tool has a helix angle of 30°, a helix length of 2 mm, an edge radius of 5 µm, a rake angle of 0°, and a clearance angle of 15°. The geometrical properties of the cutting tool were determined by measuring its features from scanning electron microscope images (Fig. 2).

Figure 2. SEM image of the diameter and edge radius of the cutting tool used in cutting experiments.

Cutting tests were conducted at various feed rates under dry cutting conditions using a triaxial test rig to determine the minimum chip thickness. The cutting tool was secured to the IMT spindle with a collet, and axis movements were set using Thorlabs software. The workpiece was held in place with a Kistler-9119AA1 mini dynamometer to measure the instantaneous cutting forces. The cutting force signals were transferred to the computer via an amplifier and converted into force measurements using Dynoware software. The cutting tool movements and cutting zone were observed using a USB microscope during the cutting process (Fig. 3).

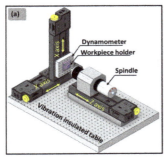 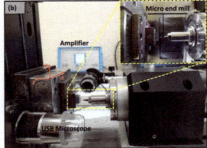

Figure 3. a) Schematic representation of the experimental setup used in cutting experiments, b) a general view and positioning of the cutting tool-workpiece on the experimental setup.

Micro milling experiments were conducted at a constant depth of cut (100 µm) and speed (30000 rpm) to determine the minimum chip thickness and observe the effect of feed rate differences per tooth on surface roughness. Ten different feed rates (0.1, 0.25, 0.5, 0.75, 1, 2, 4, 6, 8, 10 µm/tooth) were used to determine the dominant cutting process mechanism between shear and ploughing. The Nanovea 3D ST400 optical surface profilometer was used to take measurements by scanning the cutting grooves. The scanning distance was 1.2 mm x 0.6 mm with a scanning step of 3 µm. Measurements were taken from the inside of the channel at a distance of 970 µm x 600 µm. The resulting Sa (areal average surface roughness) and Sz (areal maximum surface roughness) surface roughness values were compared at different feed rates. Additionally, Ra measurements were taken parallel to the feed direction across the width of the machined channel. This revealed the impact of differences in feed rate on surface roughness for two distinct levels of roughness.

Results and Discussion
Sa and Sz values were determined at different feed rates through cutting tests and surface roughness scanning. Measurements were taken from at least three different areas of the cutting grooves, and the averages of these measurements are presented in Figure 4. A detailed graph for fz=2µm/tooth is also provided in Figure 4. The results indicate a significant increase in both Sa and Sz values for fz<0.5µm/tooth, which is defined as the Ploughing zone in both graphs. In micro milling, a negative rake angle effect occurs at the cutting edge when the feed per tooth is smaller than the minimum chip thickness. This effect makes the cutting process difficult, increases cutting forces, and causes deterioration of the machined surface quality and burr formation. Figure 4 shows a significant difference between the error bars for 0.1 and 0.25 µm/tooth, indicating a high degree of surface variability. For fz greater than 0.5µm/tooth, the milling process achieves cutting through a shear mechanism. Similar to conventional milling, increasing the feed rate results in an increase in surface roughness. As the cutting edge radius measures 2.4 µm (Fig. 2), the minimum chip thickness corresponds to 21% of the cutting edge radius, based on the surface roughness results. The Sa and Sz values at the minimum feed rate (0.1 µm/tooth) are higher than those at the maximum feed rate (10 µm/tooth), indicating that the ploughing mechanism has a significant effect on surface roughness.

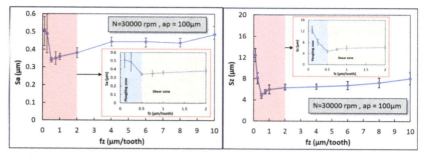

Figure 4. Variation of Sa and Sz surface roughnesses as a function of feed rate.

Figure 5 displays the surface topographies of the machined slots, illustrating the impact of the ploughing and shear mechanism on the surface in micro milling. Additionally, the two-dimensional variation plot along the slot width is presented. When the feed rate is below the minimum chip thickness (0.1 μm/tooth is provided as an example), a non-uniform surface topography is observed. The cutting marks of the cutting tool are faint and only located in the center of the slot, across its width. The surface irregularity and height difference increase towards the edges of the slot, as shown in the linear variation plot. When using fz=1 μm/tooth, the cutting marks on the surface are more pronounced, and the height difference (hpv) at the edges of the slot is smaller compared to fz=0.1 μm/tooth. Additionally, the height difference (hpv) of the surface marks between the edges and the middle region of the slot is also smaller for fz=1 μm/tooth. At a fz of 1 μm/tooth, the cutting in the middle region of the slot (100μm<ws<800 μm) is achieved through shear mechanism. However, there is a height difference between the edges and the center of the slot, indicating ploughing occurs at the edges. At a fz of 10 μm/tooth, the cutting process is also achieved through shear mechanism. The high feed rate resulted in an increase in surface roughness. Figure 5 shows that the hpv values for 0.1 and 10 μm/tooth are similar, indicating a deterioration of surface quality caused by milling with ploughing.

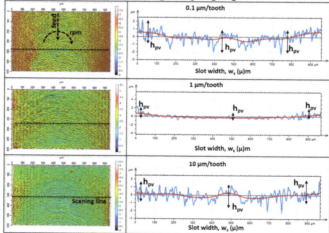

Figure 5. 3D surface topographies from the optical profilometer.

Figure 6 illustrates the variation of two-dimensional surface scanning results parallel to the feed direction with slot width. The surface roughness is high at the edges and in the middle regions in cutting processes with fz=0.1 and 0.25 μm/tooth due to the ploughing mechanism effect since h<hmin along the cutting channel. However, in cutting processes with feed rates of 0.5, 0.75 and 1 μm/tooth, Ra is minimum in the middle region of the cutting channel (100μm<ws<800 μm). In contrast, Ra increases near the starting (ws<100μm) and ending edges (ws<900μm) of the slot due to ploughing. The chip thickness is initially close to zero and increases with the rotation of the tool during cutting, even for fz=0.5 μm/tooth, where the surface roughness is minimum. However, the chip thickness at the beginning and end of the cutting process is less than the minimum chip thickness, causing ploughing. Figure 6 shows that the length at which ploughing occurs is about 100μm. At fz=0.1 and 0.25 μm/tooth, the Ra value varies across the entire width of the groove, with the maximum Ra value being at the edge and in the center of the groove. At higher feed rates (fz=8 and 10 μm/tooth), the Ra value is higher in the center of the slot (200μm<ws<800μm). However, the Ra value is low at the beginning (ws<100μm) and end edges (ws<900μm) of the slot.

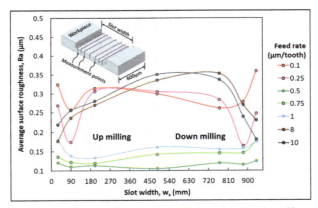

Figure 6. Two-dimensional surface topographies from the optical profilometer.

Conclusion

This study observes the size effect in the micro milling process and investigates the impact of different feed rates on surface roughness due to ploughing and shear mechanisms. The obtained results are listed below.
- Based on the surface roughness results obtained at different feed rates, it was determined that the critical feed rate was approximately fz=0.5 μm/tooth.
- It was observed that when feed rates fall below the minimum chip thickness, the ploughing mechanism significantly increases surface roughness.
- At feed rates higher than the minimum chip thickness, the surface roughness increases linearly with the feed rate because the shear mechanism is active.
- At feed rates where the shear mechanism is active, the surface roughness is higher at the starting and ending edges of the 100 μm slot than in the middle regions due to the ploughing effect.
- At high feed rates, the surface roughness in the middle regions of the slot (200μm<ws<800μm) is higher than the edge regions (ws<100μm). The Ra value increases

with increasing feed rate, which is more effective than the increase due to ploughing in the edge regions.
- At feed rates lower than the minimum chip thickness and at high feed rates, there are greater height differences in the surface topography.

Moreover, these findings hold promising implications for renewable energy applications. Understanding the interaction between feed rates and surface roughness enhances manufacturing efficiency and has significance in the fabrication of micro-components crucial for renewable energy technologies. Manufacturers can achieve smoother surfaces and improved energy conversion efficiencies in micro-devices utilized in solar panels, wind turbines, and other renewable energy systems by optimizing milling parameters to minimize plowing effects and maximize shear mechanisms. This emphasizes the pivotal role of micro-milling in advancing sustainable energy solutions.

References

[1] Aslantas, K., Demir, B., Guldibi, AS., Niinomi, M, Dikici, B., A comparative study on the machinability of β-type novel Ti29Nb13Ta4.6Zr (TNTZ) biomedical alloys under micro-milling operation, Journal of Manufacturing Processes 92 (2023) 135-146. https://doi.org/10.1016/j.jmapro.2023.02.043

[2] Boswell, B., Islam, M.N., Davies, I.J., A review of micro-mechanical cutting, Int. J. Adv. Manuf. Technol. 94 (2018) 789-806. DOI: 10.1007/s00170-017-0912-y

[3] Balazs, B.Z., Geier, N., Takacs, M., Davim, J.P., A review on micro-milling: recent advances and future trends, Int. J. Adv. Manuf. Technol. 112 (2020) 655-84. DOI: 10.1007/s00170-020-06445-w

[4] Fu, M.W., Wang, J.L., Size effects in multi-scale materials processing and manufacturing, Int. J. Mach. Tool. Manuf. 167 (2021) 103755. DOI: 10.1016/j.ijmachtools.2021.103755

[5] Dib, M.H.M., Duduch, J.G., Jasinevicius, R.G., Minimum chip thickness determination by means of cutting force signal in micro endmilling, Precis. Eng. 51 (2018) 244–62. 10.1016/j.precisioneng.2017.08.016

[6] Wan, M., Ma, Y.C., Feng, J., Zhang, W.H., Study of static and dynamic ploughing mechanisms by establishing generalized model with static milling forces, Int. J. Mech. Sci., 114 (2016) 120-31.

[7] Krajewska-Spiewak, J., Maruda, R.W., Krolczyk, G.M., Nieslony, P., Wieczorowski, M., Gawlik, J. Wojciechowski, S., Study on ploughing phenomena in tool flank face - workpiece interface including tool wear effect during ball-end milling, Tribology International 181 (2023) 108313.

[8] Cheng, K., Huo, D., Micro-cutting: fundamentals and applications (First Edition), Chichester UK, John Wiley & Sons Ltd. (2013). DOI: 10.1002/9781118536605

[9] Aslantas, K., Danish, M., Hasçelik, A., Mia M., Gupta M., Ginta T., Ijaz H., Investigations on Surface Roughness and Tool Wear Characteristics in Micro-turning of Ti-6Al-4V Alloy, Materials 13 (2020) 2998. https://doi.org/10.3390/ma13132998

[10] Park, S. S., Malekian, M., Mechanistic Modeling and Accurate Measurement of Micro End Milling Forces, CIRP Annals-Manufacturing Technology 58 (2009) 49-52. https://doi.org/10.1016/j.cirp.2009.03.060

[11] Chen, Y., Wang, T., Zhang, G., Research on Parameter Optimization of Micro-Milling Al7075 Based on Edge-Size-Effect, Micromachines 11 (2020) 197. https://doi.org/10.3390/mi11020197

[12] Wu, X., Liu, L., Du, M., et al., Experimental Study on the Minimum Undeformed Chip Thickness Based on Effective Rake Angle in Micro Milling, Micromachines 11 (2020) 924. https://doi.org/10.3390/mi11100924

[13] Chae, J., Park, S.S., Freiheit, T., Investigation of Micro-Cutting Operations, International Journal of Machine Tools and Manufacture 46 (2006) 313-332. https://doi.org/10.1016/j.ijmachtools.2005.05.015

[14] Sun, X., Cheng, K., Micro/Nano-Machining through Mechanical Cutting (2010). http://dx.doi.org/10.1016/B978-0-8155-1545-6.00002-8

[15] Sun, Z., To, S., Zhang, S., Zhang G., Theoretical and experimental investigation into non-uniformity of surface generation in micro-milling, International Journal of Mechanical Sciences, 140 (2018) 313-324. https://doi.org/10.1016/j.ijmecsci.2018.03.019

[16] Akram, S., Jaffery, S., Khan, M., et.al., Numerical and experimental investigation of Johnson–Cook material models for aluminum (Al 6061-T6) alloy using orthogonal machining approach, Advances in Mechanical Engineering 10 (2018) 1-14. https://doi.org/10.1177/1687814018797794

[17] Wakchaure, K., Chaudhari, R., Thakur, A., Fuse, K., Lopez de Lacalle, L.N., Vora, J., The Effect of Cooling Temperature on Microstructure and Mechanical Properties of Al 6061-T6 Aluminum Alloy during Submerged Friction Stir Welding, Metals 13 (2023) 1159. https://doi.org/10.3390/met13071159

From smart soles to green goals, interlacing sustainable innovations in the age of smart health: An exploratory search

Fahd KORAICHE[1,a *], Amine DEHBI[2,b] and Rachid DEHBI[1,c]

[1]Research Laboratory of Computer Science (LIS), Faculty of Science Ain Chock, Casablanca, Morocco

[2]Research Laboratory of Mathematics and Computer Science (LTI), Faculty of Science Ben M'Sick, Casablanca, Morocco

[a]fahd.koraiche@gmail.com, [b]dehbiamine1@gmail.com, [c]dehbirac@yahoo.fr

Keywords: Energy Consumption, IoT, IoMT, Healthcare, Smart Health

Abstract. Health goes beyond the mere absence of illness. It's arguably a state that encompasses a whole spectrum of physical, social, and mental well-being. It involves not only curing sickness but also promoting healthy lifestyles, behaviors, and environments that enable individuals to thrive. Health interventions can be formalized through medicine, but they can also take on a broader spectrum of approaches, including prevention, early detection, and management of diseases. By acknowledging this broader vision of health, we can see that it influences various aspects of our lives, including our work, education, relationships, food, clothes, and recreational activities. It also impacts our communities and cultures, affecting social norms, policies, and practices that shape our health outcomes. Therefore, health is not just an individual concern but also a collective one. Furthermore, the integration of health into technological practices can have a significant impact on our well-being. It can facilitate access to health information, resources, and services, enable remote monitoring and diagnosis, and enhance communication and social support networks.

Introduction

In an era where technology and health converge, the paradigm of smart health not only reshapes how we monitor well-being but also holds the potential to redefine our relationship with the environment. Smart health technologies, in addition to revolutionizing well-being management, present auspicious avenues for ameliorating our environmental footprint.
Concurrently, the urgency of mitigating climate change compels us to explore innovative solutions that optimize resource consumption and minimize environmental impact. This article delves into the intricate nexus of these two critical domains, exploring how smart health advancements can foster both individual well-being and environmental responsibility.

This article encapsulates a journey into the synergy between cutting-edge healthcare technologies and a commitment to environmental sustainability. This exploratory search navigates through the intricate tapestry of interconnected themes. We will first delve into the profound connections between sustainability practices and individual well-being. From exploring how smart shoes contribute to holistic health monitoring to the impact of programming languages on carbon emissions, we dissect the intricate threads that weave together a narrative of interconnected health and environmental consciousness.

Literature Study

Zahid et al. [1] responded to the primary weakness of Body Sensors Area Networks (BSNs) which is energy consumption which is due to their small size and limited lifetime batteries, they wrote a paper that presents two contributions. First, they propose an adaptive duty-cycle optimization algorithm (ADO) that enhances the devices' active time by taking into consideration the exact power level that saves more energy, unlike traditional methods that increase the sleep period. Second, they propose a joint Sustainable and Green smart-health framework. They conducted a thorough experimental and theoretical analysis through real-time data. The results showed that the algorithm enhances energy and reliability savings by 36.54% and 24.43%, respectively. Therefore, they concluded that the algorithm is more promising for energy-limited sensor devices in healthcare-connected platforms.

Abdellatif et al. [2] have proposed an architecture that uses wireless technologies and sensors to connect patients with medical healthcare. This framework differs from previous work in this field by using various wireless networks to optimize medical data delivery. The team aims to reduce the size of transmitted data while maintaining reliable real-time healthcare services. They have developed an energy-efficient s-Health system by incorporating wireless network components with application characteristics and using the spectrum across multiple radio access technologies to fulfill the applications' Quality of Service (QoS). The proposed MEC-based system architecture meets all s-Health requirements. The team has used intelligent data processing techniques at the network edge to achieve this. They have also introduced some edge computing techniques, such as adaptive in-network compression, event detection, and network-aware optimization, which enable MEC-based system architecture to fulfill all s-Health requirements. In addition, the team has discussed the challenges and open issues for utilizing the MEC paradigm in s-Health. This includes the use of cooperative edges to improve energy and spectrum efficiency, as well as the need and benefit of combining heterogeneous data sources at the edge.

D. Laxma Reddy et al. [4] mainly focused on finding the best Cluster Head for an energy-efficient routing protocol in Wireless Sensor Networks (WSN). The paper proposes a new hybrid algorithm called Ant Colony Optimization (ACO) integrated Glowworm Swarm Optimization (GSO) approach (ACI-GSO) as a solution for Cluster Head Selection (CHS). The main objective of CHS is to reduce the distance between the selected Cluster Head nodes. To achieve this, the fitness function uses multiple objectives such as distance, delay, and energy. Finally, the proposed work was evaluated and proved to be more effective compared to other conventional methods.

Intertwining Sustainability and Well-being
In this chapter, we delve into the intricate connections between sustainability and well-being, exploring how sustainable innovations can significantly impact and promote overall health and wellness. By examining the interplay between sustainable practices and well-being, we aim to uncover the profound implications of sustainability on fostering a healthier and more environmentally conscious future. For example, hospitals have started installing solar energy systems to meet part of their energy needs. Others have implemented waste management programs to minimize their impact on the environment.

A. Sustainable Waste Management Optimization
Sustainable waste management (SWM) optimization in the context of smart health involves leveraging technology to minimize waste generation, maximize resource efficiency, and reduce environmental impact within healthcare settings. Here's how it works:
1) Reducing Medical Waste:
    Smart health technologies can help streamline healthcare processes, leading to reduced medical waste generation. For example, digital health records and telemedicine platforms eliminate the need for paper-based records and unnecessary in-person visits, thereby reducing paper and plastic waste.

2) Optimizing Resource Utilization:
Smart sensors and IoT devices can monitor resource usage in healthcare facilities, such as energy, water, and medical supplies. By analyzing real-time data, healthcare providers can identify inefficiencies and implement strategies to optimize resource utilization, thereby reducing waste [1]. In 2017, the Republic of Türkiye's Ministry of Environment, Urbanization, and Climate Change launched the "Zero Waste Project." The project aims to prevent waste generation, reduce waste, recycle waste at its source, prevent wastage, and utilize natural resources more efficiently [5].

3) Recycling and Reuse:
Smart health technologies can facilitate the recycling and reuse [19] of medical equipment and supplies. For instance, smart inventory management systems can track the usage of medical supplies and identify items that can be recycled or sterilized for reuse, reducing the need for single-use items.

4) Waste Segregation and Disposal:
Smart waste management systems can automate the segregation and disposal of different types of waste in healthcare facilities. By using sensors and AI algorithms, these systems can identify recyclable, hazardous [18], and non-recyclable waste, ensuring proper disposal and minimizing environmental contamination.

Fig 1. Planetary Health pathway

The Green Grid of Health with Smart Technologies
We'll break down the intricate connections between smart health and sustainability, demystifying the technologies and highlighting their potential to revolutionize both our well-being and the health of our planet. In this context, it is evident that green technologies have been implemented to enhance the quality of healthcare, or at least support for energy consumption measurement should be provided [12]. Hospitals are now utilizing air quality monitoring systems to ensure a healthy environment for both patients and staff. Additionally, energy-efficient medical equipment is being used to reduce energy consumption and costs. There are many other initiatives as well [7][8].

A. Harvesting the Planet for a healthier life
  1) How Smart Shoes are Paving the Way for a More Active and Earth-Conscious Future
• **Promoting Physical Activity:** Embedded sensors can track steps, distance, and calories burned, motivating users to embrace healthier lifestyles and reduce reliance on carbon-heavy transportation.
• **Encouraging Sustainable Choices:** Gamified fitness apps linked to smart shoes can incentivize walking and cycling over car travel, reducing individual carbon footprints [12].
• **Generating Green Energy:** Kinetic energy from footsteps can be harvested to power the shoes themselves or even contribute to a microgrid for other smart health devices, eliminating reliance on batteries [8][17]. Also, wave generators and turnstiles are innovative ways to generate renewable and sustainable energy. Wave generators use natural resources like oceans' wave movement, while turnstiles [25] generate electricity each time someone passes through them, they are usually put in a population-dense area.
  2) Solar Panels as a Sustainable Powerhouse:
• **Green Energy Source:** Embedded solar panels can power healthcare facilities (Points Of Care) and health devices, minimizing dependence on conventional electricity [11], and reducing greenhouse gas emissions.
• **Decentralized Power:** Distributed solar energy collection in smart shoes and wearable devices lowers reliance on centralized grids, increasing resilience and sustainability.

• **Personal Empowerment**: Individuals gain autonomy over their energy needs, promoting environmental consciousness and fostering a sense of self-reliance.

In other words, smart shoes [17] play a multifaceted role in promoting physical activity, tracking environmental data, and contributing to sustainable transportation choices. Additionally, smart panels offer a renewable energy source for smart health monitoring systems, with the challenge lying in finding the right balance between device performance and energy consumption through thoughtful optimization strategies.

B. Invisible Footprint: Green Programming for a Healthier Planet
The digital world might seem intangible, but the lines between its code and our planet's health are not. Programming languages, software architectures, and middleware, though seemingly technical aspects of software development, hold hidden implications for CO_2 emissions and ultimately, the well-being of our planet and its inhabitants [1].

1) The Impact on Planetary and Human Health:
• **Climate Change:** Increased CO_2 emissions contribute to global warming, rising sea levels, extreme weather events, and ecosystem disruption, impacting food security, water resources, and public health.

• **Air Pollution:** Data centers rely on energy sources that often result in air pollution, exacerbating respiratory illnesses and cardiovascular diseases.

• **Mental and Social Impacts:** The relationship between our health and that of the planet is very tight, any disturbance in the planet's ecosystem is a disturbance to humanity, for instance, Climate change and its consequences pose mental health [16] challenges as well as displacement and social unrest, creating a chain reaction of suffering.

Fig 2. Illustration of Green Technologies impact

• **Data Centers and Cloud Services Energy Consumption:** Large-scale data centers that host software applications and cloud services consume significant amounts of energy. Inefficiently designed software can contribute to higher energy needs in data centers, amplifying the environmental footprint and impacting planetary health.

• **Electronic Waste (E-Waste) & Short Lifecycle:** Poorly designed software may lead to faster hardware obsolescence, contributing to the generation of e-waste. E-waste disposal has environmental consequences, with improper disposal methods leading to soil and water pollution, negatively affecting planetary health.

2) Programming languages for a sustainable future:
Different programming languages have varying levels of energy efficiency. For instance, languages that demand more computational resources may contribute to higher energy consumption during program execution. The choice of programming language affects the energy requirements of software systems and subsequently influences CO_2 emissions.

• **Software Architectures:** The design and architecture of software systems determine how efficiently resources, including processing power and memory, are utilized. Poorly optimized software architectures can result in excessive resource usage, leading to higher energy consumption and increased CO_2 emissions [1][3].

• *Middleware and communication efficiency(Green middleware):* Opting for lightweight, energy-efficient middleware solutions can streamline data communication and management while maximizing sustainability. Inefficient middleware can lead to higher data transfer loads, requiring more energy for communication processes and contributing to increased CO2 emissions [2][3].
• *Code-Level Awareness:* Educating developers about the environmental impact of their code choice.
• *Algorithmic Efficiency:* Optimizing algorithms for reduced processing power and minimizing unnecessary computations can significantly decrease energy consumption [2] [3][4].
• *Lightweight Architectures:* Choosing leaner frameworks, optimizing server usage, and minimizing geographic distribution can help reduce the carbon footprint of software systems [3].

By acknowledging the link between programming and planetary health, we can weave a new narrative. By optimizing code, architectures, and middleware, we can rewrite our digital footprint and contribute to a future where technology complements, not threatens, the well-being of our planet and its inhabitants(Global Health). Let's remember, that every line of code has the potential to leave a mark, not just on a screen, but on the very fabric of our planet's health.

C. The Tradeoff of Performance vs. Energy Consumption
1) Optimizing the Performance-Energy Trade-off:
The quest for powerful smart devices often comes at the cost of high energy consumption. However, several strategies can strike a balance:
• *Sensor Optimization:* Employing energy-efficient sensors and utilizing them strategically can significantly reduce power consumption without compromising data accuracy.
• *Low-Power Processors:* Implementing specialized, low-power processors specifically designed for wearables can further enhance energy efficiency.
• *Cloud-Based Processing:* Offloading data processing to the cloud reduces on-device energy needs, enabling more powerful features while extending battery life.
• *Adaptive Algorithms:* Developing algorithms that adjust processing power based on activity level can optimize performance while ensuring efficient energy use [2].

2) Addressing Concerns about Smart Health Device Electromagnetic Frequency Radiation
Smart devices, including wearables, emit non-ionizing radiation in the radiofrequency (RF) range. Unlike ionizing radiation (e.g., X-rays) which can damage DNA, RF radiation is generally considered safe at low levels. However, concerns remain about the potential long-term health effects of chronic exposure, particularly with the increasing number of devices we interact with daily.
• *Safety Standards and Research:* International safety guidelines, such as those set by the International Commission on Non-Ionizing Radiation Protection (ICNIRP), exist to limit exposure to RF radiation. Device manufacturers must adhere to these standards, ensuring emissions remain within safe thresholds. Ongoing research continues to investigate the potential health effects of chronic RF exposure. While some studies haven't found conclusive evidence of harm, others suggest a weak link between certain types of RF radiation and increased risk of certain cancers, like brain tumors. However, it's important to note that these studies are often complex, with confounding factors and limitations, making it difficult to establish definitive causal relationships.
• *Exposure Levels and Health Effects:* Prolonged exposure to electromagnetic frequency (EMF) radiation from smart devices, including wearables, raises concerns about its impact on human health. While EMF is a fundamental aspect of technology, addressing concerns involves understanding potential risks associated with extended exposure. Some studies suggest a possible link between long-term exposure to EMF radiation and health issues such as headaches, sleep

disturbances, and potential implications for fertility. Ongoing research aims to clarify these potential health effects and establish clear causal relationships.

• *Navigating the Uncertainty:* While uncertainty persists, responsible use of smart devices can help mitigate potential risks like minimizing Exposure; and limit the time spent close to the device, particularly when it's actively transmitting data. For example, don't sleep with your phone under your pillow or keep it directly against your body for extended periods. Opt for Airplane Mode; When not in use, turn off wireless features like Bluetooth and Wi-Fi to minimize unnecessary exposure. Choose Eco-Friendly Devices; Some devices offer lower-power modes or settings that can reduce radiation emissions. Stay Informed; Follow reputable sources, like international health organizations, for updates on research findings and safety recommendations.

• *Vulnerability and Specific Absorption Rate(SAR):* SAR measures the rate at which the body absorbs radiofrequency (RF) energy from a device. Regulatory bodies set SAR limits to prevent excessive RF energy absorption and minimize potential health risks. Also, vulnerable populations, such as pregnant individuals and children, are of particular concern due to potential sensitivity to radiation. Recognizing the need for caution, especially for those more susceptible, underscores the importance of safety standards.

Discussion
In a tapestry woven from the threads of science, technology, and environmental consciousness, our article has explored the intricate connections between individual well-being, sustainable practices, and the health of our planet. In exploring the realm of smart health and sustainable innovations, we've delved into a diverse array of topics. From the foundational concerns surrounding electromagnetic frequency radiation emitted by smart devices to the intricate interplay between programming languages, software architectures, and CO2 emissions, the journey has been comprehensive. Our study extended to the pivotal role of smart shoes in promoting physical activity and sustainable transportation choices, not forgetting their potential to generate green energy [11]. However, the reviewed literature showcases commendable efforts to address key challenges in smart health systems, particularly focusing on energy consumption efficiency and sustainable energy. While each work contributes significantly to the field, further collaborative research could explore synergies between these approaches, potentially offering holistic solutions for the evolving landscape of smart health systems. For instance, a combination of these approaches applied for a Wearable Area Sensors Network, that can rely on self-energy generation from human motion would be the most efficient, like energy-harvesting shoes.

Conclusion and Future Perspective
Our exploration of the intricate nexus between smart health technologies, green solutions, and language processing reveals a promising avenue toward a sustainable future. Intertwined threads of smart health innovation, from nuanced personal well-being monitoring to the interplay of connectivity protocols, frequency radiations, and language processing, unveil a tapestry of profound interconnectedness. The symbiotic relationship between these technological advancements and environmental stewardship offers a blueprint for a future where technological progress and ecological responsibility coexist harmoniously.
Future work may include a novel approach that follows this study to implement a greener and more sustainable Smart Health Monitoring System that monitors both human and planetary health.

References
[1] Zahid, N., Sodhro, A. H., Al-Rakhami, M. S., Wang, L., Gumaei, A., & Pirbhulal, S. (2021). An Adaptive Energy Optimization Mechanism for Decentralized Smart Healthcare Applications. 2021 IEEE 93rd Vehicular Technology Conference (VTC2021-Spring). https://doi.org/10.1109/vtc2021-spring51267.2021.9448673

[2] Abdellatif, A. A., Mohamed, A., Chiasserini, C. F., Erbad, A., & Guizani, M. (2020). Edge computing for energy-efficient smart health systems. Energy Efficiency of Medical Devices and Healthcare Applications, 53–67. https://doi.org/10.1016/b978-0-12-819045-6.00003-0

[3] Natarajan, Rajesh & H L, Gururaj & Flammini, Francesco & Premkumar, Anitha & Kumar, V Vinoth & Gupta, Dr-Shashi. (2023). A Novel Framework on Security and Energy Enhancement Based on Internet of Medical Things for Healthcare 5.0. Infrastructures. 8. 22. https://doi.org/10.3390/infrastructures8020022

[4] Reddy, D.L.; Puttamadappa, C.; Suresh, H.N. Merged glowworm swarm with ant colony optimization for energy-efficient clustering and routing in the wireless sensor network. Pervasive Mob. Comput. 2021, 71, 101338.

[5] Aykal, Güzin. (2023). Green transformation in the health sector and medical laboratories, adaptation to climate change in Türkiye. Turkish Journal of Biochemistry. https://doi.org/10.1515/tjb-2023-0207

[6] EFLM Green Labs [Internet]. https://greenlabs.eflm.eu/ [Accessed Jan 2024].

[7] Ghernouk, Chaimae & Marouane, Mkik & Dalili, Saad & Boutaky, Soukaina & Hebaz, Ali & Mchich, Hamza. (2023). The Attractive Determinants of Green Technologies: The Case of The Health Sector in Morocco. 21. 1759-1774. https://doi.org/10.55365/1923.x2023.21.192

[8] Sadegh Seddighi, Edward J. Anthony, Hamed Seddighi, Filip Johnsson, The interplay between energy technologies and human health: Implications for energy transition, Energy Reports, Volume 9, 2023, Pages 5592-5611, ISSN 2352-4847, https://doi.org/10.1016/j.egyr.2023.04.351

[9] Prajapati, Sunil & Dayal, Parmeswar & Kumar, Vipin & Gairola, Ananya & Sustain, Agri. (2023). Green Manuring: A Sustainable Path to Improve Soil Health and Fertility. 01. 24-33. https://doi.org/10.5281/zenodo.10049824

[10] Cardinali, Marcel & Balderrama, Alvaro & Arztmann, Daniel & Pottgiesser, Uta. (2023). Green Walls and Health: An umbrella review. Nature-Based Solutions. 3. https://doi.org/100070. https://doi.org/10.1016/j.nbsj.2023.100070

[11] Li, Meng & Geng, Yong & Zhou, Shaojie & Sarkis, Joseph. (2023). Clean energy transitions and health. Heliyon. 9. e21250. https://doi.org/10.1016/j.heliyon.2023.e21250

[12] Aripriharta, Aripriharta. (2023). Performance Analysis Smart-Shoes To Measure the Pulse in Dorsalis Pedis Artery. Fidelity : Jurnal Teknik Elektro. 5. 21-30. https://doi.org/10.52005/fidelity.v5i2.146

[13] Li, Qiangyi & Liu, Yangqing & Yang, Lan & Ge, Jiexiao & Chang, Xiaona & Zhang, Xiaohui. (2023). The impact of urban green space on the health of middle-aged and older adults. Frontiers in Public Health. 11. https://doi.org/10.3389/fpubh.2023.1244477

[14] Revich, B.A.. (2023). The significance of green spaces for protecting health of urban population. Health Risk Analysis. 168-185. https://doi.org/10.21668/health.risk/2023.2.17.eng

[15] Rathnayke, Saumya & Amofah, Seth. (2023). Health and Wellbeing Implications of Urban Green Exposure on Young Adults in a European City. Journal of Advanced Research in Social Sciences. 6. 53-70. https://doi.org/10.33422/jarss.v6i4.1136

[16] Zhang, Jun & Jin, Jinghua & Liang, Yimeng. (2024). The Impact of Green Space on University Students' Mental Health: The Mediating Roles of Solitude Competence and Perceptual Restoration. Sustainability. 16. 707. https://doi.org/10.3390/su16020707

[17] Kurita, Hiroki & Katabira, Kenichi & Yoshida, Yu & Narita, Fumio. (2019). Footstep Energy Harvesting with the Magnetostrictive Fiber Integrated Shoes. Materials. 12. 2055. https://doi.org/10.3390/ma12132055

[18] Ugoeze, Kenneth & Alalor, Christian & Ibezim, Chidozie & Chinko, Bruno & Owonaro, Peter & Anie, Clement & Okoronkwo, Ngozi & Mgbahurike, Amaka & Ofomata, Chijioke & Alfred-Ugbenbo, Deghinmotei & Ndukwu, Geraldine. (2024). Environmental and Human Health Impact of Antibiotics Waste Mismanagement: A Review. Advances in Environmental and Engineering Research. 05. 1-21. https://doi.org/10.21926/aeer.2401005

[19] Mohd Nawi, Mohd Nasrun & Mohd Nasir, Najuwa & Abidin, Rahimi & Salleh, Nurul & Harun, Aizul & Osman, Wan & Ahmad, M.. (2018). Enhancing construction health and safety through the practices of reuse and recycle in waste management among Malaysian contractors. Indian Journal of Public Health Research & Development. 9. 1521. https://doi.org/10.5958/0976-5506.2018.01664.9

[20] Haustein, Sonja & Koglin, Till & Nielsen, Thomas & Svensson, Åse. (2019). A comparison of cycling cultures in Stockholm and Copenhagen. International Journal of Sustainable Transportation. 14. 1-14. https://doi.org/10.1080/15568318.2018.1547463

[21] Wang, Junlin & Mukhopadhyaya, Phalguni & Valeo, Caterina. (2023). Implementing Green Roofs in the Private Realm for City-Wide Stormwater Management in Vancouver: Lessons Learned from Toronto and Portland. Environments. https://doi.org/10.3390/environments10060102

[22] Zhu, Wenhao. (2023). Vertical Farms: A Sustainable Solution to Urban Agriculture Challenges. Highlights in Science, Engineering and Technology. 75. 80-85. https://doi.org/10.54097/4n06rw70

[23] Marques, Bruno & Mcintosh, Jacqueline & Popoola, Tosin. (2018). Green Prescriptions and Therapeutic Landscapes: A New Zealand Study. International Journal of Behavioral Medicine. 25. 21.

[24] Taylor, Bron. (2013). Kenya's green belt movement: Contributions, conflict, contradictions, and complications in a prominent environmental non-governmental organization (ENGO). https://doi.org/10.1515/9780857457578-009

[25] Penagos, Hernán. (2020). Electric power generation from a turnstile. Dyna (Medellin, Colombia). 87. 156-162. https://doi.org/10.15446/dyna.v87n215.86789

Water and electricity consumption management architectures using IoT and AI: A review study

Oumaima RHALLAB[1,a*], Amine DEHBI[2,b*] and Rachid DEHBI[1,c]

[1] Research laboratory of computer science (LIS) Faculty of Sciences Aïn Chock Casablanca, Morocco

[2] Research laboratory of mathematics and computer science (LTI) Faculty of Sciences Ben M'Sick Casablanca, Morocco

[a]oumaimarhallab2@gmail.com, [b]dehbiamine1@gmail.com, [c] dehbirac@yahoo.fr

Keywords: Water and Electricity Management, Internet of Things (IoT), Artificial Intelligence (AI), Resource Management

Abstract. This in-depth article delves into the implemented architectures aimed at optimizing water and electricity consumption through the integration of the Internet of Things (IoT) and artificial intelligence (AI). It provides a detailed analysis of various developments, trends, and key technologies shaping this rapidly evolving field. The article meticulously examines background research and scrutinizes the architecture, thus offering profound insights into the technical challenges, potential benefits, and implementation obstacles of leveraging IoT and AI in resource management. By exploring these architectures, the article highlights significant advancements in terms of efficiency, resource utilization, and predictive capabilities within integrated systems. Convincing results demonstrate the positive impact of this technological convergence on environmental sustainability, waste reduction, and resource optimization, thus offering a promising vision for the future of resource management. Furthermore, an extensive discussion section critically evaluates the discussed approaches, pinpointing the strengths and weaknesses of each method and proposing avenues for improvement and development. This nuanced analysis provides a solid foundation for future research and continuous innovation in the field of resource management. This article serves as an essential resource for professionals and researchers working in the fields of water, energy, and IoT/AI. It offers an in-depth understanding of the challenges and opportunities associated with integrating these technologies and provides strategic guidance for effective and sustainable resource management in the digital age.

Introduction

In a context where the challenges of managing resources such as water and electricity are becoming increasingly pressing, advancements in the fields of the Internet of Things (IoT) and artificial intelligence (AI) offer intriguing prospects for addressing these challenges in innovative and efficient ways. This article delves deep into this technological convergence, exploring various strategies and architectures implemented to optimize water and electricity consumption. The meticulous examination of existing methodologies reveals a wealth of possibilities and complexities. Approaches such as smart telemetry, interconnected sensor networks, and AI-based predictive systems hold considerable promise for real-time monitoring, precise control, and proactive prediction of consumption patterns. These technological advancements pave the way for more efficient and sustainable resource management, with potentially revolutionary implications for industries, municipalities, and households. However, this landscape of innovation is not without its challenges. Data privacy and cybersecurity issues emerge as systems become more interconnected and data volumes increase. Moreover, the need to ensure equitable access to these technologies raises concerns about digital divides and social exclusion. Beyond technological considerations, it is crucial to examine the socio-economic and environmental implications of these advancements. How do these technologies affect the livelihoods of populations, local economies, and the health of our planet as a whole? How can we ensure that the benefits of these innovations accrue to all, without compromising fundamental rights and long-term sustainability? In this quest to address these challenges, a collaborative and interdisciplinary approach is indispensable. Policymakers, scientists, engineers, sociologists, and environmental advocates must come together to design integrated and equitable solutions. It is imperative to adopt a long-term perspective, balancing current needs with the ability of future generations to meet their own.

Literature Review
Fuentes, H[1], proposes an innovative IoT framework for water consumption management. It emphasizes secure data capture through encryption, local preprocessing of consumption data, physical security of devices, Recording and displaying water usage patterns, along with identifying leaks through the implementation of the Water Leak Algorithm. This system involves five primary elements for the collection, retention, examination, and representation of water consumption data. In the "House Data Collection" module, a smart meter captures water usage data at each T1 moment, transmitted to the "Edge Gateway" for archival, with a security feature to detect unauthorized manipulations. Cumulative consumption is sent to the Cloud at broader intervals (T2), stored with user location data obtained from their phone's GPS. The data undergoes analysis for leak detection at the Cloud, and users can visualize real-time water consumption through a web portal. This framework provides a comprehensive solution for gathering, safeguarding, storing, analyzing, and presenting water consumption data, integrating security measures for data integrity and proactive leak detection. The study offers a detailed technological overview, emphasizing the five essential elements, specifying software, programming languages, databases, and operating systems associated with each component.

Paramasivan's [2] proposed system introduces an inventive resolution for implementing prepaid energy supply while consolidating the oversight of all energy meters. The primary aim is to thwart unauthorized tampering with electricity at consumer premises while concurrently curbing labor expenses associated with billing. The GSM unit establishes a connection with the intelligent meter installed in individual residences, with each meter being allocated a unique quantity by electricity suppliers. The intelligent meter consistently records electricity usage and displays the consumed units on an LCD screen linked to the meter. The microcontroller diminishes the unit quantity in response to consumption. Upon a request from the electricity provider's server, the GSM modem is prompted to instruct the microcontroller to take necessary actions. The MAX232 module facilitates communication between the microcontroller and the GSM modem. The GSM plays a vital role in furnishing information about electricity consumption to both the electricity provider and the consumer, in real-time or as required. The GSM assumes a pivotal role in disseminating information about energy consumption to the application management and the consumer, as needed. The aerial container, affixed on or near the meter, heightens GSM communication for effective energy monitoring.

The advanced system, introduced by Ramadhan and Ali [3], offers a sophisticated solution for the cost-effective surveillance of water quality in a minimum of five water treatment stations. Its design ensures prolonged and uninterrupted operation through low energy consumption, facilitated by solar panels. Precise monitoring of water quality parameters is achieved through the utilization of ten specific sensors within the system. These parameter values are promptly displayed in real-time on a dedicated web page, providing immediate insights into the water quality across the monitored areas. The system possesses the capability to issue real-time alerts to relevant personnel via SMS and emails when abnormal or problematic values are detected. The recorded data is also archived for subsequent statistical analysis. The sensors employed in this system have undergone rigorous laboratory testing, demonstrating exceptional performance in terms of accuracy and reliability. The monitoring system is composed of three primary components: the detection node (SN), the data router (DR), and the website (WS), seamlessly integrated. Each water station is represented by a detection node (SN), and each node comprises four fundamental elements: the control unit, sensors, transceiver, and power unit. The control unit is facilitated by a programmable logic controller (PLC) of the CONTROLLINO MEGA type.

Segun O. Olatinwo [4] introduces a pioneering system architecture that integrates a Wireless Sensor Network (WSN) with multiple sensors categorized into two groups: Group 1 and Group 2. The classification is based on their prioritized schedule for information transmission, determined by a sequential data transmission method. Group 1, consisting of sensors i, becomes active during the initial cycle of the Up-Link (UL) transmission phase, while Group 2, comprising sensors q, is designated for transmitting data regarding the quality of water. in the subsequent cycle of the UL phase. Consequently, sensors from both Group 1 and Group 2 engage in data communication. To efficiently oversee the concurrent transfer of water quality data from every sensor group to a designated receiving node., a Successive Interference Cancellation (SIC) mechanism is implemented at the receiver. This mechanism acts as a congestion control measure, facilitating the separation of data emitted simultaneously by the sensors. The sensor nodes utilized in this system are Water Quality Sensors (WQS) designed to measure essential microbiological and chemical properties of water at treatment stations. These nodes are powered by Distributed Power Sources (DPS) equipped with omnidirectional antennas. The system controller possesses extensive knowledge of network resources, sensors in Groups 1 and 2, and a scheduler for activating Group 1 or Group 2 based on a predefined priority.

The architecture of the IoT application has been organized to facilitate sequential sharing of functions among all components, as suggested by Khan, M.A [5], starting from current IoT devices to the Managed-Cloud, in which data undergoes processing for making insightful decisions. The smart circuit is designed to enable the sequential measurement of current and voltage for three household appliances, utilizing ACS712 current sensors and ZMPT101B voltage sensors. Once the measured data is acquired, it is transmitted towards inputs of a Wi-Fi access controller, such as an ESP8266 module. This module, known for its online monitoring and control capabilities, is chosen for its power efficiency and affordability. Following data processing, it is forwarded to an MQTT server in the cloud, serving as a intermediary between the user and the loads. This cloud server facilitates multiple clients/users to connect and access the data. Users can oversee and control the data through a HumanMachine-Interface using either a personal computer and/or a mobile app. Furthermore, there are two additional loads included to indicate power and detect overloads. Initially linked to the power using a 4- channel relay, these loads can function in two modes: "normally open (NO)" and "normally closed (NC)." To prioritize safety, the initial configuration is set to "NO," and the voltages are consistently maintained at 220 V.

Liu, Yi [6] introduces a system proposing A framework for energy management built upon the Internet of Things (IoT), integrating advanced computing technologies and a Deep Reinforcement Learning (DRL) network. The structure is composed of three primary elements: energy devices, energy edge servers, and energy cloud servers. In this design, data is processed locally by energy edge servers and then sent via the central network to the cloud server. Agents for Deep Reinforcement Learning (DRL) are installed on edge and cloud servers. When a computing task is needed, an energy device sends it to the closest edge server, where the edge DRL agent does the task computation. Deep neural network (DNN) weight pre-training can be done in the energy cloud server to improve energy consumption in the edge server. After training, the DNN advances to the boundary, where Deep Q-Learning assumes control. Edge servers collect data from devices and transmit it to the cloud for effective processing, hence reducing energy consumption.

Background study
1. Data Sources for Adaptive Water-Electricity Management

Figure 1: Data Sources for Adaptive Water-Electricity Management

Leveraging diverse data sources is vital for adaptive water and electricity system management to optimize processes, enhance efficiency, and enable informed decision-making. Common data sources for adaptive approaches in water and electricity management include:
- **IoT Sensors and Monitoring Instruments**: Real-time data on consumption, quality, temperature, water levels, and pressure [7, 8] can be obtained using sensors deployed in distribution infrastructures. Smart Sensor Networks creating distributed monitoring systems with interconnected smart sensor networks allows for extensive data collection from multiple points [9].
- **Geographic Information Systems (GIS)**: GIS provides spatial information to analyze geographical distribution, plan maintenance routes, and optimize resource management [10].
- **Meteorological Data:** Crucial for anticipating electricity demand, especially for renewables [11], and predicting weather conditions impacting water resources.
- **Historical Consumption Data:** Analyzing past consumption patterns is vital for forecasting future demand and planning capacity [12].
- **Communication Networks:** Wired and wireless networks facilitate reliable data transmission between system components [13].
- **Operational and Maintenance Data:** Information on infrastructure operations, maintenance reports, downtime, and equipment performance is crucial for optimizing processes [14].
- **Demographic and Socio-economic Data:** Understanding population factors helps tailor management strategies to changing water and electricity demand [15].

2. Data transmission protocols

The data pathway in adaptive management of water and electricity systems can traverse various channels, with certain methods evolving in prominence [16]. Contemporary implementations predominantly leverage Bluetooth and Wi-Fi transmissions. Some systems embrace standard wireless technologies like ZigBee, XBee, and ZWave. While GSM has historically offered stability in mobile communication, it has witnessed a decline in use, being superseded by advanced cellular standards such as 3G, 4G, and 5G, offering enhanced data transfer speeds. It's noteworthy that Table 1 illustrates a comparison between some protocols [17], highlighting their distinct characteristics and functionalities.

3. Data Storage

In IoT systems dedicated to monitoring water and electricity consumption in Morocco, the data aggregation phase assumes a pivotal role. This stage involves consolidating information from multiple sensors through intelligent IoT gateways, creating a unified representation for simplified processing and analysis [18]. Utilizing buffers or local databases ensures data availability during temporary network disruptions. This approach optimizes data transmission to the central platform, maintaining efficient information flow despite potential connectivity fluctuations. Upstream of in-depth analysis, data preprocessing becomes instrumental. This stage involves processes like noise filtering, normalization of measurement units, and error correction, refining raw data quality for consistency and readiness in advanced analyses [19]. Some systems employ local storage at the sensor or gateway level, ensuring immediate data availability even without connectivity. Simultaneously, a cloud-first architecture allows direct transmission of other data to the cloud [20]. This hybrid strategy accommodates the specific constraints of each system, facilitating effective information management aligned with infrastructure needs and capabilities [21].

Table 1: Comparison between data transmission protocols

Specifications	Wi-Fi	Bluetooth	GSM	Zigbee
Network type	Point-to-point, WLAN, WAN, Mesh, Point-to- multipoint	PAN	Cellular network	Mesh network
Communication	Radio Frequency, Protocols (IEEE 802.11, 802.11b, 802.11g...)	Wireless	Wireless	Wireless
Security	Protocol (WEP, WPA, and WPA2/WPA3),	Encryption methods	Encryption methods	Encryption methods
Range	Up to 100 meters	Up to 10 meters	Several kilometers	from 10 to 100 meters
Frequency	2.4 GHz and 5 GHz	2.4 GHz	900 MHz and 1800 MHz	2.4 GHz
Bit rate	From 11Mpbs to up than 9,6 Gbps depending on the Wi-Fi standard being used	Depends on the Bluetooth version, ranging from 1 Mbps to 3 Mbps.	Up to 9.6 kbps	Between 20 and 250 kbps
Continuous sampling	Yes	Yes	NO	Yes
Interoperability	Different and same devices	Compatible devices	GSM compatible devices	Same devices

4. **AI Processing:**

Table 2: The processing AI steps

Phase	Description
Retrieval and Preparation	Following the storage of data, the AI-driven processing phase commences with the retrieval of stored information. The data is subsequently prepared for analysis, encompassing tasks such as formatting, structuring, and ensuring compatibility with AI algorithms [22].
Preprocessing with AI	AI is utilized for intricate preprocessing tasks, where machine learning models can autonomously identify patterns, outliers, and anomalies within the stored data. This step guarantees that the data is refined and well-prepared for more sophisticated analyses [23].
Predictive Analytics	AI algorithms, encompassing machine learning and predictive modeling [24], come into play for forecasting future trends based on historical data [25] [26] . This facilitates proactive decision-making and resource planning in response to anticipated consumption patterns.
Real-time Insights	AI transcends traditional batch processing limitations, enabling real-time analytics [27]. Continuously analyzing incoming data, AI algorithms furnish immediate insights into consumption patterns, potential inefficiencies, or abnormalities.

Continuous Improvement	In the realm of data processing, AI systems are engineered for continuous improvement [28]. Employing feedback loops and continuous learning mechanisms, these models evolve over time, progressively enhancing their accuracy and efficacy in dealing with a myriad of data scenarios.

Conclusion

In conclusion, this comprehensive study delved into water and electricity management architectures integrating AI and IoT, providing insights into current trends and key technologies. The background research highlighted the diverse data sources used in these architectures, emphasizing IoT sensors, communication networks, and other advanced technologies. Transmission protocols were scrutinized, comparing multiple options to inform implementation choices. It is crucial to acknowledge that this study presents findings based on the examination of six specific architectures and the comparison of four transmission protocols. The conclusions indicate significant advancements in terms of efficiency, resource utilization, and predictive capabilities in these integrated systems. However, certain limitations need to be emphasized. The restriction to six architectures and four protocols may not cover all possible variants. For a more comprehensive and nuanced understanding, future research could explore a broader range of architectures and protocols. As for future perspectives, an in-depth bibliometric study on water and electricity consumption topics, integrating AI and IoT, could provide a more holistic view of the field. This would help identify emerging trends, gaps in current research, and guide future work towards areas of innovation and particular importance.

References

[1] Fuentes, H., & Mauricio, D. (2020). Smart water consumption measurement system for houses using IoT and cloud computing. Environmental Monitoring and Assessment, 192, 1-16. https://doi.org/10.1007/s10661-020-08535-4

[2] Rajesh, M. (2021). GSM Based Smart Energy Meter System.

[3] Ramadhan, A. J., Ali, A. M., & Kareem, H. K. (2020). Smart waterquality monitoring system based on enabled real-time internet of things. J. Eng. Sci. Technol, 15(6), 3514-3527.

[4] Olatinwo, S. O., & Joubert, T. H. (2020). Energy efficiency maximization in a wireless powered IoT sensor network for water quality monitoring. Computer Networks, 176, 107237. https://doi.org/10.1016/j.comnet.2020.107237

[5] Khan, M. A., Sajjad, I. A., Tahir, M., & Haseeb, A. (2022). IoT application for energy management in smart homes. Engineering Proceedings, 20(1), 43. https://doi.org/10.3390/engproc2022020043

[6] Liu, Y., Yang, C., Jiang, L., Xie, S., & Zhang, Y. (2019). Intelligent edge computing for IoT-based energy management in smart cities. IEEE network, 33(2), 111-117. https://doi.org/10.1109/MNET.2019.1800254

[7] Ibrahim, S. N., Asnawi, A. L., Abdul Malik, N., Mohd Azmin, N. F., Jusoh, A. Z., & Mohd Isa, F. N. (2018). Web based Water Turbidity Monitoring and Automated Filtration System: IoT Application in Water Management. International Journal of Electrical & Computer Engineering (2088-8708), 8(4). https://doi.org/10.11591/ijece.v8i4.pp2503-2511

[8] Li, J., Herdem, M. S., Nathwani, J., & Wen, J. Z. (2023). Methods and applications for Artificial Intelligence, Big Data, Internet of Things, and Blockchain in smart energy management. Energy and AI, 11, 100208. https://doi.org/10.1016/j.egyai.2022.100208

[9] Jaradat, M., Jarrah, M., Bousselham, A., Jararweh, Y., & AlAyyoub, M. (2015). The internet of energy: smart sensor networks and big data management for smart grid. Procedia Computer Science, 56, 592-597. https://doi.org/10.1016/j.procs.2015.07.250

[10] Zulkifli, C. Z., Garfan, S., Talal, M., Alamoodi, A. H., Alamleh, A., Ahmaro, I. Y., ... & Chiang, H. H. (2022). IoT-based water monitoring systems: a systematic review. Water, 14(22), 3621. https://doi.org/10.3390/w14223621

[11] Saheb, T., Dehghani, M., & Saheb, T. (2022). Artificial intelligence for sustainable energy: A contextual topic modeling and content analysis. Sustainable Computing: Informatics and Systems, 35, 100699. https://doi.org/10.1016/j.suscom.2022.100699

[12] Rochd, A., Benazzouz, A., Abdelmoula, I. A., Raihani, A., Ghennioui, A., Naimi, Z., & Ikken, B. (2021). Design and implementation of an AI-based & IoT-enabled Home Energy Management System: A case study in Benguerir—Morocco. Energy Reports, 7, 699-719. https://doi.org/10.1016/j.egyr.2021.07.084

[13] Rahmadya, B., Zaini, Z., & Muharam, M. (2020). Iot: A mobile application and multi-hop communication in wireless sensor network for water monitoring. https://doi.org/10.3991/ijim.v14i11.13681

[14] Krishnan, S. R., Nallakaruppan, M. K., Chengoden, R., Koppu, S., Iyapparaja, M., Sadhasivam, J., & Sethuraman, S. (2022). Smart water resource management using Artificial Intelligence—A review. Sustainability, 14(20), 13384. https://doi.org/10.3390/su142013384

[15] Ansari, S., Ayob, A., Lipu, M. S. H., Saad, M. H. M., & Hussain, A. (2021). A review of monitoring technologies for solar PV systems using data processing modules and transmission protocols: Progress, challenges and prospects. Sustainability, 13(15), 8120. https://doi.org/10.3390/su13158120

[16] Kanellopoulos, D., Sharma, V. K., Panagiotakopoulos, T., & Kameas, A. (2023). Networking Architectures and Protocols for IoT Applications in Smart Cities: Recent Developments and Perspectives. Electronics, 12(11), 2490. https://doi.org/10.3390/electronics12112490

[17] Dehbi, A., Bakhouyi, A., Dehbi, R., & Talea, M. (2024). Smart Evaluation: A New Approach Improving the Assessment Management Process through Cloud and IoT Technologies. International Journal of Information and Education Technology, 14(1), 107-118. https://doi.org/10.18178/ijiet.2024.14.1.2030

[18] Bedi, P., Goyal, S. B., Rajawat, A. S., Shaw, R. N., & Ghosh, A. (2022). Application of AI/IoT for smart renewable energy management in smart cities. AI and IoT for Smart City Applications, 115-138. https://doi.org/10.1007/978-981-16-7498-3_8

[19] Goudarzi, S., Anisi, M. H., Soleymani, S. A., Ayob, M., & Zeadally, S. (2021). An IoT-based prediction technique for efficient energy consumption in buildings. IEEE Transactions on Green Communications and Networking, 5(4), 2076-2088. https://doi.org/10.1109/TGCN.2021.3091388

[20] Lakshmikantha, V., Hiriyannagowda, A., Manjunath, A., Patted, A., Basavaiah, J., & Anthony, A. A. (2021). IoT based smart water quality monitoring system. Global Transitions Proceedings, 2(2), 181-186. https://doi.org/10.1016/j.gltp.2021.08.062

[21] Praveena, D., Thanga Ramya, S., Gladis Pushparathi, V. P., Bethi, P., & Poopandian, S. (2021). Hybrid Cloud Data Protection Using Machine Learning Approach. In Advanced Soft

Computing Techniques in Data Science, IoT and Cloud Computing (pp. 151- 166). Cham: Springer International Publishing. https://doi.org/10.1007/978-3-030-75657-4_7

[22] Bourechak, A., Zedadra, O., Kouahla, M. N., Guerrieri, A., Seridi, H., & Fortino, G. (2023). At the Confluence of Artificial Intelligence and Edge Computing in IoT-Based Applications: A Review and New Perspectives. Sensors, 23(3), 1639. https://doi.org/10.3390/s23031639

[23] Himeur, Y., Elnour, M., Fadli, F., Meskin, N., Petri, I., Rezgui, Y., ... & Amira, A. (2023). AI-big data analytics for building automation and management systems: a survey, actual challenges and future perspectives. Artificial Intelligence Review, 56(6), 4929-5021. https://doi.org/10.1007/s10462-022-10286-2

[24] Rajawat, A. S., Mohammed, O., Shaw, R. N., & Ghosh, A. (2022). Renewable energy system for industrial internet of things model using fusion-AI. In Applications of AI and IOT in Renewable Energy (pp. 107-128). Academic Press. https://doi.org/10.1016/B978-0-323-91699-8.00006-1

[25] Jiang, D., Lian, M., Xu, M., Sun, Q., Xu, B. B., Thabet, H. K., ... & Guo, Z. (2023). Advances in triboelectric nanogenerator technology—Applications in self-powered sensors, Internet of things, biomedicine, and blue energy. Advanced Composites and Hybrid Materials, 6(2), 57. https://doi.org/10.1007/s42114-023-00632-5

[26] Olatinwo, S. O., & Joubert, T. H. (2023). Resource Allocation Optimization in IoT-Enabled Water Quality Monitoring Systems. Sensors, 23(21), 8963. https://doi.org/10.3390/s23218963

[27] Nasiri, F., Ooka, R., Haghighat, F., Shirzadi, N., Dotoli, M., Carli, R., ... & Sadrizadeh, S. (2022). Data analytics and information technologies for smart energy storage systems: A state-of-the-art review. Sustainable Cities and Society, 84, 104004. https://doi.org/10.1016/j.scs.2022.104004

[28] Kamyab, H., Khademi, T., Chelliapan, S., SaberiKamarposhti, M., Rezania, S., Yusuf, M., ... & Ahn, Y. (2023). The latest innovative avenues for the utilization of artificial Intelligence and big data analytics in water resource management. Results in Engineering, 101566. https://doi.org/10.1016/j.rineng.2023.101566

A comprehensive review on computing methods for the prediction of energy cost in Kingdom of Saudi Arabia

Nayeemuddin MOHAMMED[1,a], Andi ASIZ[1,b], Mohammad Ali KHASAWNEH[1,c], Feroz SHAIK[2,d], Hiren MEWADA[3,e], Tasneem SULTANA[4,f*]

[1]Department of Civil Engineering, Prince Mohammad Bin Fahd University, Al Khobar, Kingdom of Saudi Arabia

[2]Department of Mechanical Engineering, Prince Mohammad Bin Fahd University, Al Khobar, Kingdom of Saudi Arabia

[3]Department of Electrical Engineering, Prince Mohammad Bin Fahd University, Al Khobar, Kingdom of Saudi Arabia

[4]Artificial Intelligence and Machine Learning, Godutai Engineering College for Women, Sharnbasva University, Kalaburagi, India

mnayeemuddin@pmu.edu.sa[a], aasiz@pmu.edu.sa[b] mkhasawneh@pmu.edu.sa[c], fshaik@pmu.edu.sa[d], hmewada@pmu.edu.sa[e], tasneemsultana841@gmail.com[f*]

Keywords: Energy, Artificial Neural Network, Prediction Models, Regression, Statistical, Deep Learning

Abstract. Addressing the increasing demand for energy in the Kingdom of Saudi Arabia (KSA) poses challenges and opportunities. This necessitates effective energy planning, diversification of energy sources, and implementation of energy-efficient technologies. This study presents the energy scenario in the KSA. Later, various technical algorithms used for energy prediction from past data, including regression models, statistical models, machine learning and deep learning networks, are presented. The present study revealed that learnable models, specifically neural networks, outperformed statistical and regression networks in predicting energy demands. In addition, statistical models lack predictability and lack adoption with new data.

Introduction

Power is an exceptional resource. There must be an ongoing equilibrium between production and consumption for the power system to be stable, and this equilibrium is economically unstorable. Concurrently, the amount of economic and daily activity during off-peak and peak hours, on weekdays and weekends, etc., and weather-related factors, e.g., temperature, wind speed, precipitation, etc., determine energy demand. The forecasting of electricity is a fundamental need at the government and corporate levels for the decision-making process. Market players hedge against both volume risk and price fluctuations due to the extreme volatility of prices, which can be two orders of magnitude greater than that of any other financial asset or commodity. Thus, forecasting energy costs for a single country is valuable for several reasons. Accurate energy cost predictions enable governments and policymakers to plan for the economic development of a country. Energy costs have a significant impact on various sectors, such as manufacturing, transportation, and agriculture. By forecasting energy costs, governments can assess the competitiveness of industries, attract investments, and develop strategies to ensure an affordable and reliable energy supply. By understanding future energy costs, governments can assess the availability and affordability of energy sources. This knowledge helps in diversifying the energy supply, reducing dependence on volatile or geopolitically unstable sources, and developing strategies to maintain a stable and reliable energy infrastructure.

Predictions of energy costs are essential for developing energy legislation and policy. With the help of these forecasts, governments can create energy-saving policies, incentives, or subsidies that encourage the use of renewable energy sources and sustainable energy practices. Accurate predictions can help policymakers understand the potential impact of policy changes on energy costs for consumers and businesses. By predicting energy costs for a specific country, stakeholders can gain valuable insights for economic planning, energy security, policy formulation, consumer decision-making, and market analysis. These insights contribute to efficient energy management, sustainable development, and the overall well-being of the country and its citizens. There are several methods for predicting the energy cost. A few popular methods are regression analysis, time-series analysis, machine learning approaches, data mining and optimization algorithms. The accuracy and effectiveness of these methods depend on the availability and quality of the data, as well as the specific characteristics of the energy system being analyzed. In practice, a combination of multiple methods or an ensemble approach may be employed to improve the prediction accuracy.

In regard to global power consumption, the Kingdom of Saudi Arabia (KSA) ranks fourteenth. There has recently been remarkable growth in every sector of the Saudi Arabian economy, but the generation and consumption of electrical power have been particularly impressive. The government is currently working on strategies to improve this industry in the future. This is because it is crucial to sustainable development goals and because Vision 2030 requires the use of renewable energy to power the nation. Saudi Arabia generated an estimated 374 tera watt hours (TWh) of electricity in 2022, up 2% from 367 TWh in 2021 [1].

This paper presents recent algorithms used for energy prediction. Initially, an energy scenario in the KSA and various developed countries was presented. This paper focuses on various algorithms presented in the literature for accurate energy forecasting. Finally, limitations were presented.

Energy scenario in KSA
Resources for renewable and sustainable energy (RnSE) have gained prominence recently as being essential to the stability of economies throughout the world. Recent research has identified renewable and sustainable energy (RnSE) resources as a critical component of a healthy global economy, especially in industrialized nations like the Kingdom of Saudi Arabia (KSA) [2]. Saudi Arabia, as the biggest economy in the GCC, is heavily dependent on non-renewable resources for its economic growth. As such, it is imperative that the country look into alternate energy sources. In the region, using solar applications especially photovoltaics is seen to be the most cost-effective way to supply basic energy services [3]. Residential structures have received the majority of attention, although commercial and educational building construction is rapidly increasing [4]. The levelized cost of energy and the net present cost are used to compare the photovoltaic (PV) energy outputs of the Kingdom of Saudi Arabia with those of potential PV energy customers, such as European countries, China, India, and Pakistan [5]. Compared to the Mass Burn with recycling scenario, the Mass Burn scenario may yield twelve times as much. To compare the two situations in terms of economic, social, technological, and environmental factors need required another studies [6].

Future substantial expansion in the country's energy consumption is predicted due to a number of variables, including cheap energy prices, high economic growth, and a growing population [7]. After the construction and industrial sectors in Saudi Arabia, the electricity industry as a whole had the second-highest carbon emissions in 2018 [8].

Energy Cost Prediction Models
The Gaussian process creates a prediction function for the energy consumption with confidence bounds by modelling the intricate interactions between the input machining parameters and the

output energy consumption. Prediction models that consider various operations and process characteristics may be created using sophisticated data collecting and processing techniques to estimate a machine tool's energy usage [9]. Predicting energy usage in office buildings in cold climates was done using neural network prediction models based on evolutionary algorithms and back propagation. The highest RMSE value of the enhanced GA-BP neural network was 0.36, and the maximum MAPE value was 0.29%; both evaluation indicators were less than those of the BP prediction technique [10].

The prediction algorithm was built using two available datasets of residential and commercial structures. Accurate assessment of the energy consumption in buildings is essential for both energy policy and building energy management. The main difference between this model and the shallow machine learning (ML) model is the quantity of linear or non-linear transformations applied to the input data. The deep neural network model often makes numerous modifications to the input data before generating an output. The planned design model for the ANN-Levenberg Marquardt tool, which is employed in ANN. 3 hidden layers, with 3 inputs employed to design the model for regression and optimization using artificial neural network Levenberg Marquardt tool in MATLAB application. Each layer has different neurons such as 35, 20 and 10 were suggested to get more accuracy of prediction as shown in Figure 1 [11]. Cao et al., created and evaluated an integrated learning method that uses data permutation to gauge the significance of features in order to reduce the instability issue with building management systems [12], [13].

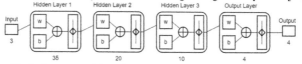

Fig 1. Design modeling for ANN-LM tool in ANN [13]

Kim et al., investigated and used the response surface approach in conjunction with the statistical method to create a prediction model for residential building energy consumption. In Seoul and Busan apartment buildings, higher window and wall thermal transmittances and infiltration rates led to higher heating energy consumption; conversely, higher SHGC was linked to reduced cooling energy consumption [14]. The suggested deep learning technique was used to forecast the energy usage of a particular building for which data on energy use over a 12-month period was gathered. The results of the experiment and comparison show that the deep learning approach performs better than a number of well-liked conventional machine learning techniques [15], [16].

Limitations in Energy Cost Prediction Models
When comparing the ANN to the BP neural network, the ELM significantly enhanced it. In contrast to typical feedforward neural networks, the ELM randomizes the network weight of a single hidden layer. The inverse matrix of Moore-Penrose is used to determine the outputs. As a result, the output calculation speed and generalization precision are rather high, and it is difficult to reach a local maximum [17]. Support vector machines (SVM), artificial neural networks (ANN), decision trees, and other statistical methods are examples of learning algorithms [18],[19]. Building and region scales are examples of spatial scales. Both short-term and long-term temporal granularities are available. There are several types of energy consumption forecast, such as total, heating, cooling, and lighting. Real and simulated datasets are two examples of dataset types [20].

Energy Cost Prediction Models Applied to KSA Scenario
The Kingdom of Saudi Arabia, an expanding nation, is seeing remarkable developments in a range of fields, such as the medical, educational, engineering, and urban sectors, particularly in the economic and industrial realms. Planning for capacity, transmission, and price all depend on the

ability to forecast energy use. The aspects of power consumption forecasting vary depending on the prediction perspective. Almuhaini's research forecasted annual TEC using statistical and machine learning techniques, namely ARIMAX, BOA–SVR, and BOA–NARX models [21]. ANFIS combines fuzzy logic and neural networks, it usually consists of five layers: fuzzification, fuzzy rule evaluation, normalization, defuzzification, and output. Each layer performs a specific function in the inference process [22]. Training includes the ANFIS model using a hybrid learning algorithm that combines gradient descent and least squares estimation. Figure 2 shows the design modelling steps involved in ANFIS method of prediction [23], [24]. These tactics function even in the absence of prior understanding of the previously outlined systems. By correctly analyzing a dataset that includes obtained output and input parameters, they try to understand the link between outputs and inputs [25], [26].

Fig 2. Design modeling for ANFIS tool in ANN [26]

A government agent may utilize the generated energy prediction model to help them plan corrective steps for future school building, design with minimal energy consumption, and make the most use of their limited finances [27]. Model accuracy and computing capacity are increased by new technological advancements including parametric modelling, simulation, and artificial neural networks (ANNs). Elbeltagi et al., used ANN application to forecast the energy demand for residential structures. Artificial neural networks (ANNs), simulation, and parametric modelling are examples of recent technology developments that improve model accuracy and processing power [28], [24]. Table 1 shows the renewable energy cost aspects and prediction of models applied in Kingdom of Saudi. Table 2 shows the study of various ANN and machine learning network used successfully in forecasting the energy consumption within different zone of Saudi Arabia. Table 2 shows deep network [29] performed best in compared to machine learning with least error in the prediction over large-scale dataset.

Table 1 Displays the detailed renewable energy cost aspects and prediction

	Aspects	Description	Ref.
Energy Scenario	Energy Production	Significant investments in renewable energy, particularly solar energy	[30]
	Renewable Energy	Ambitious plans to develop renewable energy sector, with a focus on solar and wind energy	[31]
Energy Cost Prediction models	Artificial Neural Networks (ANN)	Machine learning models that can capture complex relationships in energy cost data	[32]
	Support Vector Machines (SVM)	Machine learning models that can analyze and forecast energy costs based on historical data	[33]

Limitations in energy cost prediction models	Data Availability	Limited availability of high-quality historical data, especially for newer technologies and renewable sources	[34]
Prediction models applied in K. S. A	Hybrid Models	Combine multiple approaches (e.g., regression, machine learning) for more accurate energy cost predictions	[36]

Table 2 Analysis of various prediction algorithm adopted for electricity cost analysis in KSA

Ref	Goal	Analysis
[39]	Energy const prediction using ANN/BIM model for residential building in KSA	Design Builder was used to create a 3D model, and the energy usage was determined. The dataset covers building area, type of air conditioning, glazing system, and envelope system.
[40]	Energy consumption in school building in Riyadh, KSA	To find the best network model, several neural network (NN) design topologies were tested. The developed model had an accuracy of roughly 87.5%.
[41]	Photovoltaic power output prediction in Jubail Industrial City, KSA.	Optimal database finding suitable merit indicator can enhance the performance.
[29]	Forecasting annual electricity consumption in Saudi Arabi	A Bayesian optimized non-linear autoregressive network was developed for forecasting energy consumption.

Conclusion

Advancements in infrastructure and laying down new large projects and increasing populations in the KSA led to increased energy demands. Vision 2030 is nominated with a large number of new projects; therefore, it is necessary to forecast future energy demands. We presented a scenario in which energy requirements are met in the KSA. Later, the various algorithms presented in the literature for energy forecasting were presented. The algorithms were categorized into statistical models, i.e., ARIMAX and BOA; machine learning algorithms, including regression networks; KNN and SVM; and ANN networks, which include deep CNNs, BPNs, ANNs, and long short-term memory (LSTM). The study revealed that deep CNN performance is better than that of statistical models, which lack accuracy and adaptivity.

References

[1] "U.S. Energy Information Administration," International Energy Statistics. Accessed: Feb. 20, 2024. [Online]. Available: Energy Institute, Statistical Review of World Energy, 2023

[2] Y. H. A. Amran, Y. H. M. Amran, R. Alyousef, and H. Alabduljabbar, "Renewable and sustainable energy production in Saudi Arabia according to Saudi Vision 2030; Current status and future prospects," *Journal of Cleaner Production*, vol. 247, p. 119602, Feb. 2020. https://doi.org/10.1016/j.jclepro.2019.119602

[3] M. A. Salam and S. A. Khan, "Transition towards sustainable energy production – A review of the progress for solar energy in Saudi Arabia," *Energy Exploration & Exploitation*, vol. 36, no. 1, pp. 3–27, Jan. 2018. https://doi.org/10.1177/0144598717737442

[4] M. Abdul Mujeebu and O. S. Alshamrani, "Prospects of energy conservation and management in buildings – The Saudi Arabian scenario versus global trends," *Renewable and*

Sustainable Energy Reviews, vol. 58, pp. 1647–1663, May 2016. https://doi.org/10.1016/j.rser.2015.12.327

[5] M. Zubair, A. B. Awan, R. P. Praveen, and M. Abdulbaseer, "Solar energy export prospects of the Kingdom of Saudi Arabia," *Journal of Renewable and Sustainable Energy*, vol. 11, no. 4, p. 045902, Jul. 2019. https://doi.org/10.1063/1.5098016

[6] O. K. M. Ouda, H. M. Cekirge, and S. A. R. Raza, "An assessment of the potential contribution from waste-to-energy facilities to electricity demand in Saudi Arabia," *Energy Conversion and Management*, vol. 75, pp. 402–406, Nov. 2013. https://doi.org/10.1016/j.enconman.2013.06.056

[7] F. Alrashed and M. Asif, "Prospects of Renewable Energy to Promote Zero-Energy Residential Buildings in the KSA," *Energy Procedia*, vol. 18, pp. 1096–1105, Jan. 2012. https://doi.org/10.1016/j.egypro.2012.05.124

[8] S. Alsulamy, A. S. Bahaj, P. James, and N. Alghamdi, "Solar PV Penetration Scenarios for a University Campus in KSA," *Energies*, vol. 15, no. 9, Art. no. 9, Jan. 2022. https://doi.org/10.3390/en15093150

[9] R. Bhinge, J. Park, K. H. Law, D. A. Dornfeld, M. Helu, and S. Rachuri, "Toward a Generalized Energy Prediction Model for Machine Tools," *Journal of Manufacturing Science and Engineering*, vol. 139, no. 041013, Nov. 2016. https://doi.org/10.1115/1.4034933

[10] J. Huang, H. Lv, T. Gao, W. Feng, Y. Chen, and T. Zhou, "Thermal properties optimization of envelope in energy-saving renovation of existing public buildings," *Energy and Buildings*, vol. 75, pp. 504–510, Jun. 2014. https://doi.org/10.1016/j.enbuild.2014.02.040

[11] N. Mohammed, P. Palaniandy, and F. Shaik, "Solar photocatalytic biodegradability of saline water: Optimization using RSM and ANN," presented at the AIP Conference Proceedings, AIP Publishing, 2022.

[12] W. Cao et al., "Short-term energy consumption prediction method for educational buildings based on model integration," *Energy*, vol. 283, p. 128580, Nov. 2023. https://doi.org/10.1016/j.energy.2023.128580

[13] N. Mohammed, P. Palaniandy, and F. Shaik, "Optimization of solar photocatalytic biodegradability of seawater using statistical modelling," *Journal of the Indian Chemical Society*, vol. 98, no. 12, p. 100240, Dec. 2021. https://doi.org/10.1016/j.jics.2021.100240

[14] D. D. Kim and H. S. Suh, "Heating and cooling energy consumption prediction model for high-rise apartment buildings considering design parameters," *Energy for Sustainable Development*, vol. 61, pp. 1–14, Apr. 2021. https://doi.org/10.1016/j.esd.2021.01.001

[15] C. Li, Z. Ding, D. Zhao, J. Yi, and G. Zhang, "Building Energy Consumption Prediction: An Extreme Deep Learning Approach," *Energies*, vol. 10, no. 10, Art. no. 10, Oct. 2017. https://doi.org/10.3390/en10101525

[16] N. Mohammed, P. Palaniandy, F. Shaik, B. Deepanraj, and H. Mewada, "Statistical analysis by using soft computing methods for seawater biodegradability using ZnO photocatalyst," *Environmental Research*, vol. 227, p. 115696, Jun. 2023. https://doi.org/10.1016/j.envres.2023.115696

[17] Z. Geng, Y. Zhang, C. Li, Y. Han, Y. Cui, and B. Yu, "Energy optimization and prediction modeling of petrochemical industries: An improved convolutional neural network based on cross-feature," *Energy*, vol. 194, p. 116851, Mar. 2020. https://doi.org/10.1016/j.energy.2019.116851

[18] N. Mohammed, "Characterization of sustainable concrete made from wastewater bottle caps using a machine learning and RSM-CCD: towards performance and optimization," presented at

the AToMech1-2023 Supplement, Nov. 2023, pp. 38–46. https://doi.org/10.21741/9781644902790-4

[19] N. Mohammed, P. Palaniandya, F. Shaik, and H. Mewada, "EXPERIMENTAL AND COMPUTATIONAL ANALYSIS FOR OPTIMIZATION OF SEAWATER BIODEGRADABILITY USING PHOTO CATALYSIS," *IIUM Engineering Journal*, vol. 24, no. 2, pp. 11–33, Jul. 2023. https://doi.org/10.31436/iiumej.v24i2.2650

[20] K. Amasyali and N. El-Gohary, "Machine learning for occupant-behavior-sensitive cooling energy consumption prediction in office buildings," *Renewable and Sustainable Energy Reviews*, vol. 142, p. 110714, May 2021. https://doi.org/10.1016/j.rser.2021.110714

[21] S. H. Almuhaini and N. Sultana, "Forecasting Long-Term Electricity Consumption in Saudi Arabia Based on Statistical and Machine Learning Algorithms to Enhance Electric Power Supply Management," *Energies*, vol. 16, no. 4, Art. no. 4, Jan. 2023. https://doi.org/10.3390/en16042035

[22] N. Mohammed, P. Palaniandy, and F. Shaik, "Pollutants removal from saline water by solar photocatalysis: a review of experimental and theoretical approaches," *International Journal of Environmental Analytical Chemistry*, vol. 103, no. 16, pp. 4155–4175, Dec. 2023. https://doi.org/10.1080/03067319.2021.1924160

[23] A. Abubakar Mas'ud, "Comparison of three machine learning models for the prediction of hourly PV output power in Saudi Arabia," *Ain Shams Engineering Journal*, vol. 13, no. 4, p. 101648, Jun. 2022. https://doi.org/10.1016/j.asej.2021.11.017

[24] N. Mohammed, A. Asiz, M. A. Khasawneh, H. Mewada, and T. Sultana, "Machine learning and RSM-CCD analysis of green concrete made from waste water plastic bottle caps: Towards performance and optimization," *Mechanics of Advanced Materials and Structures*, pp. 1–9, Aug. 2023. https://doi.org/10.1080/15376494.2023.2238220

[25] N. Mohammed, P. Palaniandy, and F. Shaik, "Optimization of solar photocatalytic biodegradability of seawater using statistical modelling," *Journal of the Indian Chemical Society*, vol. 98, no. 12, p. 100240, Dec. 2021. https://doi.org/10.1016/j.jics.2021.100240

[26] N. Mohammed, P. Palaniandy, F. Shaik, H. Mewada, and D. Balakrishnan, "Comparative studies of RSM Box-Behnken and ANN-Anfis fuzzy statistical analysis for seawater biodegradability using TiO2 photocatalyst," *Chemosphere*, vol. 314, p. 137665, Feb. 2023. https://doi.org/10.1016/j.chemosphere.2022.137665

[27] A. Alshibani, "Prediction of the Energy Consumption of School Buildings," *Applied Sciences*, vol. 10, no. 17, p. 5885, Aug. 2020. https://doi.org/10.3390/app10175885

[28] E. Elbeltagi and H. Wefki, "Predicting energy consumption for residential buildings using ANN through parametric modeling," *Energy Reports*, vol. 7, pp. 2534–2545, Nov. 2021. https://doi.org/10.1016/j.egyr.2021.04.053

[29] S. H. Almuhaini and N. Sultana, "Forecasting Long-Term Electricity Consumption in Saudi Arabia Based on Statistical and Machine Learning Algorithms to Enhance Electric Power Supply Management," *Energies*, vol. 16, no. 4, Art. no. 4, Jan. 2023. https://doi.org/10.3390/en16042035

[30] J. B. Smith and J. Ming, "Renewable energy research leases: Prospects and opportunities on the Hawaiian Outer Continental Shelf (OCS)," in *OCEANS'11 MTS/IEEE KONA*, Sep. 2011, pp. 1–5. https://doi.org/10.23919/OCEANS.2011.6107285

[31] W. Strielkowski, L. Civín, E. Tarkhanova, M. Tvaronavičienė, and Y. Petrenko, "Renewable Energy in the Sustainable Development of Electrical Power Sector: A Review," *Energies*, vol. 14, no. 24, Art. no. 24, Jan. 2021. https://doi.org/10.3390/en14248240

[32] B. Bush, N. Brunhart-Lupo, B. Bugbee, V. Krishnan, K. Potter, and K. Gruchalla, "Coupling visualization, simulation, and deep learning for ensemble steering of complex energy models," in *2017 IEEE Workshop on Data Systems for Interactive Analysis (DSIA)*, Oct. 2017, pp. 1–5. https://doi.org/ 10.1109/DSIA.2017.8339087

[33] J. Zhang, Q. Liu, Q. Wang, R. Tang, Y. He, and N. Tai, "Data-Driven Intelligent Fault Diagnosis Technology for Transmission Lines of Wind Power Renewable Energy System," in *2023 3rd International Conference on Energy, Power and Electrical Engineering (EPEE)*, Sep. 2023, pp. 246–251. https://doi.org/ 10.1109/EPEE59859.2023.10351976

[34] X. Yang *et al.*, "Optimal Distribution Network Planning with CVaR Model Considering Renewable Energy Integration," in *2018 2nd IEEE Conference on Energy Internet and Energy System Integration (EI2)*, Oct. 2018, pp. 1–6. https://doi.org/ 10.1109/EI2.2018.8582256

[35] K. Kampouropoulos, F. Andrade, E. Sala, A. G. Espinosa, and L. Romeral, "Multiobjective Optimization of Multi-Carrier Energy System Using a Combination of ANFIS and Genetic Algorithms," *IEEE Transactions on Smart Grid*, vol. 9, no. 3, pp. 2276–2283, May 2018. https://doi.org/10.1109/TSG.2016.2609740

[36] M. M. Samy, H. H. Sarhan, S. Barakat, and S. A. Al-Ghamdi, "A Hybrid PV-Biomass Generation Based Micro-Grid for the Irrigation System of a Major Land Reclamation Project in Kingdom of Saudi Arabia (KSA) - Case Study of Albaha Area," in *2018 IEEE International Conference on Environment and Electrical Engineering and 2018 IEEE Industrial and Commercial Power Systems Europe (EEEIC / I&CPS Europe)*, Jun. 2018, pp. 1–8. https://doi.org/ 10.1109/EEEIC.2018.8494543

[37] Y. Bhandari, S. Chalise, J. Sternhagen, and R. Tonkoski, "Reducing fuel consumption in microgrids using PV, batteries, and generator cycling," in *IEEE International Conference on Electro-Information Technology , EIT 2013*, May 2013, pp. 1–4. https://doi.org/ 10.1109/EIT.2013.6632692

[38] M. Patterson, N. F. Macia, and A. M. Kannan, "Hybrid Microgrid Model Based on Solar Photovoltaic Battery Fuel Cell System for Intermittent Load Applications," *IEEE Transactions on Energy Conversion*, vol. 30, no. 1, pp. 359–366, Mar. 2015. https://doi.org/10.1109/TEC.2014.2352554

[39] A. Alshibani and O. S. Alshamrani, "ANN/BIM-based model for predicting the energy cost of residential buildings in Saudi Arabia," *Journal of Taibah University for Science*, vol. 11, no. 6, pp. 1317–1329, Nov. 2017. https://doi.org/10.1016/j.jtusci.2017.06.003

[40] A. Alshibani, "Prediction of the Energy Consumption of School Buildings," *Applied Sciences*, vol. 10, no. 17, Art. no. 17, Jan. 2020. https://doi.org/10.3390/app10175885

[41] A. Abubakar Mas'ud, "Comparison of three machine learning models for the prediction of hourly PV output power in Saudi Arabia," *Ain Shams Engineering Journal*, vol. 13, no. 4, p. 101648, Jun. 2022. https://doi.org/10.1016/j.asej.2021.11.017

[42] T. Alquthami and A. Alaraishy, "Comprehensive Energy and Cost evaluation through detailed Auditing and Modeling for Mosque in Saudi Arabia," vol. 16, no. 9, 2021.

Vibration analysis of 3D printed PLA beam with honeycomb cell structure for renewable energy applications and sustainable solutions

Kubilay ASLANTAS[1,a*], Ekrem ÖZKAYA[2,b], Waleed AHMED[3,c,*]

[1] Department of Mechanical Engineering, Faculty of Technology, Afyon Kocatepe University, Afyonkarahisar, Turkey

[2] Faculty of Engineering - Department of Mechanical Engineering, Turkish-German University, 34820 Beykoz / İSTANBUL, Turkey

[2] Mechanical Engineering Department College of Engineering, UAE University, Al Ain, Abu Dhabi; UAE

[a]aslantas@aku.edu.tr, [b]ekrem.oezkaya@tau.edu.tr, [c]w.ahmed@uaeu.ac.ae

Keywords: Additive Manufacturing, Vibration Analysis, Finite Element Method

Abstract. Research shows a lack of vibration analyses of structures produced by 3D printing. This study, therefore, investigates the vibration behavior of honeycomb structures made of polylactic acid (PLA) using the additive manufacturing process fused deposition modeling (FDM) in a beam element. Based on an experimental modal analysis and the determination of the damping rate, a FEM reference simulation model is created, and the results are validated with the data from the experiment. The original honeycomb structure was numerically varied in its density, i.e., in the thickness (t) of the cell wall, in the length (L) of the regular hexagons, and in its degree of filling. The results showed that the density of the honeycombs at a filling level of 19% has a marginal influence on the vibration behavior. The vibration behavior was reduced only when the filling level was increased to 30%. This study has implications for many areas of research in which vibrations play a significant role in technical applications. These findings highlight the potential for integrating renewable energy applications with sustainable solutions, emphasizing the importance of vibration dynamics in advancing environmentally friendly technologies.

Introduction

Digital manufacturing technology, also known as 3D printing or additive manufacturing, has revolutionized the field of manufacturing. It allows for the production of complex structures with high precision and customization by successively adding materials based on three-dimensional computer-aided design (CAD) data [1]. Various techniques are available for 3D printing, with the most common being powder bed fusion (PBF) and fused deposition modeling (FDM) [2]. The ability to use various materials, including thermoplastics, ceramics, graphene-based materials, and metals, offers numerous application possibilities [3]. An overview of significant advances, typical applications, current challenges, advantages, and disadvantages of 3D printing processes, and a comprehensive description of various materials can be found in [4-5]. The use of plastics is becoming increasingly popular due to their excellent properties (high strength-to-mass ratio, low cost, durable, relatively impermeable, sterilizable) with controllable flexibility. In particular, there is growing interest in the use of biodegradable polymers. The most extensively investigated (bio)degradable thermoplastic polymers, which are also suitable for additive manufacturing technologies, are polyesters such as polylactic acid (PLA), polycaprolactone (PLC), and polyhydroxyalkanoates (PHA) [6]. Polylactic acid (PLA), a biodegradable and environmentally friendly thermoplastic, has emerged as a popular choice for 3D printing due to its ease of use and versatility [7]. The properties of PLA can also be chemically altered through synthesis or physical modifications [8-9]. The integration of honeycomb geometries into structural components has been found to be applicable in many industries due to its potential to improve strength-to-weight ratios and overall mechanical efficiency [10]. However, the addition of internal honeycomb cell structures (which have good energy absorption properties) can add a layer of complexity to the material, influencing its dynamic response to external stimuli, especially vibrations. Geometric revisions to the extruded honeycomb structures will also significantly affect the vibration characteristics of the structure. The modal analysis technique is often used to determine the dynamic characterization of a structure or a machine element. Modal analysis determines a system's natural frequencies, mode shapes, and damping characteristics. The basic concept of modal analysis lies in representing the vibration response of a time-invariant linear dynamic system as a linear combination of a series of simple harmonic motions called natural modes of vibration. Modal analysis is an experimental approach that a hammer impact test or a vibration shaker can perform. Modal analysis of any structure or component must not be performed experimentally. When appropriate boundary conditions are defined, modal analysis can be performed using the finite element technique (FEM) to obtain the mode shapes and frequencies of the structure. With the widespread use of additive manufacturing in the last decade, porous structures' vibration-damping properties have been modeled experimentally and numerically. To increase the impact or vibration-damping properties of a structure, the pore structures shown in Figure 1 are widely used. Honeycomb geometry is frequently used in sandwich panels, and these structures' vibration behavior has been the subject of many studies [5].

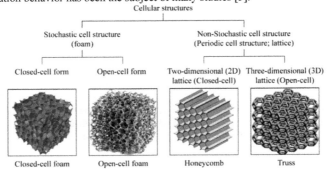

Fig. 1. Categories of cellular structures [12].

There is a lack of vibration analyses of structures produced using 3D printing in research. Understanding such structures' vibration characteristics is essential to ensure optimum performance in practical applications. This study investigates the effect of the honeycomb lattice structure on the vibration behaviour of a beam made of PLA material. In the study, firstly, a 3D printed PLA beam was produced by taking the edge length (L) and thickness (t) of the honeycomb lattice geometry as constant, and the damping ratio and resonant frequencies were found by performing a hammer test. The mode frequencies and vibration amplitudes obtained by numerical modeling are compared. FE analyses are extended for different values of L and t of the honeycomb geometry.

Material and Method
In the first stage of the study, a sample of PLA material was produced using the FDM method. The hammer test was performed on this sample, and the damping ratio of the material was determined. The damping ratio was used as input for harmonic analysis.

Sample Preparation

In this part of the study, the computer-aided design of the specified honeycomb geometry and the whole beam was carried out first. SpaceClaim software was used for the design. The 25 mm part of the beam (A region) is modeled as solid and is used to stabilize the beam (Fig. 2). The B region of the beam has a wall thickness of 0.75 mm and is filled with honeycomb. The dimensions of the honeycomb used are also given in Figure 2.

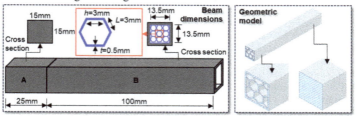

Fig.1. Dimensions and geometrical modeling of the beam

Experimental analysis

Figure 3 shows a schematic representation of the experimental procedure used in the study. The 3D-printed specimen was fixed from the A region (Fig. 2) employing a vice. An accelerometer was connected to the tooltip to measure the natural frequencies of the specimen. A Dytran brand hammer with model number 5800B4 was used for tap testing. A Dytran accelerometer (322F1) was mounted at the relevant position to measure the frequency response. A four-channel data acquisition system (Novian, model number S04) was used in conjunction with the Tap Testing measurement module of the CutPRO software to obtain the results of the vibration characteristics of the specimen. After the measurement, force-time, acceleration-time graphs and real-frequency and imaginary-frequency changes were obtained. In addition, the amplitude value occurring in the sample at the first mode frequency was also obtained. For numerical modeling, the damping ratio value of the sample must be known. The damping ratio of the sample was calculated using the real-frequency change obtained after the tap testing (Fig. 4).

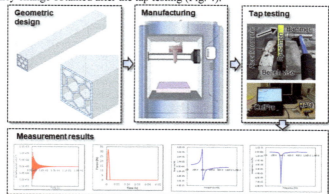

Fig. 3. Schematic representation of the experimental procedure used in the study

The specimen's real and imaginary part curves were obtained after the hammer test. The peak points (ω_1 and ω_2) in the real part curve and the values corresponding to the natural frequency (ω_n) were used. The damping ratio value of the specimen was calculated using Eq. 1.

$$\zeta = \frac{\omega_2 - \omega_1}{2.\omega_n} \tag{1}$$

Numerical modeling
Ansys software was used for FE analysis, and SpaceClaim software was preferred for geometric design. After the geometric model was imported to Ansys software [13], modal analysis was performed to find the mode frequencies. Afterward, the frequency response of the sample was obtained by performing harmonic analysis. In both analyses, part A of the specimen was fixed as anchored. In the harmonic analysis, the load was applied in the direction of the hammer load. Prior to numerical modeling, mesh convergence analysis was performed to determine the sufficient number of elements and nodes. Figure 5 shows the steps of the numerical modeling process used in the study. The modulus of elasticity of the PLA material used was taken as 2350 MPa, Poisson's ratio as 0.39, and density as 1.25 g/cm^3. In addition, the damping ratio was obtained as 0.022 using the method mentioned in the previous section. In the modeling, 31752 elements (tetrahedrons with 10 nodes) and 56764 nodes were used.

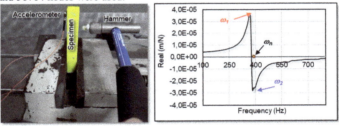

Fig. 4. Tap testing test setup and real part graph used to calculate the damping ratio.

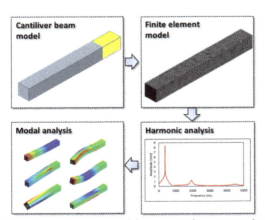

Fig. 5. Schematic representation of the steps used in numerical modeling

Results and Discussion
Tap testing results and verification of the FE model

In the study, experimental studies were first completed, and the value of the damping ratio required for numerical modeling was determined. Frequency values in each mode were obtained by hammer test. The frequency values for each mode obtained after numerical modeling were compared with the experimental results (Table 1). As can be seen from Table 1, there is a good agreement in the other frequency values except the 2nd mode value. The high margin of error in the 2nd mode frequency value may be due to two reasons. The first is that the point where the load is applied during the hammer test is not the same every time. The second is the minor dimensional errors caused during the production of the specimen. In particular, even slight differences in the wall thickness of the honeycomb may have caused this error.

Table 1. Comparison of experimental and numerical results and mode shapes

Mode No	Experimental	Modelling	Error (%)	Mode shape
1	363	364.8	0.5	
2	447.6	376.6	-18.8	
3	1687.9	1768.8	4.6	
4	1998.1	1952.4	-2.3	
5	2046.5	1986.1	-3	
6	3494.4	3435.5	-1.7	

After numerical modeling, the amplitude-frequency variation obtained for the 1st mode was compared with the experimental result (Fig. 6). As seen from Figure 6, the experimental and numerical modeling results agree very well. This shows the accuracy of the FE model and the boundary conditions used. It is seen that the maximum amplitude value occurs as 0.165 mm at 363 Hz in the experimental study, while it occurs as 0.166 mm at 364.8 Hz in the FE analysis.

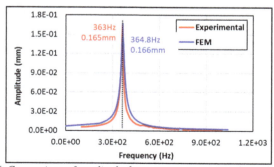

Fig. 6. Comparison of amplitude-frequency variation results in the beam

Vibration behavior of the revised honeycomb geometry

In the previous section, the vibration behavior of the honeycomb-filled PLA beam was obtained by hammer-impactor test, and the FE model was also validated. In this section, geometrical revisions were made to the honeycomb geometry, and the effects of the honeycomb's edge length, thickness, and filling rate were investigated. In lattice structures, changing the dimensions of the cell geometry causes a change in the filling rate. The FE analyses carried out in this section were performed for three different cell sizes, two different filling rates, and solid and hollow cases of the beam. Modal and then harmonic analyses were performed for the geometrical models. Thus, it was possible to calculate how much the revision in the honeycomb geometry affects the vibration amplitude and 1st mode frequency. Figure 7 shows different sections of the beam geometry used in the FE analysis. The effect of the change in cell edge length (L), cell edge thickness (t), and filling rate was investigated.

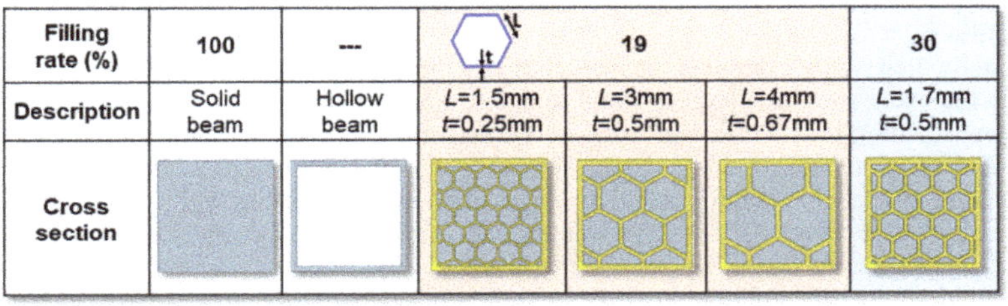

Filling rate (%)	100	---	19			30
Description	Solid beam	Hollow beam	L=1.5mm t=0.25mm	L=3mm t=0.5mm	L=4mm t=0.67mm	L=1.7mm t=0.5mm
Cross section						

Fig. 7. Cross sections of the beam geometry used in FE analysis

Figure 8 shows the amplitude-frequency variation obtained as a result of the harmonic analysis. Since the maximum amplitude occurs in the 1st mode in the beam, only 1st mode values are given in Figure 8. The hollow beam has the lowest stiffness among the analyzed sections. As expected, the maximum amplitude occurred in the hollow beam. All three sections' 1st mode frequency values with a filling rate of 19 are very close. In the inner graph in Figure 8, the graph for the % filling rate of 19% is detailed. As can be seen, the maximum amplitude was realized at L = 4 mm and t = 0.67 mm.

On the other hand, the maximum frequency was obtained for L = 1.5 mm and t = 0.25 mm. According to these three results, where the filling rate is constant, the variation of L and t does not affect the vibration amplitude and frequency much. When the filling rate ratios are compared, the amplitude value obtained for 30% is lower than 19%. On the other hand, the frequency value for 30% is lower than 19%. The amplitude value is minimum since the solid beam is the section geometry with the highest stiffness. However, the solid beam is the cross-section with the lowest 1st mode frequency value.

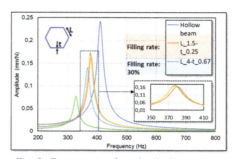

Fig. 8. Comparison of amplitude-frequency

Conclusion

This study thoroughly examined the impact of utilizing honeycomb geometry as a filling element within the beam structure. Experimental samples, crafted from PLA material with L=3 mm and t=0.5 mm (filling rate: 19%), were rigorously analyzed and validated through FE analysis. The validated FE simulation model enabled the exploration of varying densities (different L and t values) at 19% and 30% filling rates, shedding light on critical findings:

• The resonance frequency values obtained experimentally and numerically exhibited strong agreement, validating the reliability of the analysis.
• Maximal resonance amplitudes occurred at the 1st mode frequency, consistently verified through experimental and numerical approaches.
• Variations in L and t ratios, while holding the filling rate constant, minimally affected both amplitude and frequency.
• Increasing the filling rate resulted in decreased vibration amplitudes and frequencies, highlighting the role of density in vibration behavior.

The validated FE simulation model serves as a robust foundation for future investigations. Subsequent studies will explore diverse materials, filling structures, and structural densities while examining the influence of various pressure parameters. These efforts aim to advance sustainable practices and renewable energy applications, contributing to the ongoing evolution of environmentally conscious engineering solutions.

The research highlights the potential of 3D-printed honeycomb structures for renewable energy applications. Their vibration behavior, characterized through experimental analysis, offers opportunities for enhancing efficiency in energy harvesting devices and damping mechanisms. The structures' adaptability and lightweight nature make them suitable for components in wind turbines, solar panels, and energy storage systems. Integrating these structures into renewable energy technologies holds promise for advancing sustainable solutions and improving overall system performance.

References

[1] A. M. T. Syed, P. K. Elias, B. Amit, B. Susmita, O. Lisa, & C. Charitidis, "Additive manufacturing: scientific and technological challenges, market uptake and opportunities," Materials today, Vol. 1, pp. 1-16, 2017

[2] Shulman, H.; Spradling, D.; and Hoag, C.: Introduction to additive manufacturing, Ceram. Indus. 162:15-21, 2012.

[3] N. Shahrubudin, T.C. Lee, R. Ramlan, An Overview on 3D Printing Technology: Technological, Materials, and Applications, Procedia Manufacturing, Volume 35, 2019, Pages 1286-1296. https://doi.org/10.1016/j.promfg.2019.06.089

[4] Ji-chi Zhang, Kuai He, Da-wei Zhang, Ji-dong Dong, Bing Li, Yi-jie Liu, Guo-lin Gao, Zai-xing Jiang, Three-dimensional printing of energetic materials: A review, Energetic Materials Frontiers, Volume 3, Issue 2, 2022, Pages 97-108,7. https://doi.org/10.1016/j.enmf.2022.04.001

[5] Anketa Jandyal, Ikshita Chaturvedi, Ishika Wazir, Ankush Raina, Mir Irfan Ul Haq, 3D printing – A review of processes, materials and applications in industry 4.0, Sustainable Operations and Computers, Volume 3, 2022, Pages 33-42. https://doi.org/10.1016/j.susoc.2021.09.004

[6] Jennifer Gonzalez Ausejo, Joanna Rydz, Marta Musioł, Wanda Sikorska, Henryk Janeczek, Michał Sobota, Jakub Włodarczyk, Urszula Szeluga, Anna Hercog, Marek Kowalczuk, Three-dimensional printing of PLA and PLA/PHA dumbbell-shaped specimens of crisscross and transverse patterns as promising materials in emerging application areas: Prediction study, Polymer Degradation and Stability, Volume 156, 2018, Pages 100-110. https://doi.org/10.1016/j.polymdegradstab.2018.08.008

[7] Iftekar, S.F.; Aabid, A.; Amir, A.; Baig, M. Advancements and Limitations in 3D Printing Materials and Technologies: A Critical Review. *Polymers* **2023**, *15*, 2519. https://doi.org/10.3390/polym15112519

[8] Rasal RM, Janorkar AV, Hirt DE. Poly(lactic acid) modifications. Prog Polym Sci 2010;35:338-356. https://doi.org/10.1016/j.progpolymsci.2009.12.003

[9] Arrieta MP, López J, Hernández A, Rayón E. Ternary PLA-PHB-limonene blends intended for biodegradable food packaging applications. Eur Polym J 2014;50:255-270. https://doi.org/10.1016/j.eurpolymj.2013.11.009

[10] Hossein Mohammadi, Zaini Ahmad, Michal Petrů, Saiful Amri Mazlan, Mohd Aidy Faizal Johari, Hossein Hatami, Seyed Saeid Rahimian Koloor, An insight from nature: honeycomb pattern in advanced structural design for impact energy absorption, Journal of Materials Research and Technology, Volume 22, 2023, Pages 2862-2887. https://doi.org/10.1016/j.jmrt.2022.12.063

[11] Zippo, A.; Iarriccio, G.; Pellicano, F.; Shmatko, T. Vibrations of Plates with Complex Shape: Experimental Modal Analysis, Finite Element Method, and R-Functions Method. *Shock. Vib.* **2020**, 8882867. https://doi.org/10.1155/2020/8882867

[12] Park, K.-M.; Min, K.-S.; Roh, Y.-S. Design Optimization of Lattice Structures under Compression: Study of Unit Cell Types and Cell Arrangements. *Materials* **2022**, *15*, 97. https://doi.org/10.3390/ma15010097

[13] ANSYS Inc. (2022). ANSYS Software. Retrieved from https://www.ansys.com

Potential use of reject brine waste as a sustainable construction material

Seemab TAYYAB[1,a], Essam ZANELDIN[2,b*], Waleed AHMED[3,c], and Ali AL MARZOUQI[4,d]

[1]Research Assistant, Department of Civil and Environmental Engineering, United Arab Emirates University, Al-Ain, United Arab Emirates

[2]Associate Professor, Department of Civil and Environmental Engineering, United Arab Emirates University, Al-Ain, United Arab Emirates

[3]Assistant Professor, Engineering Requirements Unit, United Arab Emirates University, Al-Ain, United Arab Emirates

[4]Professor, Department of Chemical and Petroleum Engineering, United Arab Emirates University, Al-Ain, United Arab Emirates

[a]tayyabseemab14@gmail.com, [b]essamz@uaeu.ac.ae, [c]w.ahmed@uaeu.ac.ae, [d]hassana@uaeu.ac.ae

Keywords: Reject Brine, Desalination, Sustainable Construction, Resource Utilization, Environmental Impact

Abstract. In countries near the ocean, the majority of the water used for household, agricultural, and industrial purposes is attained through seawater desalination. Desalination produces highly salty water, commonly known as reject brine, which can have many drastic, negative effects on the environment. The waste results in both an environmental challenge and an opportunity for sustainable resource utilization. This research work is a literature study to investigate the feasibility and potential benefits of utilizing reject brine waste as a sustainable construction material. The results revealed that reject brine has a prodigious possibility to be used as a binder, and in place of water in concrete. The use of reject brine in cementitious composites decreases CO_2 emissions and makes them economical. Also, reject brine is fruitful in the stabilization of soil by increasing the mechanical properties and enhance the strength of soil. In essence, the use of reject brine from water desalination in construction is a sustainable and environment-friendly approach.

Introduction
Water is life. Fresh water is essential for living organisms to survive on earth and is increasingly depleted. The increase in population and industrialization lead to an increase in demand for water. About 1.8 billion people around the world will face water shortage by the end of the year 2025 [1]. The countries near the ocean (e.g., gulf countries) lack fresh water and frequently use desalination techniques to produce drinkable water. The produced water is used by human beings and for construction practices as well. Desalination methods include reverse osmosis, multistage flash evaporation, multi-effect distillation, and electrodialysis [2]. These techniques produce pure water but at the same time result in the production of a by-product waste with a high concentration of salt called reject brine. Only 35-45% of freshwater is recovered from the sea, while the remaining 55-65% of gross feed comes out of the desalination plant as waste brine [3]. It is estimated that Gulf countries produce more than 60% of the world's desalinated water with UAE's contribution alone around 13% [4]. The waste is disposed of in open sea and in valuable lands which is a major threat to aquatic life and sustainable development.

Concrete is the 2nd most widely used material after water on Earth. It is an old material consisting of cement, sand, aggregate, water, and suitable admixtures that have been in use for many centuries. The large-scale consumption of ordinary Portland cement (OPC) as a binding agent in concrete results in serious environmental challenges and issues because of the significant CO_2 emissions associated with its production. Every 1kg of cement production results in 0.9kg production of CO_2 to the environment [5-7]. The natural fine and coarse aggregates (sand, gravel, etc.) used in concrete synthesis result in their depletion with time. Concrete is a thirsty behemoth, sucking up almost 1/10th of the total world's industrial water usage. By 2050, 75% of infrastructure water demand is expected to occur in areas expected to experience water stress [8]. Therefore, in the past, various research efforts have been devoted to producing substitute construction materials in place of natural ingredients without compromising the strength, durability, and economy to promote sustainable development. A brief description of some of the studies is given in Table 1.

Table 1: A literature review of the use of different alternatives partially in place of cement, sand, and water in concrete

Material to be replaced	Material replaced with	Procedure	Result	Reference
Cement	Waste brick powder	Prepared concrete specimens using 5% and 10% of WBP in place of cement	Workability and compressive strength increased due to the shape and particle size of WBP	[9]
Cement	Marble dust	Studied the use of marble dust collected from marble blocks in concrete mixtures by 5, 10, 15, and 20%	The mechanical properties of concrete increased significantly compared to the control sample	[10]
Cement	Rice husk ash	Addressed the strength characteristics of cement mortar containing 0, 2.5, 5, 7.5, 10, 12.5, 15% RHA	Compressive strength of hardened concrete decreases with increasing RHA percentage	[11]
Sand	Coal bottom ash	Studied the effect of CBA as a replacement of sand in concrete with dosage at 20, 30, 40, 50, 75, and 100%	Workability and bleeding decreased, compressive and splitting strength of concrete did not change significantly	[12]
Sand	Recycled plastic waste	Evaluated the performance of concrete with RPW as a partial replacement of sand in different proportions	Replacing 10% of sand by volume is the best solution, saving 0.820 billion tons of sand every year	[13]
Sand	Fly ash	Concrete samples containing FA in amounts 20, 40, 60, 80, and 100% were cast and tested	Compressive strength, split tensile strength, and modulus of elasticity increased up to an optimum dosage of 40%	[14]
Water	Polyvinyl acetate resins waste water	Used the industrial wastewater discharged to replace the water in concrete completely	While compressive strength and density increased slightly, the values of slump decreased	[15]
Water	Treated Water and Waste Water	Assessed the strength of concrete using treated water and wastewater	The strength of concrete samples decreased using wastewater but the value is above the standard requirement	[16]

Desalination of Seawater

The amount of calcium, chloride, magnesium, potassium, sodium, and sulphates in seawater are 412, 19500, 1290, 380, 10770, and 905mg/L as compared to potable water containing these minerals in compositions of 75, 250, 50, 10, 200, and 400mg/L respectively. Also, the amount of total dissolved solids in seawater is 33387pm as compared to 500pm in drinkable water [17]. Desalination is the process of removing minerals from salt water. Saltwater is desalinated in order to produce water suitable for human household consumption or irrigation purposes. The formation of reject brine takes place as a byproduct of desalination. There are different methods used for the desalination process all over the world as shown in Figure 1. They meet the needs of more than 300 million people by producing 87 million cubic meters of clean water every day [18]. The schematic procedure of reject brine formation is shown in Figure 2.

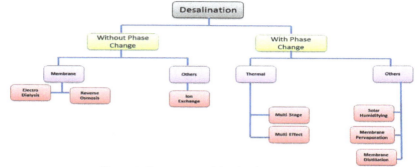

Figure 1: Processes used for desalination [19]

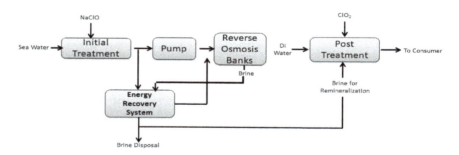

Figure 2: Schematic diagram of reject brine formation [20]

Reject Brine Use in Construction

The study of the feasibility of using desalination reject brine in construction materials has become of growing interest in this modern era. The use of reject brine in making construction materials like concrete, mortar, bricks, blocks, soils, etc. will serve two main purposes. First, natural ingredients consumption in the construction industry will decrease, releasing some pressure on the natural resources which are limited and depleting with time. Second, the brine's harmful ecological and physiochemical influences on the receiving aquatic bodies will be reduced.

Reject Brine as a Binder in Concrete
Cement is the binding agent used in construction projects obtained through the chemical process of limestone, clay, magnesia, silica, alumina, etc. [21]. It gives stability, strength, and durability to materials to guarantee they stay longer. The presence of an optimum percentage of MgO in cement ensures the setting time and strength of the composites [22]. Due to its high magnesium content (Mg^{2+}), waste brine has the character to be used as an eco- friendly and sustainable cause for cement production. S. Ruan et al. (2021) highlighted problems shown with increased emissions of magnesium oxide and carbon dioxide from brine waste. They examined the possibility of producing cement containing MgO (from brine waste) and analyzed its use as a binding agent compared to commercial MgO. The mechanical properties of the specimens were evaluated along with microstructural analysis using X-ray diffraction (XRD), thermogravimetric analysis/differential scanning calorimetry (TGA/DSC), etc. Samples containing higher reactivity magnesium oxide products were found to be stronger than samples containing commercial magnesium oxide products. The increased solubility of synthetic magnesium oxide leads to better hydration and carbonation, leading to a denser structure and structures with improved properties [23].

Reject Brine in Place of Water in Concrete
The process of hydration of cement in cementitious composites is due to the presence of water. Using salt water to prepare concrete is not a new procedure. MS. Islam et al. (2010) emphasized the effect on the setting time of concrete by partially using saline water in place of ordinary portable water [24]. Mori et al. (1981) prepared concrete samples containing fresh water and saline water. They observed a comparatively minor alteration between the mechanical strength of concrete prepared with fresh water and concrete produced with saline water [25]. However, Yamamoto et al. (1980) found that concrete prepared with salty or saline water confirms higher strength compared to concrete containing fresh water [26]. To prevent the chances of metal erosion, V. Kumar et al. (1998) suggested avoiding the use of salty water for reinforced concrete (RC) [27]. However, Dang et al. (2022) studied the long-term concrete exposure to chloride (Cl-) ions and exhibited that the effect of chloride (Cl^{-1}) ions in concrete production from seawater is quite small or insignificant [28]. In another study, F. Qu et al. (2021) determined that corrosion of steel in reinforced concrete (RC) structures is because of sulphate attack occurs not due to the presence of chloride ions (Cl^{-1}) in seawater, but due to the harsh marine environment in which the sample is located [29].

Reject Brine to Minimize CO_2 Emissions from Cementitious Composites
Cement is an important building binding material and its production accounts for 5% of the world's CO2, a powerful greenhouse gas. With the dedication of reducing the carbon footprint of the process of concrete preparation using reject brine, Fattah et al. (2017) conducted research to determine the impact of CO_2 emissions by using brine waste as water and ground- granulated-blast-furnace slag (GGBS) as a cement substitute. Concrete specimens containing different percentages of cement were prepared using ordinary portable water and wastewater containing brine. The results exhibited that the use of reject brine and GGBS enhanced the mechanical strength of concrete formed, because of the high filling character of GGBS. Consuming reject brine as the potential water source and replacing 50% of the cement with GGBS can reduce 3.74–7.5 lbs. of CO_2/cubic meter of concrete, and 388 lbs. of CO_2/cubic meter of concrete respectively. The waste reject brine usage in cementitious composites is an economical approach as it can save about AED 625–1250 per cubic meter of concrete prepared [30].

Reject Brine to Enhance the Strength of Soil
Soil is an important engineering material that serves a major role in the construction of foundations, roadbeds, dams, and buildings [31]. Mathew et al. (2012) studied the effect of increasing brine concentration on soil-bearing capacity. The samples with and without reject brine were prepared and their shear strength parameters were monitored for 364 days. The soil's bearing capacity was calculated using numerical equations. The more the contamination of soil with brine, the more decrease in the bearing capacity [32]. Kuriakose et al. (2022) researched to make use of the brine waste generated from the desalination of water for the steadiness of soil. The brine sludge was mixed with marine clay at various percentages by the dry weight of the moist clay. It was found that brine sludge can be used as a substitute in the stabilization of soft clays as it improves the geotechnical properties of soil [33]. S.L. Barbour et al. (1993) evaluated the geotechnical properties of two Ca-montmorillonite clayey soils having brine adulteration. The changes in mechanical properties (shear strength, bearing capacity, etc.), index properties (liquid limit, plastic limit, etc.), and hydraulic properties by the incorporation of reject brine were determined. A significant increase in shear strength and bearing capacity was reported [34].

Conclusion
In conclusion, this study proved that reject brine has great potential to be used in construction materials, in the formation of cement, mortar, concrete, bricks, soils, etc. It will be a great source to enhance the mechanical and durability properties of cementitious composites and also make them economical. This will in turn reduce the consumption of natural water and get rid of waste brine which is a serious concern in this modern era for the countries near the ocean. The use of reject brine in the construction industry will surely be a sustainable and environment-friendly approach in the future. Future studies by the research team are directed at defining the long-term performances of reinforced cementitious materials through carbonation tests, freeze and thaw tests, sulphate attack tests, rapid chloride ion penetration tests, sorption tests, etc. There is a great possibility of checking the feasibility of reject brine in the manufacture of bricks and also the treatment of soil with brine might enhance its properties.

References
[1] Chakraborti, R.K., Kaur, J. and Kaur, H., 2019. Water Shortage Challenges and a Way Forward in India. Journal: American Water Works Association, 111(5). https://doi.org/10.1002/awwa.1289
[2] Thimmaraju, M., Sreepada, D., Babu, G.S., Dasari, B.K., Velpula, S.K. and Vallepu, N., 2018. Desalination of water. Desalination and water treatment, pp.333-347. https://doi.org/10.5772/intechopen.78659
[3] Khan, M. and Al-Ghouti, M.A., 2021. DPSIR framework and sustainable approaches of brine management from seawater desalination plants in Qatar. Journal of Cleaner Production, 319, p.128485. https://doi.org/10.1016/j.jclepro.2021.128485
[4] Jones, E., Qadir, M., van Vliet, M.T., Smakhtin, V. and Kang, S.M., 2019. The state of desalination and brine production: A global outlook. Science of the Total Environment, 657, pp.1343-1356. https://doi.org/10.1016/j.scitotenv.2018.12.076
[5] Imbabi, M.S., Carrigan, C. and McKenna, S., 2012. Trends and developments in green cement and concrete technology. International Journal of Sustainable Built Environment, 1(2), pp.194-216. https://doi.org/10.1016/j.ijsbe.2013.05.001
[6] Ibeto, C.N., Obiefuna, C.J. and Ugwu, K.E., 2020. Environmental effects of concrete produced from partial replacement of cement and sand with coal ash. International Journal of Environmental Science and Technology, 17, pp.2967-2976. https://doi.org/10.1007/s13762-020-02682-4
[7] Ummi, R.K., 2017, July. A Comparative Study of Green Technology in Cement Industry. In ASEAN/Asian Academic Society International Conference Proceeding Series.

[8] Miller, S.A., Horvath, A. & Monteiro, P.J.M. Impacts of booming concrete production on water resources worldwide. Nat Sustain 1, 69–76 (2018). https://doi.org/10.1038/s41893-017-0009-5

[9] Arif, R., Khitab, A., Kırgız, M.S., Khan, R.B.N., Tayyab, S., Khan, R.A., Anwar, W. and Arshad, M.T., 2021. Experimental analysis on partial replacement of cement with brick powder in concrete. Case Studies in Construction Materials, 15, p.e00749. https://doi.org/10.1016/j.cscm.2021.e00749

[10] Vaidevi, C., 2013. Study on marble dust as partial replacement of cement in concrete. Indian journal of engineering, 4(9), pp.14-16.

[11] Bawankule, S.P. and Balwani, M.S., 2015. Effect of partial replacement of cement by rice husk ash in concrete. Int. J. Sci. Res, 4, pp.1572-1574.

[12] Singh, M., 2018. Coal bottom ash. In Waste and Supplementary Cementitious Materials in Concrete (pp. 3-50). Woodhead Publishing. https://doi.org/10.1016/B978-0-08-102156-9.00001-8

[13] Thorneycroft, J., Orr, J., Savoikar, P. and Ball, R.J., 2018. Performance of structural concrete with recycled plastic waste as a partial replacement for sand. Construction and Building Materials, 161, pp.63-69. https://doi.org/10.1016/j.conbuildmat.2017.11.127

[14] Majhi, R.K., Nayak, A.N. and Mukharjee, B.B., 2020. An overview of the properties of sustainable concrete using fly ash as replacement for cement. International Journal of Sustainable Materials and Structural Systems, 4(1), pp.47-90. https://doi.org/10.1504/IJSMSS.2020.106418

[15] Ismail, Z.Z. and Al-Hashmi, E.A., 2011. Assessing the recycling potential of industrial wastewater to replace fresh water in concrete mixes: application of polyvinyl acetate resin wastewater. Journal of Cleaner Production, 19(2-3), pp.197-203. https://doi.org/10.1016/j.jclepro.2010.09.011

[16] Mahasneh, B.Z., 2014. Assessment of replacing wastewater and treated water with tap water in making concrete mix. Electron. J. Geotech. Eng, 19, pp.2379-2386.

[17] Malek, A., Hawlader, M.N.A. and Ho, J.C. Large -scale seawater desalination: a technical and economic review. ASEAN J. Sci. Technol. Development Vol. 9 No. 2. pp 41-61, 1992

[18] Alix, Alexandre; Bellet, Laurent; Trommsdorff, Corinne; Audureau, Iris, eds. (2022). Reducing the Greenhouse Gas Emissions of Water and Sanitation Services: Overview of emissions and their potential reduction illustrated by utility know-how. https://doi.org/10.2166/9781789063172

[19] IDA 2004: Desalination Business Stabilized on a High Level, Int. Desal.Water Reuse, Vol. 14 (2), pages.14–17.2004.

[20] Burke, L., Chen, C., Jamil, O. and Majewska, N., 2016. Desalination-Team B.

[21] Gartner, E. and Sui, T., 2018. Alternative cement clinkers. Cement and concrete research, 114, pp.27-39. https://doi.org/10.1016/j.cemconres.2017.02.002

[22] Kara, S., Erdem, S. and Lezcano, R.A.G., 2021. MgO-based cementitious composites for sustainable and energy efficient building design. Sustainability, 13(16), p.9188. https://doi.org/10.3390/su13169188

[23] Ruan, S., Yang, E.H. and Unluer, C., 2021. Production of reactive magnesia from desalination reject brine and its use as a binder. Journal of CO2 Utilization, 44, p.101383. https://doi.org/10.1016/j.jcou.2020.101383

[24] Islam, M.S., Mondal, B.C. and Islam, M.M., 2010. Effect of sea salts on structural concrete in a tidal environment. Australian Journal of Structural Engineering, 10(3), pp.237-252. https://doi.org/10.1080/13287982.2010.11465048

[25] Mori, Y. et. al. (1981). "10 years exposure test of concrete mixed with seawater under marine environment", Journal of Cement Association, Vo.35, pp.341-344 (in Japanese).

[26] Mehta, P.K. and Malhotra, V.M., 1980. Performance of concrete in marine environment. ACI SP-65, pp.1-20.

[27] Kumar, V., 1998. Protection of steel reinforcement for concrete-A review. Corrosion Reviews, 16(4), pp.317-358. https://doi.org/10.1515/CORRREV.1998.16.4.317

[28] Dang, V.Q., Ogawa, Y., Bui, P.T. and Kawai, K., 2022. Effects of chloride ion in sea sand on properties of fresh and hardened concrete incorporating supplementary cementitious materials. Journal of Sustainable Cement-Based Materials, 11(6), pp.439-451. https://doi.org/10.1080/21650373.2021.1992683

[29] Qu, F., Li, W., Dong, W., Tam, V.W. and Yu, T., 2021. Durability deterioration of concrete under marine environment from material to structure: A critical review. Journal of Building Engineering, 35, p.102074. https://doi.org/10.1016/j.jobe.2020.102074

[30] Fattah, K.P., Al-Tamimi, A.K., Hamweyah, W. and Iqbal, F., 2017. Evaluation of sustainable concrete produced with desalinated reject brine. International Journal of Sustainable Built Environment, 6(1), pp.183-190. https://doi.org/10.1016/j.ijsbe.2017.02.004

[31] Dauncey, P.C., Bates, A.D., Poole, A.B. and Engineering Group Working Party, 2012. Chapter 10 Engineering design and construction. Geological Society, London, Engineering Geology Special Publications, 25(1), pp.347-392. https://doi.org/10.1144/EGSP25.10

[32] AYININUOLA, G.M. and AGBEDE, O.A., 2012. EFFECT OF BRINE INTRUSION ON SOIL BEARING CAPACITY.

[33] Kuriakose, M., Athira, K.N., Abraham, B.M. and Cyrus, S., Utilization of Brine Sludge to Improve the Strength and Compressibility Characteristics of Soft Clays.

[34] Barbour, S.L. and Yang, N., 1993. A review of the influence of clay–brine interactions on the geotechnical properties of Ca-montmorillonitic clayey soils from western Canada. Canadian Geotechnical Journal, 30(6), pp.920-934. https://doi.org/10.1139/t93-090

Bidding optimization for a reverse osmosis desalination plant with renewable energy in a day ahead market setting

Ebaa Al Nainoon[1,a*] and Ali Al Awami[2,b]

[1]Electrical Engineering Department, KFUPM, Dhahran, Saudi Arabia

[2]K.A.CARE Energy Research & Innovation Center, Dhahran, Saudi Arabia

[a]g202202180@stu.kfupm.sa, [b]aliawami@kfupm.edu.sa

Keywords: Generation, PV, RO, Desalination, Electricity Market, Bidding, Strategy, Renewable Energy, Maximizing Profits, Optimization, Coordination, Flexibility, Revenue, Production, Water Power Nexus, Day Ahead Market

Abstract. This study explores the relationship between power system electrical generation with PV and grid-connected reverse osmosis (RO) water desalination plants in an electricity market setting. It aims to optimize bidding strategies for renewable energy companies in this market, maximizing profits by optimizing coordination and utilizing the flexibility of RO desalination facilities. The research aims to improve economic efficiency, revenue production, and promote renewable energy resource exploitation through better coordination and integration of water desalination processes.

Introduction

Water scarcity presents a global challenge, necessitating the development of sustainable water production solutions. Grid-connected RO desalination, powered by renewable energy sources like PV, offers a promising approach. This research aims to investigate the interdependent relationship between a power system with PV generation and grid-connected reverse osmosis (RO) water desalination plants in an electricity market setting. It will examine how to best bid for a renewable energy generation company and reverse osmosis (RO) desalination plant in a future market setting with an objective to maximize profits by optimizing the coordination of PV as well as leveraging the inherent flexibility of RO plants within the bidding strategy.

Hence, this research aims to present an optimization case study with an objective to improve economic efficiency in the electrical market, maximize revenue production, and promote the exploitation of renewable energy resources through better coordination and enhanced integration of water desalination processes within an electricity market.

Literature Review

Water scarcity is a major worldwide crisis as per the United Nations as reported in [1] that around 2 billion individuals which comprise 26% of the global population didn't have access to clean, readily available drinking water on-site. Even basic access to drinking water was unavailable to 771 million people. Eight in ten were from rural areas. The majority of them were in least developed nations. In order to ensure that everyone has access to safely managed drinking water by 2030, current rates of progress must be quadrupled. Pollution and climate change are the main reasons why freshwater resources are insufficient. This imposes the development of efficient and sustainable water production technologies. Among these technologies Reverse Osmosis (RO) desalination has become a well-known technology solution to address the growing need for freshwater in coastal and dry locations. In order to guarantee economical and ecologically sustainable water production, it is crucial to investigate the operational elements of RO desalination plants and their integration into the energy market as the demand for freshwater rises [2,3].

This literature review explores RO desalination plant operational features, energy consumption, cost structures, and flexibility, and their integration into renewable energy generation companies' bidding strategies, aiming to understand system operation and market dynamics. [3]. The primary sources of cost in any of these processes that produce water are energy, operating and maintenance expenses, and capital investments. This is the case in many countries around the world where water production process consume hefty amounts of electricity [4,5]. Thermal and membrane water desalination facilities are the main types of water desalination facilities. Thermal WDP uses steam to produce saline water, which is then condensed to create freshwater. Some steam is fed into steam turbines to generate energy, enabling simultaneous generation of electricity and desalinated water. The energy for heating steam can be produced off-grid using renewable resources or fossil fuels. [6].

Reverse Osmosis (RO) water desalination uses a semi-permeable membrane filter to filter out salt, resulting in concentrated water on the membrane's high-pressure side. Electricity powers the pump, which produces the pressure needed to push water through the membrane. The desalination pressure ranges from 17 to 27 bars for brackish water and 55 to 82 bars for seawater. [7,8] Renewable energy sources like solar thermal, photovoltaic, wind, and geothermal technologies can be used as energy suppliers in water desalination plants, especially in remote areas with acute water shortages where public electricity grid connections are not practical or cost-effective. Ghaithan et.al [3] offered a multi-objective model for a grid-connected photovoltaic-wind system that would supply energy to a Saudi Arabian RO desalination plant. The model aims to reduce life cycle costs and greenhouse gas emissions by considering economic and non-economic factors. It selects three Pareto-optimal solutions and provides management insights from economic and environmental perspectives. Additionally, in their study referenced as [9] The authors introduced the two-stage pricing (TSP) method for Northeast China's electric power auxiliary service market. This involves a freshwater supply and demand balance model, upper-level optimization, and low-level wind power pricing to maximize profits and enhance energy flexibility. In addition, Authors form Malaysia in [10] designed a mini-grid hybrid power system for rural communities and emergency relief situations. This system relies solely on solar power as its primary source and incorporates renewable energy applications to minimize greenhouse gas emissions. In a pool-based energy market, retailers of electricity face several uncertainties, including those related to market pricing and demand [11]. The integration of hybrid energy sources with energy storage devices in micro-grid operations is complicated by the intermittent nature of renewable energy sources. This challenge leads retailers to face difficulties in maintaining a real-time supply-demand balance, as highlighted by Chakraborty and colleagues [12]. The authors reviewed optimization approaches aimed at achieving accurate load forecasting and maximizing profits for retailers and energy users, with the goal of reducing electricity bills. In addition, in [13] Parvania and Oikonomou disscused using desalination plants to help meet electricity demand. It discusses the challenges of desalination plants' high electricity consumption and proposes a model to optimize their participation in electricity markets. Operators of water distribution systems would be able to offer the flexibility of desalination plants in energy markets thanks to the approach. This would help to offset the costs of desalination and make it more sustainable. The model also considers the hydraulic constraints of the water distribution system and the availability of freshwater resources. Elsir and colleagues [13] Also discuss coordinating the water desalination and demand response facilities' day-ahead operation scheduling in smart grids. It talks on the necessity of water desalination and the difficulties in incorporating renewable energy sources. The authors propose a market-clearing mechanism that maximizes the performance of renewable-rich power systems and grid-connected reverse osmosis water desalination plants (RO-WDPs). The study uses a mixed-integer linear programming problem to develop a market clearing method that integrates electric demands into demand response programs, improving system efficiency without compromising water supply-demand balance.

This literature review highlights the growing need for sustainable water production and the potential of integrating grid-connected RO desalination with renewable energy like PV. It goes over the energy dependence of RO plants, integration challenges in micro-grids, and existing optimization approaches for cost reduction and profit maximization. This knowledge lays the foundation for developing an effective bidding strategy that leverages the flexibility of RO desalination within an electricity market, ultimately maximizing profits for renewable energy companies while promoting sustainable water production.

Methodology and Problem Formulation
This study aims to develop the optimal bidding strategy for renewable energy companies with RO desalination plants. It focuses on how these businesses can optimize their bidding techniques to maximize profits in the electricity market and utilize the flexibility of desalination facilities. The study examines a generation company entering the power market with renewable resources and desalination facilities in a future market setting. Our example will represent the future electricity market pool of Bahrain.

This section discusses the mathematical equations for optimizing bidding in an electricity market, focusing on cost-effective and sustainable desalination operations. The formulation uses renewable energy resources and RO flexibility, and employs the General Algebraic Modeling System (GAMS) to optimize the bidding strategy, considering energy consumption, cost structures, and market dynamics.

Profit Cost function. The problem formulation involves creating a profit cost function that considers variable operation and maintenance costs to accurately reflect the desalination plant's operation viability. This function captures the interplay between energy prices, renewable energy generation, and electricity market dynamics, allowing for optimized bidding strategies. The goal is to maintain water sales revenue while minimizing variable costs to enhance the desalination plant's profitability.

$$Profit = RT - CT \quad (1)$$

Where RT in Equation. 1 is the total revenue and CT is the total cost. calculated as:

$$R_T = \sum_{t=t_0}^{t_f}(R_{Electricity}(t) + R_{Water}(t)) \quad (2) \quad C_T = \sum_{t=t_0}^{t_f}(C_{Electricity}(t) + C_{Water}(t)) \quad (3)$$

Where $R_{Electricity}(t) = E_{Surplus}(t) \times \pi_E(t)$ and $R_{Water}(t) = V_{demand}(t) \times \pi_W(t)$

And $C_{Electricity}(t) = E_{deficit}(t) \times \pi_E(t)$ and $C_{Water}(t) = V_{RO}(t) \times MC$

$\pi_E(t)$ represents the Price of electricity for hour t and $\pi_W(t)$ represents the price of water, while MC represents the variable Maintenance Cost per m³ of Water Produced.

The energy requirement will be represented by the energy balance equation:

$$E_{RO}(t) = E_{PV}(t) + E_{deficit}(t) - E_{Surplus}(t) \quad (4)$$

These equations will be further elaborated in the relevant following subsections.

RO Water Desalination Energy Consumption and Cost structure. Reverse osmosis (RO) is a process that removes impurities from water by pressurizing it through a semipermeable membrane, leaving only pure water. The energy input for RO water production is influenced by the pressure needed to overcome osmotic pressure. This text presents a straightforward mathematical generalization of energy use, cost structures, and the adaptability of RO.

In terms of Energy Consumption RO systems require energy to operate the high-pressure pumps that push water through the membrane. The energy consumption of an RO system depends on various factors such as feed water quality, system design, and recovery rate [7]. RO systems consume significant energy, especially in large-scale applications. Advancements in technology have led to more energy-efficient membranes. The cost structure of an RO system includes capital and operating expenses, including equipment, energy, membrane replacement, sanitization, and monitoring. The complexity and capacity of the system influence these costs. Capital costs are disregarded for bidding purposes. A simplified equation for RO water production accounting for operating costs and energy consumption with temperature can be obtained using the following formulation.

Operational Cost = Energy Cost + Maintenance Cost

The energy cost component is reliant on the RO system's energy consumption, which is impacted by the surrounding ambient temperature. Higher or lower temperatures may increase energy requirements due to reduced water flux and increased osmotic pressure whenever you deviate from the optimal operating temperatures [14,15,16].

The connection between the production of water in RO and energy input is referred to as specific energy consumption (SEC). Which measures the energy needed to use the RO process to generate a specific volume of purified water. Usually, it is given as (kWh/m3) or kWh/gal. [7].

$$E_{RO}(t) = SEC(t) \times V_{RO}(t) \quad (5) \qquad C_{Water}(t) = V_{RO}(t) \times MC \quad (6)$$

The calculation of energy requirement as depicted in Equation. 5 will involve multiplying the SEC (kWh/m³) by the volume of water produced in m³ to obtain the total energy input needed at a certain hour to produce the desired water output, which is part of the energy balance in Equation 4. In addition, the costs of replacing membranes, cleaning, sanitizing, and other regular maintenance tasks are included in the maintenance cost component. An estimate of the maintenance cost (MC) for a certain RO system can be made using industry standards or historical data. It can be stated as a price per generated unit of filtered water (cost/m³) for example as in Equation. 6 Above. As such, Total cost of RO Water Production was summarized in Equation. 3.

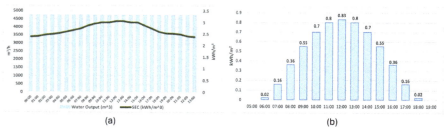

Figure 1. (a) Relationship between SEC and output of a SWRO plant (b) Average energy output for solar PV system in Bahrain [19].

We mentioned that SEC can be affected by ambient temperature, therefore for a summer day in GCC greater SEC would be required for hotter times of day to produce the same water output. this shall be taken into account while scaling SEC input parameters. For this proposal, since it is in a future market setting we will scale data for a current IPWP in Bahrain. This IPWP has a generation capacity of 50 MIGD for Sea Water RO (SWRO) which is equal to 227,304.5 m³/day and is expected to keep an almost constant water production in its two production blocks throughout the day with an average of 9,471.02 m³/hr [17] total and 4,735.51 per block. Therefor the expected water output for 1 block vs. SEC should look like the graph in Fig. (1a) for 24 hours if we take into consideration the effect of ambient temperature on SEC and an average SEC of 2.374 kWh/m³ for seawater [7].

In addition, the Maintenance cost has been approximated by averaging the cost of replacement of different kinds of membranes as stated in [18] which came to about 0.1638 $/m³ and it was rounded to 0.2 $/m³ to account for other maintenance or operational aspects.

PV Generation and Cost structure. The variable cost of PV power production requires precise data collection and analysis, tailored to the specific PV system. The capacity, efficiency, sun irradiation, and ambient temperature of a photovoltaic system are some of the elements that affect its energy generation rate. A generic formula will be examined, considering operational expenses and energy generation, for simplicity. Since the company already owns the PV panels Energy costs for PV will be very minimal and Maintenance costs (MC) are relatively very small and thus can be ignored. For this problem we will use the simulated data for the solar system which is already deployed in Bahrain as seen in [19] to more realistically tailor the optimization problem for an IPWP in Bahrain.

The Energy output of the PV will be calculated for this proposal using the previously stated PV system data in [19] which has the energy output in Fig. 3 per m². In order to calculate the Relevant Energy production for the IPWP we are studying, we will assume that PV panels are installed on 60% of the total land area of the power plant, where the total area for the IPWP in the proposal is approximately 32,400 m² according to coordinates. Therefore, we'll assume the total area of PV installation is around 19,440 m² and we'll assume a higher efficiency of PV panels (10%) since it is a future application. As observed in Figure. (1a) we can achieve higher energy output with higher ambient temperatures, since higher temperature is usually correspondent with higher solar irradiance. which is where the PV characteristics can nicely align with optimal generation for an RO facility in the summer during higher SEC hours.

Problem Definition

The above equations have been used to formulate the profit maximization objective function with constraints and relevant variables. Thus, the problem formulation will be as follows:

Maximize Equation. (1) $Profit = RT - CT$
subject to:

$Eqution. (4)\ and\ Eqution. (5)$
$E_{deficit}(t).E_{surplus}(t) \leq 0$
$V_{demand}(t) = Hourly\ Demand\ Target$
$V_{RO}(t) \leq Production\ Block\ Maximum$
$V_{tank}(t) \leq Tank\ Storage\ Capacity$
$V_{RO}(t).V_{tank}(t) \geq 0$
$V_{tank}(t) = V_{tank}(t-1) + V_{RO}(t) - V_{demand}(t)$

where t ranges between $t_o = 1$ and $t_f = 24$ for a 24 hour time period.

Since Bahrain doesn't currently have a market setting and hourly pricing, we will use Day ahead hourly market prices available in [20] and scale it with the same pattern to match the regional prices in GCC. As for the water price, a constant value of $2.012/m³ will be considered.

Table 1. Optimization Problem Input Parameters

t	EPV	$/MWh	$/m³	t	EPV	$/MWh	$/m³	t	EPV	$/MWh	$/m³
1	0	39.698	2.012	9	8.9813	75.267	2.012	17	8.9813	87.658	2.012
2	0	48.52	2.012	10	13.721	78.533	2.012	18	3.9917	105.22	2.012
3	0	47.522	2.012	11	17.464	78.753	2.012	19	0.499	111.18	2.012
4	0	39.698	2.012	12	19.958	80.718	2.012	20	0	101.36	2.012
5	0	42.268	2.012	13	20.707	78.042	2.012	21	0	92.534	2.012
6	0	54.16	2.012	14	19.958	82.669	2.012	22	0	84.075	2.012
7	0.499	67.571	2.012	15	17.464	83.19	2.012	23	0	76.469	2.012
8	3.9917	75.819	2.012	16	13.721	85.64	2.012	24	0	73.77	2.012

Table. 1 Displays the expected output of the PV panels EPV values and market day ahead prices of power and water for the time frame of 24 hours. 4 scenarios of power/water exchange dynamics have been considered as follows.

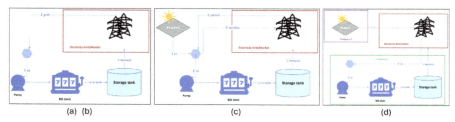

Figure 2. Modeling Scenarios

Scenario A in Fig 2a. represents the basic, classical Energy in - Water out RO system where no optimization can be done, and you have to operate the system in the same way to produce the same output regardless of energy prices or system conditions (no storage tank). On the other hand, Scenario B is an upgrade to the basic system in which a water storage tank is added to the system allowing some degree of flexibility in production.

Thirdly, Scenario C considers the full PV-RO system in which there is one main company that owns both RO facilities and PV facilities. The optimization problem will revolve around how the energy cost efficiency with the addition of PV and flexibility of RO can be maximized.

For the last case, Case D, the PV and RO facilities are owned by two separate companies that bid separately into the electricity market.

Each of the above cases will be examined in terms of profit and cost vs. revenue to observe the optimization process for each configuration.

Results and Discussion

The Four above Scenarios were simulated using GAMS software to obtain the profit maximization results while considering a 24-hour time period.

Table 2. Scenario Comparison of Cases A and B

Variable	Total Cost	Total Revenue	Total Profit
Scenario A	$28,994.91	$144,864.00	$115,869.09
Scenario B	$27,446.62	$144,864.00	$117,417.38

As can be seen from the table when the storage was added in scenario B the flexibility of the RO allowed for reduced water production during higher price periods while still maintaining the supply demand. This is done by increasing the water production during the periods of time where the electricity is cheaper and then utilizing the water storage to maintain demand during the pricey hours to buy less electricity. Hence, the cost reduced for scenario B allowing for increased profits for the same revenue and demand.

This process can be observed in Table. 3 with details about charging, discharging, electricity price and SEC for the 24 h period.

Table 3. Variation of storage tank with respect to time

t	$/MWh	SEC	in tank	out tank	Tank (t)	t	$/MWh	SEC	in tank	out tank	Tank (t)
1	39.69756	0.002374	4735.51	3000	1735.51	13	78.04188	0.003056	4735.51	3000	9000
2	48.52008	0.002388	4735.51	3000	3471.02	14	82.6686	0.003056	1264.49	3000	7264.49
3	47.52216	0.002456	4735.51	3000	5206.53	15	83.19024	0.003002	4735.51	3000	9000
4	39.69756	0.002483	4735.51	3000	6942.04	16	85.63968	0.003002	0	3000	6000
5	42.26796	0.002524	4735.51	3000	8677.55	17	87.6582	0.002865	4735.51	3000	7735.51
6	54.15984	0.002592	3322.45	3000	9000	18	105.2201	0.002729	0	3000	4735.51
7	67.57128	0.002661	3000	3000	9000	19	111.1774	0.002592	0	3000	1735.51
8	75.81924	0.002729	3000	3000	9000	20	101.3645	0.002524	1264.49	3000	0
9	75.26736	0.002865	3000	3000	9000	21	92.5344	0.00251	3000	3000	0
10	78.53328	0.002947	3000	3000	9000	22	84.07476	0.002483	3000	3000	0
11	78.75252	0.003002	3000	3000	9000	23	76.4694	0.002401	3000	3000	0
12	80.71812	0.003002	1264.49	3000	7264.49	24	73.77048	0.002374	3000	3000	0

In the (out tank) column of Table. 3 we see that the number is constant which is the water demand that should be supplied at all times, and you can notice the relationship between state of storage and electricity price. Moreover, tank capacity can also play a role in the optimization process. Therefore, Scenario B has been simulated considering 3 different water tank capacities.

Table 4. Scenario B: Capacity Comparison

Capcity	9000 m^3	18000 m^3	36000 m^3
Total Cost	$27,446.62	$26,978.74	$26,950.05
Total Revenue	$144,864.00	$144,864.00	$144,864.00
Total Profit	$117,417.38	$117,885.26	$117,913.95

As observed from Table 4 increasing the capacity can increase the profits due to the added flexibility, however at some point increasing the capacity beyond a certain volume will be unprofitable. Increasing the capacity from 18000 m^3 to 36000 m^3 will only increase the profits by $ 28.69 per day, this small increase in profit will likely only be worth it if the cost of upgrading the tank can be justified.

Meanwhile, for scenarios C and D, it can be noticed that there is an extra source of revenue. Therefor in Table. 5 for C not only did the cost reduce significantly due to utilizing energy from PV, but the revenues have also increased as the company can sell excess PV energy to the grid.

Table 5. Scenario Comparison of Cases C and D

Variable	Total Cost (W)	Total Revenue	PV Revenue	RO Revenue	Total Profit	RO Profit
Case C	$21,469.325	$150,007.47	$5,143.47	$144,864	$128,538.144	$123,394.675
Case D	$27,446.623	$155,984.77	$11,120.767	$144,864	$128,538.14	$117,417.38

Moreover, for scenario D this will be different, as PV is no longer part of the same company and is selling energy separately to the grid, the interesting thing is, that both scenarios C and D have reached the same optimal profit value for the total system. However, the profits here are divided between 2 companies, the revenues from selling PV power have increased because the output of the PV is sold entirely to the grid. But, for the RO company the costs saw an increase, as there is no longer a free energy source to use, and more must be bought from the grid. This is probably due to the balance of Energy bought and sold, as the same amount of energy is still being exchanged however it is in different directions (for different beneficiaries). Assuming, Company B were to sell Electricity to Company A directly under a predetermined contract, profits might improve because both parties would be able to take advantage of the flexibility of these resources accessible for their best interests.

Conclusion

The study examined the connection between power system electrical generation, PV, and grid-connected reverse osmosis (RO) water desalination plants in an electricity market setting, aiming to maximize profits by optimizing RO facility coordination and flexibility. The study examined four scenarios: basic RO system, RO with water storage tank, combined PV-RO system, and Seperate PV-RO system with energy optimization. It found that incorporating a water storage tank and using PV energy during daylight hours can reduce costs, increase revenues and profits. Moreover, while bidding the PV separately can result in the same combined profits it would mean compromising on some profits for the RO owner compared to the combined system. Additionally, the simulation considered different water tank capacities where the results indicated that increasing the capacity can enhance profits and earnings only up to a certain point, beyond which incremental benefits become negligible.

The study highlights the economic benefits of integrating RO desalination plants into renewable energy generation companies' bidding strategies. It emphasizes the cost reduction and revenue maximization benefits of water storage and PV energy utilization, while balancing profits and costs in a day ahead electricity Market setting.

References

[1] UN-Water, "Summary progress update 2021: Sdg 6 - water and sanitation for all," Online, 2021. [Online]. Available: https://www.unwater.org/sites/default/files/app/uploads/2021/12/SDG-6-SummaryProgress-Update-2021 Version-July-2021a.pdf

[2] M. Rouholamini, C. Wang, C. J. Miller, and M. Mohammadian, "A review of water/energy co-management opportunities," in 2018 IEEE Power & Energy Society General Meeting (PESGM), 2018, pp. 1–5. https://doi.org/10.1109/PESGM.2018.8586013

[3] A. M. Ghaithan, A. Mohammed, A. Al-Hanbali, A. M. Attia, and H. Saleh, "Multi-objective optimization of a photovoltaic-wind- grid connected system to power reverse osmosis desalination plant," Energy, vol. 251, p. 123888, 2022. https://doi.org/10.1016/j.energy.2022.123888

[4] S. Shu, D. Zhang, S. Liu, M. Zhao, Y. Yuan, and H. Zhao, "Power saving in water supply system with pump operation optimization," in 2010 Asia-Pacific Power and Energy Engineering Conference, 2010, pp. 1–4. https://doi.org/10.1109/APPEEC.2010.5449192

[5] A. Siddiqi and L. D. Anadon, "The water–energy nexus in middle east and north africa," Energy Policy, vol. 39, no. 8, pp. 4529–4540, 2011, at the Crossroads: Pathways of Renewable and Nuclear Energy Policy in North Africa. https://doi.org/10.1016/j.enpol.2011.04.023

[6] M. Elsir, A. T. Al-Awami, M. A. Antar, K. Oikonomou, and M. Parvania, "Risk-based operation coordination of water desalination and renewable-rich power systems," IEEE Transactions on Power Systems, vol. 38, no. 2, pp. 1162–1175, 2023.

[7] A. Karabelas, C. Koutsou, M. Kostoglou, and D. Sioutopoulos, "Analysis of specific energy consumption in reverse osmosis desalination processes," Desalination, vol. 431, pp. 15–21, 2018, "Desalination, energy and the environment" in honor of Professor Raphael Semiat. https://doi.org/10.1016/j.desal.2017.04.006

[8] A. Al-Karaghouli and L. L. Kazmerski, "Energy consumption and water production cost of conventional and renewable energy powered desalination processes," Renewable and Sustainable Energy Reviews, vol. 24, pp. 343–356, 2013. https://doi.org/10.1016/j.rser.2012.12.064

[9] S. Chu, S. Zhang, W. Ge, G. Cai, and Y. Li, "The pricing method for abandoned wind power contract between wind power enterprises and desalination plants in bilateral transactions," Electric Power Systems Research, vol. 214, 2023. https://doi.org/10.1016/j.epsr.2022.108918

[10] N. Baharudin, T. Mansur, R. Ali, A. Wahab, N. Rahman, E. Ariff, and A. Ali, "Minigrid power system optimization design and economic analysis of solar powered sea water desalination plant for rural communities and emergency relief conditions," in 2012 IEEE International Power Engineering and Optimization Conference Melaka, Malaysia, 2012, pp. 465–469.

[11] S. M. Mousavi, T. Barforoushi, and F. H. Moghimi, "A decision-making model for a retailer considering a new short-term contract and flexible demands," Electric Power Systems Research, vol. 192, p. 106960, 2021. https://doi.org/10.1016/j.epsr.2020.106960

[12] N. Chakraborty, N. B. D. Choudhury, and P. K. Tiwari, "Profit maximization of retailers with intermittent renewable sources and energy storage systems in deregulated electricity market with modern optimization techniques: A review," Renewable Energy Focus, vol. 47, p. 100492, 2023. https://doi.org/10.1016/j.ref.2023.100492

[13] K. Oikonomou and M. Parvania, "Optimal participation of water desalination plants in electricity demand response and regulation markets," IEEE Systems Journal, vol. 14, no. 3, pp. 3729–3739, 2020.

[14] C. Koutsou, E. Kritikos, A. Karabelas, and M. Kostoglou, "Analysis of temperature effects on the specific energy consumption in reverse osmosis desalination processes," Desalination, vol. 476, p. 114213, 2020. https://doi.org/10.1016/j.desal.2019.114213

[15] M. M. Armendáriz-Ontiveros, G. E. Dévora-Isiordia, J. Rodríguez-López, R. G. Sánchez-Duarte, J. Álvarez Sánchez, Y. Villegas-Peralta, and M. d. R. Martínez-Macias, "Effect of temperature on energy consumption and polarization in reverse osmosis desalination using a spray cooled photovoltaic system," Energies, vol. 15, no. 20, 2022. https://doi.org/10.3390/en15207787

[16] S. Yagnambhatt, S. Khanmohammadi, and J. Maisonneuve, "Reducing the specific energy use of seawater desalination with thermally enhanced reverse osmosis," Desalination, vol. 573, p. 117163, 2024. https://doi.org/10.1016/j.desal.2023.117163

[17] "NOMAC - AL DUR PHASE II IWPP," NOMAC. [Online]. Available: https://www.nomac.com/en/our-operations/nomac-globally/al-dur2-iwpp/.

[18] S. Avlonitis, K. Kouroumbas, and N. Vlachakis, "Energy consumption and membrane replacement cost for seawater RO desalination plants," Desalination, vol. 157, no. 1, pp. 151–158, 2003, desalination and the Environment: Fresh Water for all. https://doi.org/10.1016/S00119164(03)00395-3

[19] W. Alnaser, N. Alnaser, and I. Batarseh, "Bahrain's bapco 5mw pv grid-connected solar project," Int. J. Power Renew. Energy Syst., vol. 1, pp. 72–84, 01 2014.

[20] "Market data - Nord Pool," Nord Pool. [Online]. Available: https://www.nordpoolgroup.com/en/Market-data1/GB/Auction-prices/UK/Hourly/?view=table.

Adaptive neuro-fuzzy inference system for DC power forecasting for grid-connected PV system in Sharjah

Tareq SALAMEH[1,2,a], Mena Maurice FARAG[2,3,b*], Abdul Kadir HAMID[2,3,c*], Mousa Hussein[4,d]

[1]Department of Sustainable and Renewable Energy Engineering, College of Engineering, University of Sharjah, United Arab Emirates

[2]Sustainable Energy and Power Systems Research Centre, Research Institute for Sciences and Engineering (RISE), University of Sharjah, Sharjah, United Arab Emirates

[3]Department of Electrical Engineering, College of Engineering, University of Sharjah, Sharjah, United Arab Emirates

[4]Department of Electrical and Communication Engineering, College of Engineering, United Arab Emirates University, Al Ain, UAE

[a]tsalameh@sharjah.ac.ae, [b]u20105427@sharjah.ac.ae, [c]akhamid@sharjah.ac.ae, [d]mihussein@uaeu.ac.ae

Keywords: Artificial Intelligence, ANFIS, Solar Energy, PV Systems

Abstract. Solar energy forecasting is essential to maintain PV system's performance in uncertain environmental conditions. Factors such as module temperature, ambient temperature, solar irradiance, and wind speed contribute to the DC current generated by PV systems. In this study, an adaptive neuro-fuzzy inference system (ANFIS) is developed on MATLAB to study a 2.88 kW grid-connected PV system in the harsh weather conditions of Sharjah. Solar irradiance, ambient temperature, module temperature, and wind speed are considered as the input membership functions in the developed ANFIS model. The output parameter considered in this study is the current DC generation, which critically depends on the defined membership functions. The accuracy of the model was determined based on the comparison with the experimental dataset. The R^2 value has shown that the proposed model can forecast the DC current with minimal error, with a value of 99.12% and 99.13% for training and testing, respectively. Moreover, the spatial 3-D surface has shown that the optimum DC current generation is achieved at the highest solar irradiance and ambient temperature while minimizing the module temperature for enhanced electrical efficiency.

Introduction

Conventional energy resources have been the main contributors to global warming and climate change, leading to the integration of clean energy sources [1]. Solar emissions received yearly present a huge portion of clean energy resources, making solar energy a remarkable source of alternate energy [2]. Photovoltaic (PV) systems have been designed to utilize solar emissions for clean electricity generation, therefore reducing carbon emissions and maintaining a clean economy [3].

Renewable Energy Sources (RES) have been increasingly integrated into the energy mix, specifically in oil-dependent countries such as the United Arab Emirates [4]. Moreover, solar irradiance exposure globally counts to 1367 W/m², which is sufficient to meet the electrical energy demand requirements worldwide [5,6].

However, to successfully integrate solar energy within the energy mix and attain electricity demands, its accurate prediction is necessary to ensure stability in power generation and injection into the energy grid [7,8]. Power generation from PV plants is known to be dynamic due to weather conditions dependency, presenting potential effects in its coupling with electrical networks [9]. In this notion, it's essential to forecast PV power plant generation to boost solar energy use and maintain electrical network stability. In principle, forecasting based on time-series data is extracted based on numerous amounts of sensors, guaranteeing precise data production and interval-based data measurement [10].

The accurate prediction of PV plants' performance is a critical task since weather conditions drive the operating conditions of PV plants. The dynamic change in solar irradiance causes fluctuation in PV power generation, leading to inconsistent electricity production. Multiple forecasting models are proposed to accurately predict PV plant generation [11]. Precise prediction of DC power generation requires the employment of machine learning (ML) techniques to enable learning complex pattern recognition and regression analysis [12]. Multiple forecasting models have been developed in the scientific literature such as ARIMA [13], support vector regression (SVR) [14], artificial neural network (ANN) and ANN-based hybrid models [15], hybrid intelligent system (HIS), Convolutional Neural Networks (CNN) [16], Long Short-Term Memory (LSTM) [17], and numerous other forecasting models.

The amalgamation of more than one forecasting offers an efficient compromise between prediction accuracy, computation time, and direct forecasting of PV plant power generation. Recently, ANN has been designed and modeled using a Fuzzy Inference System to develop an Adaptive Neuro Fuzzy Inference System (ANFIS) [18]. ANFIS and Support Vector Machine (SVM) have been popular tools for forecasting techniques and applications. However, ANFIS has proven to be precise as compared to SVM. Several studies discussed the employment of ANFIS in many fields of sciences [19,20], and as a model for accurate PV plant generation forecasting, proving its feasibility [21].

In this study, ANFIS has been employed to predict the performance of a 2.88 kW on-grid PV system installed in the terrestrial conditions of Sharjah, UAE. The forecasting of DC current is considered the main factor affecting PV power generation and therefore is considered in this study. Moreover, the relationship between the DC current and environmental factors such as solar irradiance, ambient temperature, module temperature, and wind speed is established. Therefore, the forecasting results are compared with the experimental results measured by the system to demonstrate the accuracy of the proposed model.

Experimental Setup

As seen in Fig. 1, a 2.88 kW on-grid photovoltaic system is installed on the rooftop of the W-12 central laboratories building at the University of Sharjah's main campus (Lat. 25.34° N; Long. 55.42° E) [22,23]. For immediate access to system data, the system is powered by a real-time data capture system [24]. The system's potential to function over short and long-time spans, together with its comprehensive infrastructure and improved data recording capabilities, were previously highlighted [25,26]. Furthermore, the system functions as a comprehensive grid-connected photovoltaic system, contributing AC electrical energy to the three-phase local utility grid. In addition, the on-grid photovoltaic system logs environmental and electrical characteristics every five minutes [27]. Table 1 represents the technical specifications of the on-grid PV system state-of-the-art.

Table 1. Technical Specifications of the 2.88 kW on-grid PV system

Equipment	Specifications
On-grid PV system State-of-the-Art	- 9 PV modules of 320 W electrical capacity. - Single Axis azimuth tracking. - 3.7 kWac Grid Inverter for Electrical Network Coupling and Synchronization. - Data manager for centralization system measurements on a common online interface.
Sensor Box	Responsible for coupling environmental sensors such as: - PT1000 thermocouples for module and ambient temperature measurement. - Irradiation sensor for solar insolation measurement. - Anemometer for wind speed measurement.

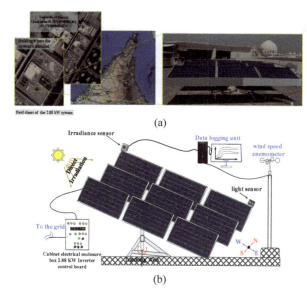

Fig 1. Illustration of 2.88 kW on-grid PV system (a) experimental setup (b) system visualization.

Adaptive network-based fuzzy inference system (ANFIS)
Deployment of fuzzy logic (FL) in control research has grown daily since the 1960s. The FL's foundation is expanding the well-known Boolean logic operations. The Fl employs binary systems (0 and 1) during the modeling procedure. Moreover, FL deals with multi-valued interpretation, which ranges from 0 to 1, just like in human perception. In the 1990s of this centuries, Jang created an Adaptive Network-based Fuzzy Inference System (ANFIS) based on the artificial neural network (ANN) and the fuzzy inference system (FIS) [28]. ANFIS can demonstrate a complicated system with extreme nonlinearity. The FL structure consists of three primary components: defuzzification, inference, and fuzzification.

In ANFIS, two different kinds of structures are combined to produce a fuzzy rule. The first kind is called Mamdani, and it was established in 1975 for the controller layout of heat transfer techniques where the Center of Gravity (COG) is the best defuzzifier. The second type, Sugeno (TSK), uses Weighted Average (Wtaver) as a defuzzifier. Eqs. (1) and (2) illustrate how these two types differ for fuzzy systems with two inputs and one output.

Should A be MF$_A$ and B be MF$_B$, then C is the MF$_C$ Mamdani form. (1)

In the event when A is MF$_A$ and B is MF$_B$, C= F (A, B) Form of Sugeno (2)

MF$_A$ and MF$_B$ represent fuzzy membership functions (MFs) for A and B, respectively. A and B are the two input (antecedent) fuzzy systems. The membership functions for the Mamdani form are MFC, the membership functions for the Sugeno form are F (A, B), and the output (consequence) fuzzy system is C. For A and B as input factors, F (A, B) can be either a linear or nonlinear function [29].

In this study, since DC current generated from PV systems is critically dependent on several environmental factors such as solar irradiance, ambient temperature, and module temperature, the compound Sugeno form (TSK) model is considered to represent a nonlinear function and is studied. The phases of fuzzification and defuzzification were translated using the membership functions, respectively. To carry out this translation, the input and output values were converted from crisp to fuzzy during the fuzzification stage and from fuzzy to crisp during the defuzzification step.

The study's data was used to extract the rules. As indicated by Eq. 3, the total output value was computed using the number of input variables and the quantification of rules implemented.

$$y(x) = \frac{\sum_{i=1}^{n} w_i y_i(x)}{\sum_{i=1}^{n} w_i}$$ (3)

where the i$_{th}$ fuzzy rule's weight, input, and output are represented by the variables y$_i$, x, and w$_i$, respectively. The weight of the ith fired rule has a range of [0 1].

A dataset of experimental measurements is used for the training and testing phases of 601 data points. With four input parameters, training and testing data were done for every fuzzy model output. Table 2 demonstrates every detail of the fuzzy model that was used in this investigation.

Table 2. Specifications of the developed fuzzy model

Fuzzy component	Type	Details
Inputs	Variables	4
Epochs	Number of Training Data	410
	Number of Testing Data	191
Rule	Number	4
	Base builder	SC
	ANDing Operation	Product
	ORing Operation	Probabilistic OR
	Implication	Min
	Aggregation	Max
	Defuzzification	Wtaver
Output	Function	Linear

Results and Discussion

A large data set of experimental data was employed for the developed ANFIS model. The prediction of DC current by the ANFIS model under the terrestrial conditions of Sharjah during harsh weather conditions is demonstrated in Fig 3. Moreover, the accuracy and precision of the proposed model according to the R^2 value, as presented in Fig 3, represent 99.1% accuracy. The ANFIS model can sufficiently forecast the DC current with respect to the experimental data that is measured from the system, as presented in Fig 2 and Fig 3.

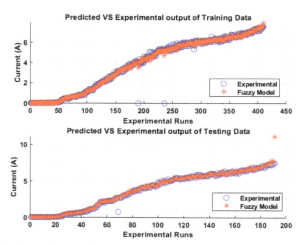

Fig 2. ANFIS model predictions for training and testing datasets for the experimental data.

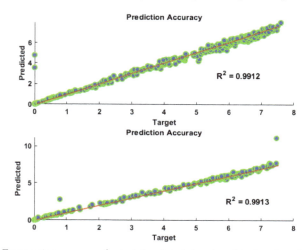

*Fig 3. Forecasting accuracy for training data (**above**) and testing data (**below**).*

The input parameter membership functions that are employed in this investigation are presented in Fig 4. These functions are typically used as rules to convert between the fuzzification and defuzzification procedures. The different colors represent the various membership functions that are utilized to connect the fuzzification and defuzzification procedures within the fuzzy inference system architecture.

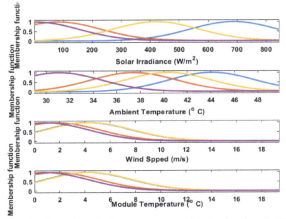

Fig 4. Membership Functions for input parameters for the developed ANFIS model.

Figs. 5 (a)-(e) represent the spatial 3-D surface for the DC Current (A) for the given PV system for each combination of the input parameters of the developed fuzzy model. The effect of ambient temperature, module temperature, and solar irradiance on the DC current generation is demonstrated in Fig. 5 (a) and (b), respectively. A direct relationship is observed with respect to ambient temperature and solar irradiance on the DC current generation. However, as module temperature increases significantly, the DC current is observed to reduce due to the reduction in the overall electrical efficiency of the system.

Moreover, Fig. 5 (c) depicts the impact of wind speed and ambient temperature on DC current generation. The spatial 3-D surface further demonstrates the direct relation of ambient temperature with DC current generation, while wind speed does not influence the DC current generation much.

Moreover, Figs. 5 (d) and (e) demonstrate the adverse impact of module temperature on the DC current generation as it increases significantly. The relative decrement in module temperature and maximizing the ambient temperature demonstrates an increase in the DC current generation as observed in Fig 5 (d). Furthermore, the minimal effect of wind speed on the DC current generation further confirms the findings presented in Fig 5 (c).

Therefore, the developed ANFIS model has demonstrated the performance of the demonstrated PV system based on four input variables. The DC current generation yield is at its highest when ambient temperature and solar irradiance are maintained at peak, while module temperature is minimized to attain maximum yield. This demonstrates the importance of incorporating cooling techniques to improve the PV system's electrical yield, sustaining its performance under harsh weather conditions.

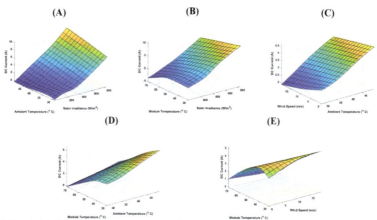

Fig 5. Spatial 3-D surface representation of DC current with respect to membership functions

Summary

PV System's DC current generation is affected by several environmental factors such as module temperature, ambient temperature, solar irradiance, and wind speed. In this study, the Adaptive Neuro-Fuzzy Inference System (ANFIS) model was developed to forecast the performance of a 2.88 kW on-grid PV system in the terrestrial conditions of Sharjah, UAE. The model adopted four input membership functions that relate to the DC current output power generation. DC current generation has observed a direct proportional relationship with respect to solar irradiance and ambient temperature, while module temperature presents an indirect relationship. This presents that DC current is improved at lower module temperatures while maintaining high solar irradiance exposure to enhance the electrical efficiency of the PV system. Moreover, the model presented a notable prediction accuracy and a significant correlation based on the least error, as the R^2 value corresponded to 99.12% and 99.13% for training and testing, respectively. Optimization techniques will be incorporated and tested in future work to improve the forecasting accuracy for larger datasets.

References

[1] J.L. Holechek, H.M.E. Geli, M.N. Sawalhah, R. Valdez, A Global Assessment: Can Renewable Energy Replace Fossil Fuels by 2050?, Sustainability. 14 (2022) 4792. https://doi.org/10.3390/su14084792

[2] T. Güney, Solar energy, governance and CO2 emissions, Renew. Energy. 184 (2022) 791–798. https://doi.org/10.1016/j.renene.2021.11.124

[3] A. Mehmood, J. Ren, L. Zhang, Achieving energy sustainability by using solar PV: System modelling and comprehensive techno-economic-environmental analysis, Energy Strateg. Rev. 49 (2023) 101126. https://doi.org/10.1016/j.esr.2023.101126

[4] M.M. Farag, R.C. Bansal, Solar energy development in the GCC region – a review on recent progress and opportunities, Int. J. Model. Simul. 43 (2023) 579–599. https://doi.org/10.1080/02286203.2022.2105785

[5] M. Jamei, M. Karbasi, M. Ali, A. Malik, X. Chu, Z.M. Yaseen, A novel global solar exposure forecasting model based on air temperature: Designing a new multi-processing ensemble deep learning paradigm, Expert Syst. Appl. 222 (2023) 119811.

https://doi.org/10.1016/j.eswa.2023.119811

[6] M.M. Farag, A.K. Hamid, T. Salameh, E.M. Abo-Zahhad, M. AlMallahi, M. Elgendi, ENVIRONMENTAL, ECONOMIC, AND DEGRADATION ASSESSMENT FOR A 2.88 KW GRID-CONNECTED PV SYSTEM UNDER SHARJAH WEATHER CONDITIONS, in: 50th Int. Conf. Comput. Ind. Eng., 2023: pp. 1722–1731.

[7] C. Scott, M. Ahsan, A. Albarbar, Machine learning for forecasting a photovoltaic (PV) generation system, Energy. 278 (2023) 127807. https://doi.org/10.1016/j.energy.2023.127807

[8] U.K. Das, K.S. Tey, M. Seyedmahmoudian, S. Mekhilef, M.Y.I. Idris, W. Van Deventer, B. Horan, A. Stojcevski, Forecasting of photovoltaic power generation and model optimization: A review, Renew. Sustain. Energy Rev. 81 (2018) 912–928. https://doi.org/10.1016/j.rser.2017.08.017

[9] M.M. Farag, N. Patel, A.-K. Hamid, A.A. Adam, R.C. Bansal, M. Bettayeb, A. Mehiri, An Optimized Fractional Nonlinear Synergic Controller for Maximum Power Point Tracking of Photovoltaic Array Under Abrupt Irradiance Change, IEEE J. Photovoltaics. 13 (2023) 305–314. https://doi.org/10.1109/JPHOTOV.2023.3236808

[10] E. Kim, M.S. Akhtar, O.-B. Yang, Designing solar power generation output forecasting methods using time series algorithms, Electr. Power Syst. Res. 216 (2023) 109073. https://doi.org/10.1016/j.epsr.2022.109073

[11] A. Alcañiz, D. Grzebyk, H. Ziar, O. Isabella, Trends and gaps in photovoltaic power forecasting with machine learning, Energy Reports. 9 (2023) 447–471. https://doi.org/10.1016/j.egyr.2022.11.208

[12] F. Pereira, C. Silva, Machine learning for monitoring and classification in inverters from solar photovoltaic energy plants, Sol. Compass. 9 (2024) 100066. https://doi.org/10.1016/j.solcom.2023.100066

[13] Y. Chen, M.S. Bhutta, M. Abubakar, D. Xiao, F.M. Almasoudi, H. Naeem, M. Faheem, Evaluation of Machine Learning Models for Smart Grid Parameters: Performance Analysis of ARIMA and Bi-LSTM, Sustainability. 15 (2023) 8555. https://doi.org/10.3390/su15118555

[14] M. Jobayer, M.A.H. Shaikat, M. Naimur Rashid, M.R. Hasan, A systematic review on predicting PV system parameters using machine learning, Heliyon. 9 (2023) e16815. https://doi.org/10.1016/j.heliyon.2023.e16815

[15] S. Pereira, P. Canhoto, R. Salgado, Development and assessment of artificial neural network models for direct normal solar irradiance forecasting using operational numerical weather prediction data, Energy AI. 15 (2024) 100314. https://doi.org/10.1016/j.egyai.2023.100314

[16] S.B. Bashir, M.M. Farag, A.K. Hamid, A.A. Adam, A.G. Abo-Khalil, R. Bansal, A Novel Hybrid CNN-XGBoost Model for Photovoltaic System Power Forecasting, in: 2024 6th Int. Youth Conf. Radio Electron. Electr. Power Eng., 2024. https://doi.org/10.1109/REEPE60449.2024.10479878

[17] M.S. Ibrahim, S.M. Gharghory, H.A. Kamal, A hybrid model of CNN and LSTM autoencoder-based short-term PV power generation forecasting, Electr. Eng. (2024). https://doi.org/10.1007/s00202-023-02220-8

[18] G. Perveen, M. Rizwan, N. Goel, An ANFIS-based model for solar energy forecasting and its smart grid application, Eng. Reports. 1 (2019). https://doi.org/10.1002/eng2.12070

[19] T.M.M. Abdellatief, M.A. Ershov, V.M. Kapustin, E.A. Chernysheva, V.D. Savelenko, T. Salameh, M.A. Abdelkareem, A.G. Olabi, Uniqueness technique for introducing high octane environmental gasoline using renewable oxygenates and its formulation on Fuzzy modeling, Sci. Total Environ. 802 (2022) 149863. https://doi.org/10.1016/j.scitotenv.2021.149863

[20] T.M.M. Abdellatief, M.A. Ershov, V.M. Kapustin, E.A. Chernysheva, V.D. Savelenko, T. Salameh, M.A. Abdelkareem, A.G. Olabi, Novel promising octane hyperboosting using isoolefinic gasoline additives and its application on fuzzy modeling, Int. J. Hydrogen Energy. 47 (2022) 4932–4942. https://doi.org/10.1016/j.ijhydene.2021.11.114

[21] T. Salameh, E.T. Sayed, A.G. Olabi, I.I. Hdaib, Y. Allan, M. Alkasrawi, M.A. Abdelkareem, Adaptive Network Fuzzy Inference System and Particle Swarm Optimization of Biohydrogen Production Process, Fermentation. 8 (2022) 483. https://doi.org/10.3390/fermentation8100483

[22] M.M. Farag, F.F. Ahmad, A.K. Hamid, C. Ghenai, M. Bettayeb, M. Alchadirchy, Performance Assessment of a Hybrid PV/T system during Winter Season under Sharjah Climate, in: 2021 Int. Conf. Electr. Comput. Commun. Mechatronics Eng., IEEE, 2021: pp. 1–5. https://doi.org/10.1109/ICECCME52200.2021.9590896

[23] F.F. Ahmad, M. Abdelsalam, A.K. Hamid, C. Ghenai, W. Obaid, M. Bettayeb, Experimental Validation of PVSYST Simulation for Fix Oriented and Azimuth Tracking Solar PV System, in: 2020: pp. 227–235. https://doi.org/10.1007/978-981-15-4775-1_25

[24] M.M. Farag, F.F. Ahmad, A.K. Hamid, C. Ghenai, M. Bettayeb, Real-Time Monitoring and Performance Harvesting for Grid-Connected PV System - A Case in Sharjah, in: 2021 14th Int. Conf. Dev. ESystems Eng., IEEE, 2021: pp. 241–245. https://doi.org/10.1109/DeSE54285.2021.9719385

[25] T. Salameh, A.K. Hamid, M.M. Farag, E.M. Abo-Zahhad, Energy and exergy assessment for a University of Sharjah's PV grid-connected system based on experimental for harsh terrestrial conditions, Energy Reports. 9 (2023) 345–353. https://doi.org/10.1016/j.egyr.2022.12.117

[26] T. Salameh, A.K. Hamid, M.M. Farag, E.M. Abo-Zahhad, Experimental and numerical simulation of a 2.88 kW PV grid-connected system under the terrestrial conditions of Sharjah city, Energy Reports. 9 (2023) 320–327. https://doi.org/10.1016/j.egyr.2022.12.115

[27] M.M. Farag, A.K. Hamid, Experimental Investigation on the Annual Performance of an Actively Monitored 2.88 kW Grid-Connected PV System in Sharjah, UAE, in: 2023 Adv. Sci. Eng. Technol. Int. Conf., IEEE, 2023: pp. 1–6. https://doi.org/10.1109/ASET56582.2023.10180880

[28] J.-S.R. Jang, ANFIS: adaptive-network-based fuzzy inference system, IEEE Trans. Syst. Man. Cybern. 23 (1993) 665–685. https://doi.org/10.1109/21.256541

[29] T. Salameh, P.P. Kumar, E.T. Sayed, M.A. Abdelkareem, H. Rezk, A.G. Olabi, Fuzzy modeling and particle swarm optimization of Al2O3/SiO2 nanofluid, Int. J. Thermofluids. 10 (2021) 100084. https://doi.org/10.1016/j.ijft.2021.100084

Robust testing requirements for Li-ion battery performance analysis

Muhammad SHEIKH[1,a *], Muhammad RASHID[1,b] and Sheikh REHMAN[2,c]

[1]WMG, The University of Warwick, Coventry, UK

[2]School of Computing, Engineering and Digital Technologies, Teesside University, Middlesbrough, UK

[a]muhammad.sheikh@warwick.ac.uk, [b]R.Muhammad.1@warwick.ac.uk, [c]S.Rehman@tees.ac.uk

Keywords: Battery Degradation, Capacity Fade, Failure Analysis, Testing and Characterization, Test Matrix Design, Design and Development

Abstract. Lithium-ion batteries are considered reliable option for Electric vehicle propulsion and portable applications. Various battery chemistries are being developed to enhance safety and performance of batteries to improve lifespan and reliability. Battery use case scenario often dictate requirements of different li-ion battery types. When target applications are fulfilled, other key considerations are implemented which include testing and characterisation to understand useful performance indicators from chosen battery type. This paper investigates current testing and characterisation needs to understand capacity fade and battery degradation with respect to temperature variations. Cycling tests followed by reference performance tests are used to analyse capacity fade. Due to limitation for the paper size only capacity fade analysis along with immersed test setup are focused to understand battery degradation with respect to various C-rates. Key findings are discussed, and comparative analysis is provided with future recommendations.

Introduction

Lithium-Ion batteries (LIBs) have attracted more attention due to their great energy storage capacity, high current density, extended lifespan, low self-discharge, lack of memory effects, and minimal environmental impact [1]. LIBs are mostly used in portable electronics, energy storage systems, and electric vehicles (EVs) throughout a variety of industries, including transportation and aerospace [2][3], such as Tesla (Model 'S' and Roadster) and Nissan (Leaf) are among the main automotive manufacturers using LIBs for their fleets [4]. LIBs are the widely used technology for energy storage applications, despite the existence of alternative technologies [5]. This is due to their ability to fast-charging capacity with higher cycle life and energy density when compared to established technologies that are available commercially [6]–[8].

Battery Safety

LIBs are an effective energy source for present electric vehicles (EVs), their safety must be taken into consideration before these batteries are large-scale deployment. One of the issues is short circuits in batteries, which can propagate fast within battery packs or modules if they are not managed at the cell level [9]. Battery failures are evident once they are exposed to abusive conditions, however, when using test and validation techniques to determine battery potential, predicting these failures before the time is extremely essential [10]. Various LIBs have been recalled in recent years as a result of explosion and fire incidents [12–15], significantly damaging LIBs reputation and causing serious economic problems for associated market sectors [11].

Although LIBs provide many advantages, it is important to carefully assess their durability and safety [12]. When a battery's capacity approaches 80% of its initial value, battery is at the end of its life. Excessive use of a battery beyond its end-of-life (EOL) specifications can result in bad system performance and occasionally catastrophic events [13][14]. Remaining usable life (RUL) prediction is therefore required to ensure battery safety and reliable operation. Battery's operation can be managed via a battery management system (BMS), based on the RUL prediction results. Accurately estimating the state of health (SOH) of lithium batteries has become difficult, because of the uncertainty and diversity of their internal side reactions and external working conditions [15].

The chemistry of the battery [16][17], its working environment, and its abuse tolerance [18] all have a significant impact on battery safety. Electrochemical system instability is the root cause of a LIBs internal failure [19]. Optimising battery design and making thoughtful selections regarding electrode materials, separators, and electrolytes can greatly increase LIB safety and performance stability. In typical circumstances, external techniques such as cell balancing and cooling can also significantly improve LIBs safety performance [11]. Therefore, it is essential to include appropriate safety precautions in the design, manufacture, and second life of LIBs, such as through the appropriate design of short circuit protection or temperature management systems [20].

Capacity degradation
The actual capacity degrades as the battery cycles, It affects the vehicle's driving range and increases "range anxiety"[21]. Repeated cycles of charging and discharging can cause LIBs to degrade over time in terms of their durability. This might affect their charge retention capacity and their lifespan. Cycle-life performance of LIBs is intrinsically correlated to the fundamental understanding of ageing mechanisms [22]. Therefore, continuous research is carried out to advancements in materials, battery management systems and electrode designs. The goal of LIB development is to increase their efficiency by using eco-friendly components [23]. Batteries can fail at any point in their life cycle for several reasons, including degradation, abuse conditions, and manufacturing errors. Battery abuse loads, both mechanical and thermal, have been simulated through the development of tests [24]. In the EV industry, new techniques that enable continuous battery condition monitoring are currently being used [12].

Literature shows that the causes of capacity fade can be classified into two groups namely calendar ageing and cycling ageing. Whereas cycling ageing is usually influenced by ambient temperature, the number of charge cycles or charge throughput, C-rate, and DoD, calendar ageing is generally primarily affected by the storing temperature, SoC, and time, which represents how long the battery placed in the storage or in resting state [25].

Battery Capacity Fading
Battery capacity fading can be divided into three stages; constant capacity fading, rapid capacity fading, and repetition between capacity increase and decrease [26]. According to Jialong et al, they used incremental capacity analysis and electrochemical impedance spectroscopy to investigate relevant aging mechanisms in their experiment work [27]. They found; The formation of solid electrolyte interface (SEI) films causes a rapid decrease in capacity during the first stage. The capacity decreases slowly due to the stable state of the lithium-ion battery in the second stage. In the third stage, the capacity decreases rapidly again due to the decrease in charge acceptance capability and damage to active materials.

To optimise battery design, management, and operation, precise measurement and prediction of LIBs performance and degradation are essential. As a result, a lot of study has been done to look at the models and testing procedures used to determine the lifetime and capacity fading of LIBs. Several tests to identify capacity fading have been suggested in the literature such as tests based on electrochemical models [28],equivalent circuit models [29], an analytical model with empirical data fitting [30], and performance-based models have been proposed. Safari et al created an electrochemical model to investigate how ageing affects impedance rise and capacity fading [31]. Wang t al proposed advanced data-driven methods for predicting the remaining useful life (RUL) and whole life cycle state of charge (SOC) of lithium-ion batteries [32]. Selcuk et al; presented a novel ageing mechanism, this mechanism improves upon the standard approach of transport limited models that incorporates (i) multi-layered SEI, (ii) lithium-plating, and (iii) reduction of anode porosity. This method attempted to represent more realistic ageing kinetics in order to obtain an understanding of linear and nonlinear capacity fading [22]. Muhammad et al developed dataset for rapid state of health estimation of lithium batteries using EIS and machine learning, this dataset encompasses all ageing statistics for commercially accessible and commonly used lithium-ion batteries. It also evaluates how increased charge throughput (ageing) affects the cell's retained energy capacity and impedance. The dataset quantifies the inter-dependency between LIB impedance's temperature and SOC at various ageing states between 100% and 80% SOH [33]. A variety of scientists and engineers working on battery-related projects can use the datasets. Truong et al. report a thorough investigation on decreased lithium-ion battery degradation, through state-of-charge pre-conditioning techniques that enable an electric car to engage in vehicle-to-grid activities when the vehicle is parked [34].

Test procedure and experimental results
This section provides details of the test procedures used on the cells under examination for ageing analysis. These include cycle ageing at various charge/discharge rates and temperatures. Preconditioning and characterisation of cells are done prior to cell testing. The experiment conditions were selected from an originally more comprehensive test matrix.
When conducting long term ageing tests, equipment selection is crucial as it has direct impact on cell performance estimation. To overcome the high-temperature issue and ensure the safety of cells, schematic of a fully immersed setup is shown in figure 1(a), where dielectric oil (Kryo-51) is used and cell fixtures are shown in figure 1(b). The initial characterisation tests proved this method of thermal management more effective, enabling improved temperature control throughout the test.

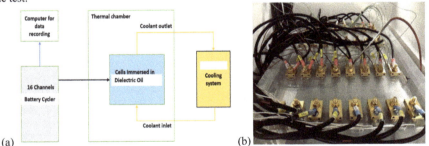

Figure 1: (a) Schematic of ully immersed test setup, (b) Cell fixtures for immersed setup

To record caapcity variations, reference performance tests (RPT) are recorded periodically, and the data is analysed according to the ageing test conditions. The results are discussed in detail in the following sections. The cell capacity is tracked periodically to analyse capacity fade to a given usage profile, with uncertainty intervals based on the four cells used per experiment. To ensure the safe running of the tests, routine monitoring of the oil-rig test setup, running programs and the live readings of the cycler data are performed.

Table 1: Summary of the cycle ageing tests performed

Temperature/Rate	Full Charge-Discharge cycling		
	0.3C Charge- 0.3C Discharg	0.5C Charge- 0.3C Discharge	0.7C Charge- 0.3C Discharge
0°C	4 cells	4 cells	4 cells
10°C	4 cells	4 cells	4 cells
25°C	4 cells	4 cells	4 cells

Table 1 provide details of ageing test conditions used where three temperature conditions are monitored with differect charge and discharge currents applied. The battery investigated was a 5Ah, 21700 cylindrical cell manufactured by LG Chem. This cell utilises nickel-rich NMC811 and SiOy-graphite active materials. For cycle ageing testing we have considered three test conditions which are 0.3C charge-0.3C discharge, 0.5C charge-0.3C discharge, 0.7C charge-0.3C discharge, and three temperature conditions at 0°C, 10°C, and 25°C. Four cells are used for each cycling ageing test condition and the same cell numbers are used throughout this work. Capacity checks are done after one week of cycling for all temperatures. The End-of-Life (EoL) for these cells are defined as 80% capacity compared to the initial capacity (5Ah). Figure 2 shows voltage vs capacity changes with respect to total number of cycles for each cycling condition underwent before reaching EoL.

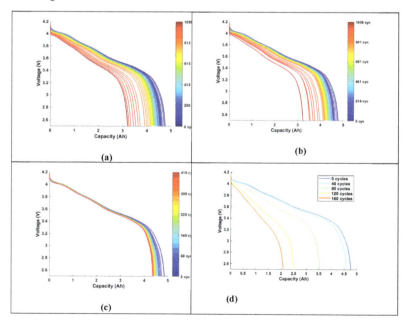

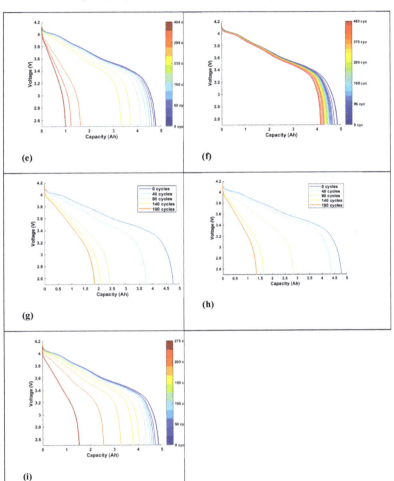

Figure 2: Capacity Test Voltage Curves (a) 0.3C/0.3C–0°C (b) 0.3C/0.3C–10°C (c) 0.3C/0.3C–25°C (d) 0.5C/0.3C–0°C (e) 0.5C/0.3C–10°C (f) 0.5C/0.3C–25°C (g) 0.7C/0.3C–0°C (h) 0.7C/0.3C–10°C (i) 0.7C/0.3C–25°C

Figures 2(a-c), show there is less affect of immersed temperature with low charge-discharge currents (1.67A). Analysis is further broaden to compare number of cycles and at the same number of cycles (418 cycles) we have same capacity fade. When charge current is increased and testing is done at 0.5C (2.5A) charge and 0.3C (1.67A) discharge as shown in figures 2(d-f), we can see less number of cycles are completed at 0°C and 10°C and capacity fade is rapid which, but at 25°C high number of cycles are achieved and capacity fade is lower which is only 20% after 485 cycles.

Figures 2(g-i) show that fewer cycles are achieved with the 0.7C charge (3.5A) and 0.3C discharge (1.67A) for all temperature cases.

Overall, it can be observed that with the low charge current (0.3C) condition, degradation has occurred more slowly compared with high charge currents (0.5C and 0.7C) for all temperature conditions except 25°C, whereby the performance becomes more comparable. However, ordinarily at lower temperatures, the diffusion kinetics for Li are slower and there is an inevitable trade-off in some performance level. Thus, higher cycling currents would not be sustainable to achieve a long lifespan.

Summary

This paper investigated current testing and characterisation needs to understand capacity fade, battery degradation and temperature dependance. Capacity fade analysis along with immersed test setup is provided to understand battery degradation with respect to various C-rates. Low temperatures in general, can induce deterioration in battery performance for a whole host of reasons; ultimately reducing the discharge voltage and accelerating capacity decay. The most severe capacity fading process has been reported to relate to effects from Li-plating on the anodes. This will result in lowered lithium inventory and reductions in accessible active material – capacity decay will thus continue. This at first may appear counterintuitive; however, this indicates that while cycling within the maximum and minimum voltage limits, the lower discharge current causes lower voltage losses in the battery and allows a higher utilisation of the electrodes. As a result, the charge throughput increases and the cells are worked harder while operating with the constant discharge, resulting in fewer cycles to reach EoL.

Author Contributions

Methodology, M.S.; formal analysis, M.S. and S.R.; investigation M.S. and M.R.; writing—original draft, M.S., M.R. and S.R., writing—review and editing, M.S. All authors have read and agreed to the published version of the manuscript.

References

[1] S. Windisch-Kern et al., "Recycling chains for lithium-ion batteries: A critical examination of current challenges, opportunities and process dependencies," *Waste Manag.*, vol. 138, pp. 125–139, 2022. https://doi.org/10.1016/j.wasman.2021.11.038

[2] Z. Cui, L. Wang, Q. Li, and K. Wang, "A comprehensive review on the state of charge estimation for lithium-ion battery based on neural network," *Int. J. Energy Res.*, vol. 46, no. 5, pp. 5423–5440, 2022. https://doi.org/10.1002/er.7545

[3] R. Gong et al., "A sustainable closed-loop method of selective oxidation leaching and regeneration for lithium iron phosphate cathode materials from spent batteries," *J. Environ. Manage.*, vol. 319, no. March 2022, p. 115740, 2022. https://doi.org/10.1016/j.jenvman.2022.115740

[4] M. Sheikh, S. Rehman, and M. Elkady, "Numerical simulation model for short circuit prediction under compression and bending of 18650 cylindrical lithium-ion battery," *Energy Procedia*, vol. 151, pp. 187–193, 2018. https://doi.org/10.1016/j.egypro.2018.09.046

[5] G. E. Blomgren, "The Development and Future of Lithium Ion Batteries," *J. Electrochem. Soc.*, vol. 164, no. 1, pp. A5019–A5025, 2017. https://doi.org/10.1149/2.0251701jes

[6] S. S. Rangarajan et al., "Lithium-Ion Batteries—The Crux of Electric Vehicles with Opportunities and Challenges," *Clean Technol.*, vol. 4, no. 4, pp. 908–930, 2022. https://doi.org/10.3390/cleantechnol4040056

[7] Y. Bai, N. Muralidharan, Y. K. Sun, S. Passerini, M. Stanley Whittingham, and I. Belharouak, "Energy and environmental aspects in recycling lithium-ion batteries: Concept of

Battery Identity Global Passport," *Mater. Today*, vol. 41, no. December, pp. 304–315, 2020. https://doi.org/10.1016/j.mattod.2020.09.001

[8] Y. Liang et al., "A review of rechargeable batteries for portable electronic devices," *InfoMat*, vol. 1, no. 1, pp. 6–32, 2019. https://doi.org/10.1002/inf2.12000

[9] M. Sheikh, A. Elmarakbi, and S. Rehman, "A combined experimental and simulation approach for short circuit prediction of 18650 lithium-ion battery under mechanical abuse conditions," *J. Energy Storage*, vol. 32, no. August, p. 101833, 2020. https://doi.org/10.1016/j.est.2020.101833

[10] M Sheikh; M Elmarakbi; S Rehman and A Elmarakbi, "Internal Short Circuit Analysis of Cylindrical Lithium-Ion Cells Due to Structural Failure," *J. Electrochem. Soc.*, 2021.

[11] Y. Chen et al., "A review of lithium-ion battery safety concerns: The issues, strategies, and testing standards," *J. Energy Chem.*, vol. 59, pp. 83–99, 2021. https://doi.org/10.1016/j.jechem.2020.10.017

[12] X. C. A. Chacón, S. Laureti, M. Ricci, and G. Cappuccino, "A Review of Non-Destructive Techniques for Lithium-Ion Battery Performance Analysis," *World Electr. Veh. J.*, vol. 14, no. 11, 2023. https://doi.org/10.3390/wevj14110305

[13] X. Li, D. Yu, V. Søren Byg, and S. Daniel Ioan, "The development of machine learning-based remaining useful life prediction for lithium-ion batteries," *J. Energy Chem.*, vol. 82, pp. 103–121, 2023. https://doi.org/10.1016/j.jechem.2023.03.026

[14] Q. Yu, C. Wang, J. Li, R. Xiong, and M. Pecht, "Challenges and outlook for lithium-ion battery fault diagnosis methods from the laboratory to real world applications," *eTransportation*, vol. 17, no. May, p. 100254, 2023. https://doi.org/10.1016/j.etran.2023.100254

[15] Y. Wang, X. Zhang, K. Li, G. Zhao, and Z. Chen, "Perspectives and challenges for future lithium-ion battery control and management," *eTransportation*, vol. 18, no. June, p. 100260, 2023. https://doi.org/10.1016/j.etran.2023.100260

[16] Q. Zhao et al., "Hierarchical flower-like spinel manganese-based oxide nanosheets for high-performance lithium ion battery," *Sci. China Mater.*, vol. 62, no. 10, pp. 1385–1392, 2019. https://doi.org/10.1007/s40843-019-9442-x

[17] X. Chen, H. Li, Z. Yan, F. Cheng, and J. Chen, "Structure design and mechanism analysis of silicon anode for lithium-ion batteries," *Sci. China Mater.*, vol. 62, no. 11, pp. 1515–1536, 2019. https://doi.org/10.1007/s40843-019-9464-0

[18] L. Wang et al., "Unlocking the significant role of shell material for lithium-ion battery safety," *Mater. Des.*, vol. 160, pp. 601–610, 2018. https://doi.org/10.1016/j.matdes.2018.10.002

[19] Y. Kang, C. Deng, X. Liu, T. Liang, Zheng; Li, Q. Hu, and Y. Zhao, "Binder-Free Electrode based on Electrospun-Fiber for Li Ion Batteries via a Simple Rolling Formation," *Scopus*, 2020.

[20] A. G. Olabi et al., Battery thermal management systems: Recent progress and challenges, vol. 15, no. June. Elsevier Ltd, 2022.

[21] R. Xiong, Z. Li, R. Yang, W. Shen, S. Ma, and F. Sun, "Fast self-heating battery with anti-aging awareness for freezing climates application," *Appl. Energy*, vol. 324, no. 5, p. 119762, 2022. https://doi.org/10.1016/j.apenergy.2022.119762

[22] S. Atalay, M. Sheikh, A. Mariani, Y. Merla, E. Bower, and W. D. Widanage, "Theory of battery ageing in a lithium-ion battery: Capacity fade, nonlinear ageing and lifetime prediction," *J. Power Sources*, vol. 478, p. 229026, 2020. https://doi.org/10.1016/j.jpowsour.2020.229026

[23] J. C. Barbosa, R. Gonçalves, C. M. Costa, and S. Lanceros-Mendez, "Recent advances on materials for lithium-ion batteries," *Energies*, vol. 14, no. 11, 2021. https://doi.org/10.3390/en14113145

[24] H. Wang, E. Lara-Curzio, E. T. Rule, and C. S. Winchester, "Mechanical abuse simulation and thermal runaway risks of large-format Li-ion batteries," *J. Power Sources*, vol. 342, pp. 913–920, 2017. https://doi.org/10.1016/j.jpowsour.2016.12.111

[25] A. Ahmadian, M. Sedghi, A. Elkamel, M. Fowler, and M. Aliakbar, "Plug-in electric vehicle batteries degradation modeling for smart grid studies : Review , assessment and conceptual framework," *Renew. Sustain. Energy Rev.*, vol. 81, no. June 2017, pp. 2609–2624, 2018. https://doi.org/10.1016/j.rser.2017.06.067

[26] B. Xu, A. Oudalov, A. Ulbig, G. Andersson, and D. S. Kirschen, "Modeling of lithium-ion battery degradation for cell life assessment," *IEEE Trans. Smart Grid*, vol. 9, no. 2, pp. 1131–1140, 2018. https://doi.org/10.1109/TSG.2016.2578950

[27] J. Liu *et al.*, "Capacity fading mechanisms and state of health prediction of commercial lithium-ion battery in total lifespan," *J. Energy Storage*, vol. 46, no. December 2021, p. 103910, 2022. https://doi.org/10.1016/j.est.2021.103910

[28] A. Khalid and A. I. Sarwat, "Fast charging li-ion battery capacity fade prognostic modeling using correlated parameters' decomposition and recurrent wavelet neural network," *2021 IEEE Transp. Electrif. Conf. Expo, ITEC 2021*, pp. 27–32, 2021. https://doi.org/10.1109/ITEC51675.2021.9490177

[29] D. Andre, C. Appel, T. Soczka-Guth, and D. U. Sauer, "Advanced mathematical methods of SOC and SOH estimation for lithium-ion batteries," *J. Power Sources*, vol. 224, pp. 20–27, 2013. https://doi.org/10.1016/j.jpowsour.2012.10.001

[30] K. M. Tsang and W. L. Chan, "State of health detection for Lithium ion batteries in photovoltaic system," *Energy Convers. Manag.*, vol. 65, pp. 7–12, 2013. https://doi.org/10.1016/j.enconman.2012.07.006

[31] M. Safari; C. Delacourt, "Simulation-Based Analysis of Aging Phenomena in a Commercial Graphite/LiFePO4 Cell," *J. Electrochem. Soc.*, vol. 158, no. 12, 2011.

[32] S. Wang, P. Takyi-Aninakwa, S. Jin, C. Yu, C. Fernandez, and D. I. Stroe, "An improved feedforward-long short-term memory modeling method for the whole-life-cycle state of charge prediction of lithium-ion batteries considering current-voltage-temperature variation," *Energy*, vol. 254, p. 124224, 2022. https://doi.org/10.1016/j.energy.2022.124224

[33] M. Rashid, M. Faraji-niri, J. Sansom, M. Sheikh, D. Widanage, and J. Marco, "Dataset for rapid state of health estimation of lithium batteries using EIS and machine learning : Training and validation," *Data Br.*, vol. 48, p. 109157, 2023. https://doi.org/10.1016/j.dib.2023.109157

[34] T. M. N. Bui *et al.*, "A Study of Reduced Battery Degradation Through State-of-Charge Pre-Conditioning for Vehicle-to-Grid Operations," *IEEE Access*, vol. 9, pp. 155871–155896, 2021. https://doi.org/10.1109/ACCESS.2021.3128774

[35] Y. Tian, Q. Wang, and J. Liu, 'Analysis of Lithium-Ion Battery through Direct Current Internal Resistance Characteristic', SAE International Journal of Electrified Vehicles, vol. 12, no. 2, pp. 173–184, Apr. 2022. https://doi.org/10.4271/14-12-02-0009

Date fruit type classification using convolutional neural networks

Abdullah ALAVI[1,a], Md Faysal AHAMED[1,b], Ali ALBELADI[1,c], Mohamed MOHANDES[1,d *]

[1]King Fahd University of Petroleum and Minerals, Department of Electrical Engineering, Dhahran, Saudi Arabia

[a]g202313690@kfupm.edu.sa, [b]g202309930@kfupm.edu.sa, [c]albeladi@kfupm.edu.sa, [d]mohandes@kfupm.edu.sa

Keywords: Date Fruit Type Classification, Convolutional Neural Network, Squeezenet, Pretrained Network, Transfer Learning

Abstract. Classification of objects is an important task for convolutional neural networks (CNNs). They have been applied to numerous fields with excellent results. In this study, we use CNNs to classify five categories of Sukkari dates, namely Galaxy, Mufattal, Nagad, Qishr, and Ruttab. Transfer learning is when a pretrained model is taken and only the final layers are trained to make a prediction. In this paper, we used the following five models: SqueezeNet, GoogLeNet, EfficientNet-b0, ShuffleNet, and MobileNet V2. The results show that SqueezeNet outperforms the other networks with a classification accuracy of 92% on the testing set. The testing accuracy for GoogLeNet, EfficientNet-b0, ShuffleNet, and MobileNet V2, on the other hand are 85.14%, 82.86%, 89.14%, and 87.43%, respectively. As this is a classification task, other metrics like precision, recall, and F1 score are also evaluated. These values for the SqueezeNet on the testing set are 92.67%, 92%, and 92.33%, respectively. ShuffleNet was second with values of 89.41%, 89.14%, and 89.28%, respectively. EfficientNet scored the lowest with 83.10%, 82.86%, and 82.98%, respectively.

Introduction

The recent breakthroughs in computer vision and artificial intelligence (AI) have led to myriad applications, ranging from facial recognition to self-driving cars. All such applications have a common theme, that it is relatively easy for people to solve with good accuracy but nearly impossible to program and implement on a computer [1]. To solve this problem, AI-based systems need to possess the ability to extract the patterns from the raw data and produce an output based on this knowledge [1, 2, 3, 4]. Agriculture is one of the areas that fall in this category and has benefitted from the developments of computer vision. Applications include machines to sort fruits, automatic fruit harvesting, and fruit scanners in markets [5, 6].

Deep Learning (DL) is a subset of AI that has gained significant popularity in recent years. It has a high level of abstraction and can automatically learn patterns from images [7]. Convolutional Neural Network (CNN) [8] is a popular architecture for applications that involve image processing [1, 4, 9, 10]. CNNs use a convolution operation in at least one of the layers [1, 11]. CNNs have started to gain popularity after 2012, when Krizhevsky et al. [12] won the ILSVRC competition on ImageNet [13]. Since then, they have found various applications in computer vision including fruit classification and detection [14, 15, 16, 17, 18].

Dates fruits are popular in the Middle East, North Africa, and Southwest Asia [19]. In the Kingdom of Saudi Arabia (KSA), date palm trees occupy nearly a quarter of the total cultivated land [20]. Various types of dates are cultivated and they vary greatly in terms of their size, color, and taste [21]. The recent success of AI and DL on inspecting a variety of fruits has inspired a number of works pertaining to date fruits. Date fruit quality classification, for instance, has gained traction among several researchers [22, 23]. Alresheedi et al. [24] compared the accuracy of several machine learning models with CNN. The dataset consisted of nine classes of date fruits. The authors found that the CNN model boasted the highest accuracy. In [25], a framework is proposed for date recognition. 500 images of three types of date fruits were used. The framework was based on a deep CNN and achieved an accuracy of 89.2%. In [20], Faisal et al. proposed a solution consisting of three estimation functions to classify date fruits based on maturity, type, and mass. The work used a Support Vector Machine (SVM) and achieved an accuracy of 99% among all the estimation functions. In [26], the authors focused on sorting dates based on their health and maturity. Four date types in different maturity stages along with defective dates were used in a CNN model based on VGG-16. The accuracy reported was 97%. In [27], Perez et al. used Medjool dates to compare the performance of eight different CNNs. The target was to sort the dates based on maturity stage. Out of all models, VGG-19 performed the best with an accuracy of 99.32%. In [26], the authors used transfer learning with fine-tuning using two pretrained networks, namely AlexNet and VGGNet. They used a dataset of 8,000 images separated into 5 classes and achieved an accuracy of 97.25%. In [28], the authors implemented a machine vision framework to deploy in a harvesting robot. The framework used three models to classify according to the type, maturity, and harvesting decision. The models are based on pretrained AlexNet and VGG-16. The VGG-16 model achieved accuracy of 99.01%, 97.25%, and 98.59% on date type, maturity, and harvesting decision classification, respectively. In [29], the authors used a dataset of 1,658 images belonging to nine categories of dates. The model used MobileNet V2 and resulted in an accuracy of 96%.

Although, numerous works have been published using individual pretrained models, there has been very little regarding the comparison of the performances of these various pretrained state-of-the-art models. Hardware and datasets for training and testing are different in published works, which make it unfair to compare performances. In this work, we present a comparison between various pretrained CNN models and observe how they perform on the same dataset. The goal of the CNN models is to classify five subclasses of dates of the same family. The paper is organized as follows. Section II describes the CNN models used. Section III explains the explains the dataset and the methodology. Section IV discusses the results. Finally, the conclusion is presented in Section V.

Overview of the CNN Models
A number of popular state-of-the-art pretrained networks like VGG-16, ResNet, Inception, and AlexNet are implemented frequently. However, due to their great number of layers, they usually require a long time to train even on powerful hardware. To mitigate this issue, researchers are looking for ways to minimize the size and training time by restructuring the CNN in various ways while maintain a comparable accuracy. In this paper, we implemented five such models and compared their performance in this classification task. The models are SqueezeNet [30], GoogLeNet [31], EfficientNet-b0 [32], ShuffleNet [33], and MobileNet V2 [34]. A brief summary is provided below about each model and the techniques they used to improve the efficiency.

In SqueezeNet, the authors achieved accuracy comparable to that of AlexNet [12] while using 1/50th of its parameters. They also reduced the size to less than 0.5 MB, which is 510 times smaller than AlexNet. They achieved this by replacing the 3x3 filters with 1x1 filters. They also decreased the number of input channels and down sampling the later layers in the network. These strategies were then incorporated with other modifications and packed in what is known as a *Fire module*.

In GoogLeNet, the authors managed to increase the depth and width of the network without increasing the computational demand. They have achieved this by implementing the *Inception module*. This module applies 1x1, 3x3, 5x5 parallel convolutions along with dimensionality reduction. This helped in capturing details of varying sizes.

In EfficientNet, the primary motivation was how to scale up a CNN. Generally, CNNs are developed with certain computational constraint in mind. If the model performs satisfactorily, then it is scaled up to further increase the accuracy. Scaling up can be done by increasing the depth, width or resolution. Instead of scaling up arbitrarily, the authors proposed a relationship between the three parameters known as *compound coefficient*. Using this technique, they have managed to reduce the size and increase the speed when compared to existing state-of-the-art CNNs.

In ShuffleNet, the primary motivation was to design a CNN that can run on mobile devices with extremely limited hardware. They used two new operations, namely pointwise group convolutions and channel shuffling. Using this, the ShuffleNet architecture achieved a superior performance, outperforming MobileNet in terms of accuracy in the ImageNet top-1 error on a computational power of 40 MFLOPS. On ARM-based computing hardware, ShuffleNet achieved a 13x speedup over AlexNet while maintaining a similar level of accuracy.

In MobileNetV2, the authors implemented what is known as an inverted residual structure. In this structure, the shortcut connections are between the thin bottleneck layers. Then, in the intermediate expansion layers, lightweight depth wise convolutions are performed to filter features in order to introduce non-linearity. Also, in the narrow layers, they discovered that it is important to remove non-linearities. This helped in maintaining representational power. On the ImageNet dataset, the MobileNet architecture improved the state-of-the-art on various performance measures in addition to reducing the model complexity.

Dataset and Methodology

The dataset used in this paper came from a real date palm plantation. It consists of images of dates that belong to five subcategories of Sukkari dates. They are known as Galaxy, Ruttab, Mufattal, Qishr, and Nagad. The original images were cropped to 500 by 500 pixels with RGB channels of size 8 bits each with white background. The images were then down sampled to 250 by 250 to reduce the size. Fig. 1 illustrates the five classes of images. As can be observed, the subclasses bear similar resemblance to each other. However, upon closer inspection, it can be seen that they have slightly different texture.

The dataset contains a total of 1,689 images that include the five classes of dates mentioned above. The images were not equally divided between the five classes, rather each class had roughly 400-440 images. For testing, a total of 175 images were set aside from the 1,689 images, and each class contained 35 images. The remaining images were used for training and validation with a split of 80% for training and 20% for validation, respectively. After splitting the dataset, the five CNNs mentioned in the previous section are loaded and trained. Fig. 2 illustrates the flowchart of the overall process of the collection of the images, cropping and down sampling, and the training phase of the CNNs.

The pretrained CNNs were originally trained on the ImageNet dataset [35]. For this study, only the final convolutional, classification and softmax layer were modified to produce the outputs of 5 classes. Also, the learning rate of the final layers were increased. This is discussed further in the next section. Apart from that, no other parameters of the CNNs were modified and the default values were used for simulation. The size of the input layers and the subsequent layers and their connections were also kept the same. During the training, validation, and testing phase, the images were automatically resized to fit the size of the default input layer of the pretrained networks.

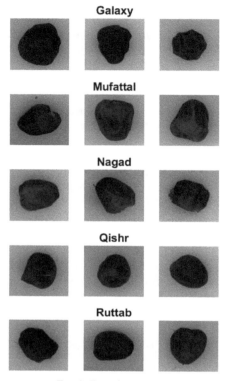

Fig. 1. Five classes of dates.

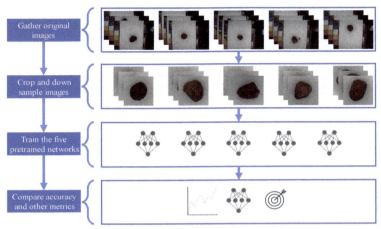

Fig. 2. Flowchart of the overall simulation.

Simulation Results
The results of the simulation provide an overview of the performances of the various networks. MATLAB 2023b along with the built-in Deep Learning Toolbox were used to train, validate, and test the networks. The networks were trained on a Lenovo IdeaPad Gaming 3 laptop. The networks were specifically trained on the GPU only. The GPU on the laptop is Nvidia GeForce RTX 4050 (6 GB). The CPU is an AMD Ryzen 7-7735HS. The RAM is 16 GB. As the networks are pretrained, the training does not need to be as extensive because the previous layers have already learned the detailed features. The epochs have been set to 10 with a batch size of 32 images. The validation frequency has been set to every two iterations. Cross-entropy loss was used with a train-test-validation split of 72-18-10, respectively. The learning rate for all layers except the last layer has been set to 0.0001. For the final layer, it has been set to 10 for both the weights and biases. For augmenting the images, random X and Y-reflections were applied. Also, random X and Y-translation has been applied varying from -30 to 30 pixels. Table 1 summarizes the parameters used for the training of the network.

Table 1. Training parameters for the CNNs.

Training Parameters	Setting
Epochs	10
Mini Batch Size	32
Total Iterations	480
Validation Frequency (Iterations)	2
Train-Val-Test Split	72-18-10
Loss Function	Cross-entropy
Overall Learning Rate	0.0001
Final Layer Weight and Bias Learning Rate	10
Image Augmentation	X-reflection
	Y-reflection
	X-translation (30 to -30 pixels)
	Y-translation (30 to -30 pixels)

Table 2 summarizes the results of the testing for all the models. The primary metric is the accuracy. However, as the task is classification, other metrics such as precision, recall and F1 score are also of great significance. The training time is also included in the table. The results on the test set indicate that SqueezeNet outperforms the other networks. The overall accuracy of SqueezeNet is 92%, with precision, recall, and F1 score of 92.67%, 92%, and 92.33%, respectively. In the second place was ShuffleNet with an accuracy of 89.14%, with precision, recall, and F1 score of 89.41%, 89.14%, and 89.28%, respectively. EfficientNet recorded the lowest accuracy at 82.86%, with precision, recall, and F1 score of 83.10%, 82.86%, and 82.98%, respectively. In terms of the training time, SqueezeNet was also the quickest, taking 9 minutes and 17 seconds to train. GoogLeNet was the second at 14 minutes and 46 seconds. EfficientNet took the longest with 34 minutes and 17 seconds.

Table 2. Summary of the results of testing and training time.

CNN	Accuracy (%)	Precision (%)	Recall (%)	F1 Score (%)	Training Time (mm:ss)
SqueezeNet	92	92.67	92	92.33	09:17
ShuffleNet	89.14	89.41	89.14	89.28	15:20
MobileNet	87.43	87.65	87.43	87.54	21:05
GoogLeNet	85.14	85.68	85.14	85.41	14:46
EfficientNet	82.86	83.10	82.86	82.98	34:17

Fig. 3 illustrates the confusion matrix for SqueezeNet in the test set. The vertical axis is the actual class. The horizontal axis is the predicted class. The diagonal elements represent the correct predictions while the off-diagonal elements represent the misclassified predictions. The outer horizontal percentages give the precision for each class, whereas the vertical percentages give the recall for each class.

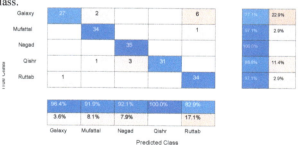

Fig. 3. Confusion matrix for SqueezeNet in the Test set.

Fig. 4 illustrates how the accuracy of the SqueezeNet evolved as the training and validation iterations progress. The number of iterations for the 10 epochs is 480 (for batch size of 32 images). As can be seen, the model converges quickly as it is a pretrained network. Fig. 5 illustrates the loss vs. the iterations. The loss will decrease as the iterations progress. Accuracy and loss are inversely related. Therefore, as accuracy increases, loss decreases.

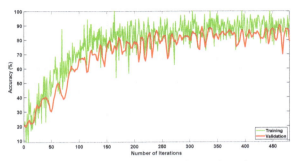

Fig. 4. Accuracy of SqueezeNet vs. the number of iterations.

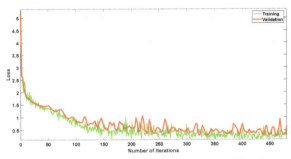

Fig. 5. Loss of SqueezeNet vs. the number of iterations.

Conclusion
Fruit classification is an important area of research for industries as they proceed towards automating the classification process. Various researches have been published in this area using various fruit datasets. However, most studies either include a single model or train a model from a scratch. Comparisons between various papers on different models are usually not homogeneous in nature as they are trained on different datasets and/or different hardware which can significantly affect the results. In this study, we used transfer learning on five popular pretrained CNNs to classify five subcategories of Sukkari dates. We used the same dataset of 1,689 regular RGB images and run the simulation on the same hardware for an even comparison. Comparing the results of the training for the CNNs, it is observed that SqueezeNet performs the best in terms of classification accuracy and the training time. The overall testing accuracy of SqueezeNet is 92% and took 9 minutes and 17 seconds to train on Nvidia GeForce RTX 4050 (6 GB) GPU. The second highest accuracy recorded was the ShuffleNet with a testing accuracy of 89.14%, whereas the second fastest in terms of training time was the GoogLeNet with a training time of 14 minutes and 46 seconds. For the future, we plan to perform further studies on how to improve the accuracy of these models and include other types of date fruits.

References
[1]   I. Goodfellow, Y. Bengio, and A. Courville, Deep Learning. MIT Press, 2016.
[2]   B. Coppin, Artificial intelligence illuminated. Jones & Bartlett Learning, 2004.

[3] M. I. Jordan and T. M. Mitchell, "Machine learning: Trends, perspectives, and prospects," Science (1979), vol. 349, no. 6245, pp. 255–260, 2015. https://doi.org/10.1126/science.aaa8415
[4] Y. LeCun, Y. Bengio, and G. Hinton, "Deep learning," Nature, vol. 521, no. 7553, pp. 436–444, 2015. https://doi.org/10.1038/nature14539
[5] E. Rachmawati, I. Supriana, and M. L. Khodra, "Toward a new approach in fruit recognition using hybrid RGBD features and fruit hierarchy property," in 2017 4th International Conference on Electrical Engineering, Computer Science and Informatics (EECSI), 2017, pp. 1–6. https://doi.org/10.1109/EECSI.2017.8239110
[6] Y. Tao and J. Zhou, "Automatic apple recognition based on the fusion of color and 3D feature for robotic fruit picking," Comput Electron Agric, vol. 142, pp. 388–396, 2017. https://doi.org/10.1016/j.compag.2017.09.019
[7] J. Naranjo-Torres, M. Mora, R. Hernández-Garcia, R. J. Barrientos, C. Fredes, and A. Valenzuela, "A review of convolutional neural network applied to fruit image processing," Applied Sciences, vol. 10, no. 10, p. 3443, 2020. https://doi.org/10.3390/app10103443
[8] Y. LeCun, L. Bottou, Y. Bengio, and P. Haffner, "Gradient-based learning applied to document recognition," Proceedings of the IEEE, vol. 86, no. 11, pp. 2278–2324, 1998. https://doi.org/10.1109/5.726791
[9] Y. Guo and others, "Deep learning for visual understanding." Neurocomput, 2017.
[10] M. D. Zeiler and R. Fergus, "Visualizing and understanding convolutional networks," in Computer Vision–ECCV 2014: 13th European Conference, Zurich, Switzerland, September 6-12, 2014, Proceedings, Part I 13, 2014, pp. 818–833. https://doi.org/10.1007/978-3-319-10590-1_53
[11] R. Singh and S. Balasundaram, "Application of extreme learning machine method for time series analysis," International Journal of Computer and Information Engineering, vol. 1, no. 11, pp. 3407–3413, 2007.
[12] A. Krizhevsky, I. Sutskever, and G. E. Hinton, "Imagenet classification with deep convolutional neural networks," Adv Neural Inf Process Syst, vol. 25, 2012.
[13] O. Russakovsky et al., "Imagenet large scale visual recognition challenge," Int J Comput Vis, vol. 115, pp. 211–252, 2015. https://doi.org/10.1007/s11263-015-0816-y
[14] Y. Lu, "Food image recognition by using convolutional neural networks (cnns)," arXiv preprint arXiv:1612.00983, 2016.
[15] Y.-D. Zhang et al., "Image based fruit category classification by 13-layer deep convolutional neural network and data augmentation," Multimed Tools Appl, vol. 78, pp. 3613–3632, 2019. https://doi.org/10.1007/s11042-017-5243-3
[16] J. Steinbrener, K. Posch, and R. Leitner, "Hyperspectral fruit and vegetable classification using convolutional neural networks," Comput Electron Agric, vol. 162, pp. 364–372, 2019. https://doi.org/10.1016/j.compag.2019.04.019
[17] S. W. Chen et al., "Counting apples and oranges with deep learning: A data-driven approach," IEEE Robot Autom Lett, vol. 2, no. 2, pp. 781–788, 2017. https://doi.org/10.1109/LRA.2017.2651944
[18] S. Bargoti and J. Underwood, "Deep fruit detection in orchards," in 2017 IEEE international conference on robotics and automation (ICRA), 2017, pp. 3626–3633. https://doi.org/10.1109/ICRA.2017.7989417
[19] K. Albarrak, Y. Gulzar, Y. Hamid, A. Mehmood, and A. B. Soomro, "A deep learning-based model for date fruit classification," Sustainability, vol. 14, no. 10, p. 6339, 2022. https://doi.org/10.3390/su14106339
[20] M. Faisal, F. Albogamy, H. Elgibreen, M. Algabri, and F. A. Alqershi, "Deep learning and computer vision for estimating date fruits type, maturity level, and weight," IEEE Access, vol. 8, pp. 206770–206782, 2020. https://doi.org/10.1109/ACCESS.2020.3037948

[21] A. Alsirhani, M. H. Siddiqi, A. M. Mostafa, M. Ezz, and A. A. Mahmoud, "A novel classification model of date fruit dataset using deep transfer learning," Electronics (Basel), vol. 12, no. 3, p. 665, 2023. https://doi.org/10.3390/electronics12030665
[22] D. Chaudhari and S. Waghmare, "Machine vision based fruit classification and grading—a review," in ICCCE 2021: Proceedings of the 4th International Conference on Communications and Cyber Physical Engineering, 2022, pp. 775–781. https://doi.org/10.1007/978-981-16-7985-8_81
[23] T. Shoshan, A. Bechar, Y. Cohen, A. Sadowsky, and S. Berman, "Segmentation and motion parameter estimation for robotic Medjoul-date thinning," Precis Agric, vol. 23, no. 2, pp. 514–537, 2022. https://doi.org/10.1007/s11119-021-09847-2
[24] K. M. Alresheedi, S. Aladhadh, R. U. Khan, and A. M. Qamar, "Dates Fruit Recognition: From Classical Fusion to Deep Learning.," Computer Systems Science & Engineering, vol. 40, no. 1, 2022. https://doi.org/10.32604/csse.2022.017931
[25] A. Magsi, J. A. Mahar, S. H. Danwar, and others, "Date fruit recognition using feature extraction techniques and deep convolutional neural network," Indian J Sci Technol, vol. 12, no. 32, pp. 1–12, 2019. https://doi.org/10.17485/ijst/2019/v12i32/146441
[26] A. Nasiri, A. Taheri-Garavand, and Y.-D. Zhang, "Image-based deep learning automated sorting of date fruit," Postharvest Biol Technol, vol. 153, pp. 133–141, 2019. https://doi.org/10.1016/j.postharvbio.2019.04.003
[27] B. D. Pérez-Pérez, J. P. Garcia Vazquez, and R. Salomón-Torres, "Evaluation of convolutional neural networks' hyperparameters with transfer learning to determine sorting of ripe medjool dates," Agriculture, vol. 11, no. 2, p. 115, 2021. https://doi.org/10.3390/agriculture11020115
[28] H. Altaheri, M. Alsulaiman, and G. Muhammad, "Date fruit classification for robotic harvesting in a natural environment using deep learning," IEEE Access, vol. 7, pp. 117115–117133, 2019. https://doi.org/10.1109/ACCESS.2019.2936536
[29] M. F. Nadhif and S. Dwiasnati, "Classification of Date Fruit Types Using CNN Algorithm Based on Type," MALCOM: Indonesian Journal of Machine Learning and Computer Science, vol. 3, no. 1, pp. 36–42, 2023. https://doi.org/10.57152/malcom.v3i1.724
[30] F. N. Iandola, S. Han, M. W. Moskewicz, K. Ashraf, W. J. Dally, and K. Keutzer, "SqueezeNet: AlexNet-level accuracy with 50x fewer parameters and< 0.5 MB model size," arXiv preprint arXiv:1602.07360, 2016.
[31] C. Szegedy et al., "Going deeper with convolutions," in Proceedings of the IEEE conference on computer vision and pattern recognition, 2015, pp. 1–9. https://doi.org/10.1109/CVPR.2015.7298594
[32] M. Tan and Q. Le, "Efficientnet: Rethinking model scaling for convolutional neural networks," in International conference on machine learning, 2019, pp. 6105–6114.
[33] X. Zhang, X. Zhou, M. Lin, and J. Sun, "Shufflenet: An extremely efficient convolutional neural network for mobile devices," in Proceedings of the IEEE conference on computer vision and pattern recognition, 2018, pp. 6848–6856. https://doi.org/10.1109/CVPR.2018.00716
[34] M. Sandler, A. Howard, M. Zhu, A. Zhmoginov, and L.-C. Chen, "Mobilenetv2: Inverted residuals and linear bottlenecks," in Proceedings of the IEEE conference on computer vision and pattern recognition, 2018, pp. 4510–4520. https://doi.org/10.1109/CVPR.2018.00474
[35] "Pretrained Deep Neural Networks." Accessed: Feb. 25, 2024. [Online]. Available: https://www.mathworks.com/help/deeplearning/ug/pretrained-convolutional-neural-networks.html

Bidding optimization for hydrogen production from an electrolyzer

Nouf M. ALMUTAIRY[1,a *], Ali T. ALAWAMI[2,b]

[1]Department of Electrical Engineering, King Fahd University of Petroleum and Minerals, Saudi Arabia

[2]Department of Electrical Engineering, King Fahd University of Petroleum and Minerals, Saudi Arabia

[a] nouf459@gmail.com, [b] aliawami@kfupm.edu.sa

Keywords: Hydrogen Production, Electrolyzer, Bidding Optimization, Sustainability

Abstract. This paper presents a comprehensive study on the bidding optimization for hydrogen production from an electrolyzer, focusing on a single day comprising 24 hours. With the rising demand for clean energy sources, the research aims to optimize profitability and efficiency in hydrogen production. The primary objective is to maximize profit while ensuring the fulfillment of the targeted hydrogen production by the end of the day. The optimization formulation incorporates electrolyzer maintenance, electricity, and water consumption costs. The model considered two cases with different electrolyzer efficiency values to optimize power usage, allowing for a comprehensive analysis of hydrogen production optimization. Ramping limits are imposed to maintain power system stability and reliability, preventing sudden fluctuations. By solving the formulated equations and considering factors such as energy and water prices, the research findings demonstrate the effectiveness of the bidding optimization approach in optimizing resource utilization and maximizing profit. Notably, the model successfully achieves the targeted hydrogen production by the end of the day while maximizing profit. This research contributes valuable insights into the bidding optimization process for hydrogen production, highlighting the potential for economic and sustainable hydrogen generation from electrolyzers.

Introduction

Climate change necessitates a shift to clean, sustainable energy sources to mitigate its environmental impact. Hydrogen, as a clean and efficient energy carrier, has garnered attention in the energy industry. However, traditional hydrogen production from fossil fuels without carbon capture contributes to greenhouse gas emissions. The shift towards hydrogen produced from fossil fuels with carbon capture, utilization, and storage (CCUS) offers a viable alternative, considering the high carbon production from natural gas and coal sources. This transition is crucial in combating climate change. Hydrogen can contribute to a resilient and sustainable energy future by utilizing alternative and cleaner production methods and diversifying its sources. Additionally, hydrogen can be utilized in new applications and complement electricity use, enhancing its potential. Today, more countries invest in hydrogen technologies since they recognize their importance for the future of energy. Austria announced that as part of the Austrian Climate and Energy Strategy for 2030, a hydrogen strategy based on renewable electricity would be developed. Even in Saudi Arabia, Saudi Aramco and Air Products announced their plans to construct Saudi Arabia's first hydrogen refueling station. Thus, this paper focuses on the bidding optimization for hydrogen production from an electrolyzer. [1]

To produce hydrogen as an energy carrier, an electrolyzer is required. This device utilizes electricity to split water into hydrogen and oxygen through an electrochemical process called electrolysis. The electrolyzer connects to an external circuit to provide the required electric current for electrolysis. This process makes water break down into its constituent elements of hydrogen and oxygen gases. The electrolyzer consists of an anode and a cathode. At the anode, a process called oxidation takes place. Water molecules near the anode lose electrons and form oxygen gas (O2) and positively charged hydrogen ions (H+). The cathode attracts hydrogen ions (H+) from the electrolyte. At the cathode, the hydrogen ions gain electrons from the external circuit and combine to form hydrogen gas (H2) in a reduction process. [2]

Bidding optimization is identifying the optimal bidding strategy in energy markets to maximize profits. Optimizing bidding for hydrogen production from an electrolyzer is crucial for a greener future and for addressing carbon emissions. Developing bidding strategies considering market prices for hydrogen, electricity, and electrolyzer costs can maximize revenue and profitability as hydrogen production gains traction. Hydrogen plants play a vital role in the clean energy sector, making bid optimization for electrolyzer-based hydrogen production a focal point. With the growing interest in hydrogen as a sustainable energy alternative, bid optimization can significantly contribute to the global transition toward a greener future [3].

Several studies have focused on hydrogen production from electrolyzer. Study [4] provides an overview of water electrolysis-based systems for hydrogen production, particularly those utilizing hybrid/solar/wind energy sources. The article emphasizes the importance of hydrogen as a clean energy carrier and discusses system configurations, electrolyzer types, catalysts, and energy sources. Study [5] delves into the role of catalysts in enhancing the efficiency and performance of electrolyzers for hydrogen production, exploring different types of catalysts and recent advancements in catalyst design. In [6], the focus is on a membrane-based seawater electrolyzer that directly splits seawater into hydrogen and oxygen using a proton-conducting membrane.

Recent studies have focused on optimization for hydrogen production from electrolyzer. In a study [7], this paper discusses optimizing a high-temperature electrolysis system for hydrogen production, considering the degradation of cell materials. It explores the factors causing degradation and investigates the influence of operating conditions on the degradation process. The paper also proposes operation strategies to balance hydrogen production efficiency and the lifespan of the stack. In the study [8], this paper's primary emphasis is modeling and enhancing an alkaline water electrolysis system employed for hydrogen generation. The study extensively covers the design and operational aspects of the electrolyzer, along with an exploration of diverse optimization methodologies. This research aims to improve hydrogen production efficiency and cost-effectiveness through electrolysis.

Optimizing bidding strategies is crucial for maximizing profits in electrolyzer-based hydrogen production, and it has gained significant research attention. In the study [9], an optimal bidding strategy is developed for hydrogen production from electrolyzers in renewable energy systems. The proposed mathematical model considers uncertainties in renewable energy generation and electricity prices to maximize profit. The results demonstrate the effectiveness of the strategy in maximizing profit. In [10], the potential participation of virtual power plants (VPPs) with hydrogen energy storage in multi-energy markets is discussed. The study highlights the role of hydrogen storage in VPPs and presents a two-layer optimization model considering resource complementarity and external market bidding strategies.

In addition to previous studies on hydrogen production from electrolyzers, it is essential to understand the motivation behind this research. Electrolyzers are electric loads that require electric energy to produce hydrogen (H2). However, hydrogen production plants' storage capacity allows them to schedule their operations based on electricity prices. The lack of an optimization framework integrating bidding prices presents a research gap in hydrogen production from electrolyzers. This aspect is crucial in optimizing the operational scheduling of electrolyzers. Therefore, the objective is to integrate bidding prices into the optimization process and identify the most effective strategy for achieving maximum profit while meeting the targeted hydrogen production quota. The study will utilize the GAMS software for accurate and reliable results.

Methodology

A. Objective function

The study utilizes a bidding optimization model to maximize the profit from hydrogen production. This bidding optimization model is achieved through (1), which explains that the optimization model will maximize the profit. In this equation, F represents the profit, R represents the revenue from hydrogen production, and C represents the cost of the electrolyzer needed to produce hydrogen.

$$\max F = R - C \tag{1}$$

The analysis assumes that the demand for hydrogen is already established, which allows for revenue calculation using (2). As reference [11] indicates, the selling price has been set at $11 per kilogram. The study target for hydrogen production is 1000 kilograms, resulting in a total revenue of 11000 dollars.

$$R = \text{selling price} * \text{quaintity of hydrogen target} \tag{2}$$

Eq. 3 represents the cost function for the electrolyzer. In this equation, C_M represents to the maintenance cost for the electrolyzer, as derived from the reference paper [10]. According to formula 4, C_M can be calculated as 2% of the electrolyzer capital cost.

Eq. 5 to 6 illustrate the cost functions for electricity and water consumption, respectively. These equations quantify the expenses associated with utilizing an electrolyzer to produce hydrogen.

$$C = C_E + C_W + C_M \tag{3}$$
$$C_M = 2\% \times \text{capital cost} \tag{4}$$
$$C_M = 2\% \times 1765 \text{ \$/kW} = 35.3 \text{ \$/kW}$$

Eq. 5 defines the cost function for electricity consumption. In this equation, P(t) represents the decision variable denoting the electricity required at each time interval, covering the 24-hour duration of the study. The variable Energy price(t) represents the corresponding energy price for each specific time interval. The data used for the energy price, expressed in $/kWh, was sourced from Norway on December 5, 2023 [12].

$$C_E = \sum_{t=1}^{24} P(t) * \text{energy price}(t) \tag{5}$$

Eq. 6 defines the cost function for water consumption. In this equation, W(t) represents the decision variable indicating the water required at each time interval within the 24-hour study period. The variable water price(t) represents the corresponding water price for each specific time interval. The water price data, equal to 0.00669 $/kg, is sourced from Norway [13].

$$C_W = \sum_{t=1}^{24} W(t) * \text{water price}(t) \tag{6}$$

B. Equality constraints:

1. Constraint for Electricity Consumption:

The model considers two cases for electricity consumption related to the efficiency of the electrolyzer. The electrolyzer efficiency is set at 70% in the first case, as referenced in [9]. Eq. 7 establishes the relationship between the power input to the electrolyzer and the resulting hydrogen production. The decision variable H(t) represents the hydrogen produced at each time interval. In the second case, the efficiency is assumed to be 80%. Analyzing these two cases allows the model to explore the impact of different electrolyzer efficiencies on electricity consumption and hydrogen production.

$$P(t) = \frac{50 * H(t)}{\text{efficiency of electrolyzer}} \qquad (7)$$

2. **Constraint for Water Consumption:**
The model includes a constraint on water consumption, which defines the relation between the water consumption and hydrogen production decision variables, shown in Eq.8.

Additionally, the water needed in kg to produce 1 kg of hydrogen is 9 kg [15].

$$W(t) = 9 * H(t) \qquad (8)$$

3. **Constraint for the hydrogen tank:**
Eq. 9 represents a constraint that governs the hydrogen storage in the tank at a specific time t. It ensures that the amount of hydrogen stored in the tank at time t is determined by the sum of the hydrogen storage in the tank at the previous time and the amount of hydrogen produced during time t. This constraint captures the dynamics of the hydrogen storage system, where the current storage level depends on the previous storage level and the hydrogen production during the current period.

$$\text{TankStore}(t) = \text{TankStore}(t-1) + H(t) \qquad (9)$$

C. **Inequality constraints:**
1. **Maximum and Minimum Consumption:**
The proposed model incorporates certain constraints to ensure electricity and water consumption remain within certain limits. Specifically, the model uses inequality constraints to set the upper and lower limits on electricity consumption as shown in Eq. 10 and water consumption as shown in Eq. 11. The model's maximum allowable electricity consumption (P_{max}) is 6000 kWh, while the minimum allowable consumption (P_{min}) is zero. Similarly, the maximum allowable water consumption (W_{max}) is assumed to be 3000 kg, while the minimum allowable consumption (W_{min}) is also zero.

$$0 \leq P(t) \leq 6000 \qquad (10)$$

$$0 \leq W(t) \leq 3000 \qquad (11)$$

2. **Constraint for the hydrogen target:**
Eq. 12 represents a target constraint that ensures the cumulative sum of H(t) over all periods is greater than or equal to the target hydrogen value. This constraint ensures that the total hydrogen produced throughout all periods meets or exceeds the desired target value.

$$\text{Target} \leq \sum_{t=1}^{24} H(t) \qquad (12)$$

3. **Ramping constraints:**
Ramping limits play a crucial role in maintaining power system stability and reliability. These limits constrain the rate of change of electricity, hydrogen production, and water consumption rates, preventing sudden and excessive fluctuations. Ramping limits help optimize resource utilization and mitigate the risk of disruptions or imbalances by ensuring a gradual and controlled adjustment of these variables over consecutive periods. The constraints from Eq. 13 to 15 define allowable changes in electricity consumption, hydrogen production rate, and water consumption

rate between periods. They set bounds on the differences between current and previous values, preventing rapid increases or decreases. Enforcing these ramping limits supports grid stability and enhances overall system performance.

$$-355.5 \leq P(t) - P(t-1) \leq 355.5 \quad (13)$$
$$-25 \leq H(t) - H(t-1) \leq 25 \quad (14)$$
$$-100 \leq W(t) - W(t-1) \leq 100 \quad (15)$$

Results and discussion
The bidding optimization model for hydrogen production from the electrolyzer was solved using the General Algebraic Modeling System (GAMS), with the solver of choice being a linear Programming (LP) solver. This approach efficiently optimized the model, determining optimal values for decision variables and constraints. The results provided insights into maximizing profit from hydrogen production. For each period, optimal values for decision variables, including electricity consumption, water consumption, and hydrogen production, were determined. To optimize power usage, the model considered two cases with different electrolyzer efficiency values.

A. Case 1: Electrolyzer efficiency equals 70%.
In this scenario, the electricity consumption formula considers an electrolyzer efficiency of 70%. Based on Table 1, the objective function aims to maximize profit and minimize total cost. Table 2 displays the results for H(t), W(t), P(t), and Tank Storage(t), satisfying all constraints. The table confirms that the hydrogen target is achieved in the hydrogen tank at t=24. The results demonstrate the successful consideration of the dynamics of the hydrogen storage system in the bidding optimization model. Eq. 10 ensures that the amount of hydrogen stored in the tank at each time TankStore(t) is accurately calculated based on the previous storage level TankStore(t-1) and the current hydrogen production (H(t)). This enables informed decision-making for optimal hydrogen production scheduling.

Table 1. Output parameters in dollars for case 1.

Profit	Revenue	Total cost	C_E	C_W	C_M
$1497.799	$11000	$9502.201	$9406.691	$60.210	$35.500

The model successfully satisfied the inequality constraints (Eq. 11-12) on electricity and water consumption, ensuring they remained within the specified limits. Additionally, the optimized hydrogen production (H(t)) adhered to the ramping constraint (Eq. 15), maintaining a controlled rate of change between consecutive periods. Figures 1 and 2, corresponding to Table 2, illustrate the 24-hour trends in electricity usage and hydrogen tank storage. Figure 1 shows that P(t) steadily increases as time progresses and then slightly decreases due to the increase in energy prices. Subsequently, hydrogen production increases, leading to simultaneous increases in P(t) and W(t). The initial hydrogen production is recorded as 4.977 kg, reaching a maximum value of 77.117 kg after 24 hours.

Table 2. Hourly output parameters for case 1.

Time [hr]	1	2	3	4	5	6	7	8	9	10	11	12
H(t)	4.98	9.95	14.93	19.91	24.89	29.86	34.84	39.82	44.79	49.77	52.23	47.26
Tank storage (t)	4.98	14.93	29.86	49.77	74.66	104.5	139.36	179.17	223.97	273.74	325.97	373.22
P(t)	355.50	711.00	1066.50	1422.00	1777.50	2133.00	2488.50	2844.00	3199.50	3555.00	3730.83	3375.33
W(t)	44.79	89.59	134.38	179.17	223.97	268.76	313.55	358.34	403.14	447.93	470.08	425.29
Time [hr]	13	14	15	16	17	18	19	20	21	22	23	24
H(t)	42.28	37.30	32.32	37.30	42.28	47.26	52.23	57.21	62.19	67.16	72.14	77.12
Tank storage (t)	415.50	452.80	485.12	522.42	564.70	611.96	664.19	721.40	783.58	850.74	922.88	1000.000
P(t)	3019.83	2664.33	2308.83	2664.33	3019.83	3375.33	3730.83	4086.33	4441.83	4797.33	5152.83	5508.33
W(t)	380.49	335.71	290.91	335.71	380.50	425.29	470.08	514.88	559.67	604.46	649.26	694.05

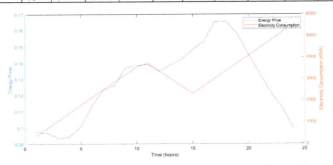

Figure 1. Electricity consumption and energy price for case 1.

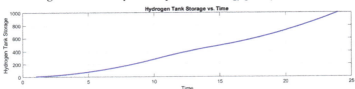

Figure 2. Hydrogen Tank Storage for case 1.

B. Case two : electrolyzer efficiency equals 80%.

In this scenario, the electricity consumption equation considers an electrolyzer efficiency of 80%. According to the results presented in Table 3, the objective function achieved in this case, considering both maximum profit and minimum total cost, shows improvement compared to case 1. The profit is higher, while the total cost of the electrolyzer is reduced.

Table 3. Output parameters in dollars for case 2.

Profit	Revenue	Total cost	C_E	C_W	C_M
$ 2743.899	$11000	$ 8256.101	$ 8160.591	$ 60.210	$35.500

The results for $H(t), W(t), P(t)$, and Tank Storage(t) are shown in Table 4. All the constraints were satisfied for all the variables. Fig. 3 visualizes the trends in electricity consumption with energy prices over time. As depicted in the figure, $P(t)$ steadily increases as time progresses until reaching a stable point. Subsequently, it decreases due to the rise in energy prices. However, when the energy prices start decreasing, $P(t)$ steadily increases again. In this scenario, hydrogen production ranges between 5.688 kg and 78.043 kg. Notably, the hydrogen production range is more significant than in case 1, highlighting the improved flexibility and variability in hydrogen production achieved through the optimization framework.

Table 4. Hourly output parameters for case 2.

Time [hr]	1	2	3	4	5	6	7	8	9	10	11	12
H(t)	5.69	11.38	17.06	22.75	28.44	34.13	39.82	45.50	51.19	55.29	49.60	43.92
Tank storage (t)	5.69	17.06	34.13	56.88	85.32	119.45	159.26	204.77	255.96	311.25	360.85	404.77
P(t)	355.50	711.00	1066.50	1422.00	1777.50	2133.00	2488.50	2844.00	3199.50	3455.67	3100.17	2744.67
W(t)	51.19	102.38	153.58	204.77	255.96	307.15	358.34	409.54	460.73	497.62	446.42	395.23
Time [hr]	13	14	15	16	17	18	19	20	21	22	23	24
H(t)	38.23	32.54	26.85	32.54	38.23	43.92	49.60	55.29	60.98	66.67	72.36	78.04
Tank storage (t)	442.99	475.53	502.38	534.92	573.15	617.06	666.67	721.96	782.94	849.60	921.96	1000.00
P(t)	2389.17	2033.67	1678.17	2033.17	2389.17	2744.67	3100.17	3455.67	3811.17	4166.67	4522.17	4877.67
W(t)	344.04	292.85	241.66	292.85	344.04	395.23	446.42	497.62	548.81	600.00	651.19	702.38

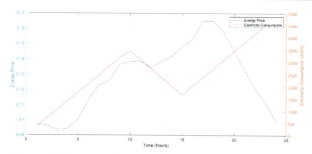

Figure 3: Electricity Consumption and Energy price for case 2.

Conclusion

In conclusion, this research paper has presented a comprehensive study on optimizing the bidding process for hydrogen production through an electrolyzer. The primary objective was to develop an optimization framework that maximizes profitability while meeting the targeted hydrogen production quota. Two cases were considered: Case 1, which incorporated an electrolyzer efficiency of 70% in the electricity consumption equation, and Case 2, where electrolyzer efficiency was 80%. For Case 1, hydrogen production started at 4.977 kg and reached 77.117kg after 24 hours. However, Case 2, which considered 80% electrolyzer efficiency, yielded improved outcomes regarding maximum profit and minimum total cost. Hydrogen production ranged from 5.688 kg to 78.043 kg., with higher profitability achieved. These results emphasize the importance of considering various factors, such as electrolyzer efficiency, in the bidding optimization process. Overall, this research emphasizes the importance of bidding optimization in hydrogen production, offering valuable insights for the energy industry's transition towards sustainable and economically viable hydrogen generation.

References

[1] IEA. (2019). The Future of Hydrogen, Report Prepared by the IEA for the G20, Japan. Seizing Today's Opportunities.

[2] What is an electrolyzer and what is it used for? | Accelera. Available at: https://www.accelerazero.com/news/what-is an-electrolyzer-and-what-is-it-used-for (Accessed: 10 November 2023).

[3] Kim, M., & Kim, J. (2016). Optimization model for the design and analysis of an integrated renewable hydrogen supply (IRHS) system: Application to Korea's hydrogen economy. International Journal of Hydrogen Energy, 41(38), 16613-16626. https://doi.org/10.1016/j.ijhydene.2016.07.079

[4] Nasser, M., Megahed, T. F., Ookawara, S., & Hassan, H. (2022). A review of water electrolysis–based systems for hydrogen production using hybrid/solar/wind energy systems. Environmental Science and Pollution Research, 29(58), 86994-87018. https://doi.org/10.1007/s11356-022-23323-y

[5] Wang, S., Lu, A., & Zhong, C. J. (2021). Hydrogen production from water electrolysis: role of catalysts. Nano Convergence, 8, 1-23. https://doi.org/10.1186/s40580-021-00254-x

[6] Xie, H., Zhao, Z., Liu, T., Wu, Y., Lan, C., Jiang, W., ... & Shao, Z. (2022). A membrane-based seawater electrolyser for hydrogen generation. Nature, 612(7941), 673-678. https://doi.org/10.1038/s41586-022-05379-5

[7] Yuan, J., Li, Z., Yuan, B., Xiao, G., Li, T., & Wang, J. Q. (2023). Optimization of High-Temperature Electrolysis System for Hydrogen Production Considering High-Temperature Degradation. Energies, 16(6), 2616. https://doi.org/10.3390/en16062616

[8] K. Stewart et al., "Modeling and Optimization of an Alkaline Water Electrolysis for Hydrogen Production," 2021 IEEE Green Energy and Smart Systems Conference (IGESSC), Long Beach, CA, USA, 2021, pp. 1-6. https://doi.org/10.1109/IGESSC53124.2021.9618679

[9] Morton, E. M., Deetjen, T. A., & Goodarzi, S. (2023). Optimizing hydrogen production capacity and day ahead market bidding for a wind farm in Texas. International Journal of Hydrogen Energy, 48(46), 17420-17433. https://doi.org/10.1016/j.ijhydene.2022.12.354

[10] Zhang, W., Shen, Y., Wang, X., Li, M., Ren, W., Xu, X., & Zhang, Y. Research on multi-market strategies for virtual power plants with hydrogen energy storage. Frontiers in Energy Research, 11, 1260251. https://doi.org/10.3389/fenrg.2023.1260251

[11] Economics SGH2 Energy. Available at : https://www.sgh2energy.com/economics (Accessed: 05 December).

[12] Market data | Nord Pool. Available at: https://www.nordpoolgroup.com/en/Market-data1/Dayahead/Area-Prices/de-lu/hourly/?view=table (Accessed: 05 December 2023).

[13] Salas, E.B. (2023) Europe: Tap water prices in select cities , Statista. Available at: https://www.statista.com/statistics/1232847/tap-water-prices-in-selected-european-cities/ (Accessed: 5 December 2023).

[14] KURRER, C. M. (2020). The potential of hydrogen for decarbonising steel production.

[15] Saulnier, R., Minnich, K., & Sturgess, P. K. (2020). Water for the hydrogen economy

Li-ion batteries life cycle from electric vehicles to energy storage

Muhammad RASHID[1,a *], Muhammad SHEIKH[1,b] and Sheikh REHMAN[3,c]

[1]WMG, The University of Warwick, Coventry, UK

[2]School of Computing, Engineering and Digital Technologies. Teesside University, Middlesborough, UK

[a]R.Muhammad.1@warwick.ac.uk, [b]muhammad.sheikh@warwick.ac.uk, [c]S.Rehman@tees.ac.uk

Keywords: Li-ion Battery, State of Health (SOH), Retired Batteries, 2nd Life Application, Energy Storage Systems (ESS)

Abstract. Vehicle electrification is an emerging solution to reduce fossil fuel dependence and the environmental pollution caused by automobile emissions. Electric vehicles (EVs) are powered by Li-ion batteries which degrade with use and time, and once their state of health (SOH, ratio of current capacity to the initial capacity) reaches 80% they retire from the EVs and need a replacement. In this study, battery degradation behaviour has been investigated and demonstrated under different electrical and thermal loading conditions. A different rate of cell degradation has been observed with different environmental and electrical loading conditions. The rate of degradation of the cells is higher at low temperatures and at high current charging conditions. Additionally, it has been demonstrated that the temperature of the cells within a battery module is different across the 6S2P battery module which would be significantly higher in the case of a bigger battery module. Hence for the potential second-life applications of the retired electric vehicle batteries, knowing the correct cell SOH is highly essential to grouping them which will lead to optimized use of this battery in 2nd life applications.

Introduction

Government and policymakers are continuously promoting vehicle electrification to reduce fossil fuel dependency and minimize carbon emissions [1]. Li-ion batteries are the key contender to power EVs and HEVs (hybrid electric vehicles) due to their high gravimetric and volumetric energy density. However, the bottleneck with LIB is its endurance with time [1-3] which is caused by the capacity degradation due to repeated charge/discharge and storage. This degradation leads to a reduction in the driving range of the vehicle and eventually makes the LIBs incompatible with EVs when they reach 80% SOH (varying with the EV manufacturer and government regulations) [2-5]. At this stage, the battery needs to be replaced which costs significantly to the EV owners and the disposal of the retired batteries is hazardous to the environment [4-5]. The replacement cost of EV batteries could be reduced by selling the batteries to the businesses involved in utilizing the retired batteries in low-power and stationery applications [3,5]. Hence the selection of batteries from the first life and their SOH investigation and finally categorizing and grouping them for the potential second life is an open challenge which needs significant researchers' attention [1-6].

Since the cycle and the shelf life of each cell within a battery pack are dictated by the rate of capacity degradation and operational conditions. In the case of EVs and HEVs, multiple cells are arranged in series, parallel and mixed connections to achieve the desired voltage and current [1,4]. Hence, (i) cells within a battery pack have variations in electrical loading based on the series-parallel architecture of the battery systems and (ii) they would have different temperature conditions based on their location within the pack thermal management system [5-8]. These two variations lead to the uneven degradation of the cells across the battery pack [5-8]. However, the SOH of the pack is specified by the weakest cell [6,7] which has faster degradation due to the nonhomogeneous electrical loading or due to the manufacturing defect caused by slower ionic/charge transport [9,10]. Hence, an understanding of the cell behaviour in parallel to the battery pack is highly needed to pinpoint the root causes of the pack failure. Identifying weak cells within the battery pack can facilitate the replacement of that module/cell containing dead cells and can improve the capacity utilization of the battery pack.

Battery End of Life, Reuse and Recycling
End-of-life (EOL) of EVs and HEVs battery systems are considered once they hit 80% SOH, however, these retired batteries have significant leftover energy [11] which must be utilized by some means otherwise will go to waste [9-13]. An EOL of the EV batteries will strongly impact the economic, environmental, and political dynamics of the nation because the battery raw materials are not evenly accessible across the globe [1-3]. Hence, reuse of the batteries for Energy storage systems (ESS) can minimize that impact by offering significant economic and environmental benefits [9-10]. ESS is an essential factor for the smart grid which helps store energy at times of load shifting and low energy demands and supply at times of high energy demand and saves the cost for the consumers and suppliers. For instance, reusing these batteries in utilities can provide energy at the time of the fluctuations since the cost of the retired batteries ranges from \$38-\$147/kWh as compared to new LIBs ~\$209/kWh [14]. However, the application of the retired batteries needs to be encouraged by the policymakers by providing funds to the new businesses [7,9,14]. Major areas of battery 2nd life application are:
- Low-speed EVs
- Mobile power
- Energy storage for home use
- Backup power and energy storage sites

2nd life application of the batteries can encourage business model innovation which will link transportation with other energy applications [1,4]. These applications could reduce the effective price of EVs and reduce its life cycle impacts on the EV owners. Reuse of the batteries will promote the EVs uses on the road which can also greatly reduce CO_2 emissions and improve air quality using retired batteries in the grid application and UPSs [10]. Since batteries are not simple waste which can be disposed of anywhere, they are electrochemical systems which if not disposed of safely might cause accidents and fire [12]. Therefore, its disposal is also expensive which generates various harmful gases which pollute the environment and atmosphere and this can be minimized by the reuse of the battery systems [2,13-14].

Battery SOH and Capacity Estimation
Battery pack SOH and rate of capacity degradation estimation are essential for the appropriate 2nd life applications or subsequent recycling. For 2nd life applications of the batteries, it is highly desirable to have a similar rated capacity and comparable SOH and, also have a capacity greater than 50% of the original capacity [7-8]. A dissimilarity in cell SOH within a pack leads to incomplete utilization of the battery pack since the battery capacity is defined by the capacity of the weakest cell connected in a series [9]. Since the SOH of the battery pack is determined by the SOH of the weakest cell it is inadequate to maximize the energy utilization of the entire pack [4,6]. Therefore, identification of the cell SOH is highly desirable for the best possible uses, because the parallelly connected cells undergo different currents due to varying internal resistance which leads to different degradation rates [10,11]. Even cells in a series connection can have different SOH and degradation rates. Therefore, the SOH and capacity estimation of the pack without knowing the SOH of the cell is misleading information about the battery systems. For the 2nd life application, cells from the same SOH and internal resistance are recommended even in case of parallel connections [12,13]. There are a variety of SOH assessment techniques reported in the literature, however, either they are time-consuming or are not accurate enough [6-8, 15-18]. The SOH

methods must be cheap and fast enough to reduce the SOH testing apparatus time as well as cost which makes it economical and benign [1,8,9]. Battery SOH is widely assessed by a series of charge/discharge cycles which take several hours to conduct and are hence time-consuming and costly [4,8-9]. However, because of 2nd life application, a fast-screening method with high accuracy is highly desirable to make the reuse of EV batteries sustainable. In this study, battery degradation behavior has been investigated and demonstrated under different electrical and thermal loading conditions. Additionally, a simulation of thermal gradients for a 6S2P battery module has been conducted to demonstrate temperature variation within the cells and across the module.

Test Methodology and Data Collection
To conduct the test, fresh batteries are aged under various loading and operational conditions up to the EOL. This process involves initial electrochemical milling/formation of the cells, followed by a reference capacity test (RCT) within a thermal chamber using a battery cycler at 25°C. To accurately assess the cells' performance, we aged them through mild and aggressive electrochemical cycling and various drive cycles, simulating temperature conditions ranging from 0-25°C. To achieve the required thermal environment, the batteries are placed within an incubator to imitate the real-use scenarios. By regularly conducting RCTs, we monitored the SOH and performance of the cells throughout their first life until the cells reached their EOL. With this approach, battery ageing data was collected and analyzed, which are discussed in the results section.

Results and Discussion
In this section, discussion of the cycling behavior of LGM50 commercial cell is presented. Figure 1 shows the variation in cell capacity with respect to cycling at different temperature conditions which is 0, 10 and 25°C. As can be seen, the cycling behavior of the cell is impacted by the environmental temperature. Three different charging conditions have been used which are 0.3C, 0.5C and 0.7C, however the discharge current is 0.3C for all the cases. Observing the discharge behaviour of the cell it is evident that cell degradation is highly dependent on temperature conditions as well as applied current which indicates the loading condition of the cell. Additionally, for similar cycling/environmental conditions there is significant cell-to-cell variation. Even in similar test conditions cell performance and cell degradation would be different, which indicates that even within the same battery module in a real use case scenario the cells within a battery module will degrade differently just because of different loading conditions. Hence talking about Figure 1(a) cells at 0.3C charge condition for all the temperature conditions have shown less degradation even after 800 cycles compared to 1000 cycles. After 800 cycles cell has lost only 15% of its capacity. Furthermore, before the occurrence of the knee point cell-to-cell variation for the capacity degradation is minimal however after the knee point significant variation between cell-to-cell is observed which is highest at 0°C and 10°C.

After further analyzing the capacity variation of the cell with cycling at 0.7C charge condition for the similar temperature conditions we can see that cells lose their capacity rapidly for all the temperature conditions. It is also observed that the rate of degradation is highest in case of 0°C as can be seen in the figure 1(b). For this condition the degradation behavior at 25°C is significantly less as compared to 0°C and 10°C. Cell lose 20% of SoH after 200 cycles however in case of 10°C and 0°C cell lose 50% of SoH within 70 and 80 cycles respectively. Aging these cells further even after losing 20% of its SoH, we can see the cells lose around 60% and 70% of SoH within 200 cycles for 0°C and 10°C, however at 25°C cell can go up to 300 cycles before it loses 70% SoH. This degradation at low temperature is attributed to the slower reaction kinetics and transport limitations for ions and electrons and lower diffusivity of lithium ions within the cathode and anode.

Further analysing the resistance rise in the cells with cycling condition at 0.3C charge for all the temperature conditions we can see that the rate of rise of cell resistance is quite consistent for all the temperature conditions up to 800 cycles. Additionally, the cell-to-cell variation in internal resistance of the cells is quite minimal up to 800 cycles, however after 800 cycles the rate of rise of the cell resistance is significantly higher, which is also a variation in rate of rise for different temperature conditions as can be seen in the figure 2. For capacity variation with cycling, we can see that the rate of rise in cell resistance at 10°C is significantly higher than the other two temperature conditions including higher cell-to-cell variation. Rapid resistance increases after 800 cycles support the previous discussion and the plots shown in Figure 1(a).

In Figure 1(d) we see variation in cell resistance with cycling at 0.7 C charging current and with 0.3C discharging current at 0, 10 and 25° C temperature conditions, all other conditions are the same for all the cells. The rate of resistance rise in the cell at 25°C is very minimal as compared to 0°C and 10°C up to 170 cycles however after that a significant rise in resistance can be observed even at 25°C. Looking for the resistance rise at 0 and 10° C we can see a significant rise even in less than 50 cycles with 40% and 50% as compared to 1.5% in the case of 25° C. With further cycling the rate of rise of the cell resistance is high in cases of 0 and 10°C and within 180 cycles the cell resistance is 2.6 and 2.8 times the initial resistance of the cell. However, in the case of 25°C cell resistance is up to 2.6 times which shows a rapid increase after 250 cycles. By analyzing the cell resistance, it is observed that a variable rate of cell degradation for different temperature conditions is shown in Figure 1(c).

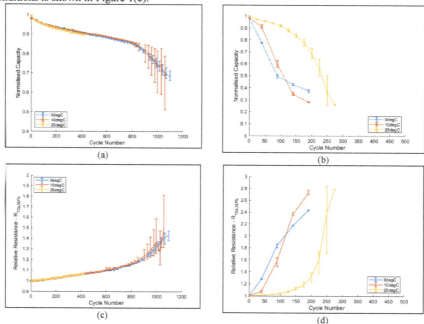

Figure 1: Discharge capacity vs cycle Number (a) 0.3C and (b) 0.7C charging current, cell resistance with cycle number (c) 0.3C and (d) 0.7C charging current at 0, 10, 25°C. Discharge current for all the cases is 0.3C.

In figure 2 variation in thermal condition across a battery module with 6S2P configuration has been shown. As can be seen, the temperature of the individual cells depends on the location within a battery module. Hence, for a same thermal management system of the battery module we can see a significant variation between cell temperature within the cell as well as across the battery module. Even for a small battery module (figure 2) this much variation in temperature can be seen, hence, in case of a commercial module/pack significant variation in temperature can happen which may lead to different rate of degradation of the cells for similar electrical loading. Further on, cells within a module or a battery pack would have different electrical loading conditions due to different SP architects which may lead to different rate of heat generation within the cells which will degrade cells differently. As discussed in the results shown in Figure 1 that the rate of degradation of the cells are highly dependent of temperature conditions as well as electrical loading. And figure 2 dominates a significant temperature variation which will lead to significantly different rate of degradation of the cells within a battery module. Hence at the rate retired stage of the battery happens when battery reaches 80% SOH because of the weakest cells. These retired modules would have a significant number of cells with higher SOH. With this study, it has been demonstrated the importance of identification of the SOH of the cells for potential 2nd life applications. For an effective utilization of the remaining capacity of the retired cells, cells with capacity need to be grouped and utilized in secondary applications.

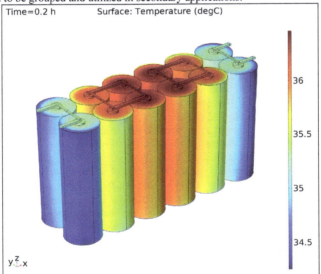

Figure 2: Demonstration of thermal gradient for 6S2P module across the cells.

Conclusion
In conclusion, a different rate of cell degradation has been observed with different environmental are electrical loading conditions. The rate of degradation of the cells is higher at low temperature and at high current charging conditions. In case of 0.3C charging conditions the rate of degradation is slower and after that and after the knee point is reached, the rate of degradation is steep. However, in case of 0.7C charging condition the rate half degradation is steep from the early stage of the cycling. Additionally, it has been demonstrated that the temperature of the cells within a

battery module is different across the 6S2P battery module which would be significantly higher in case of a bigger battery module. This temperature condition will lead to a different rate of degradation of the cells within a battery module. Hence for the potential second life applications of the retired electric vehicle batteries, knowing correct cell SOH is highly essential to group them which will lead to optimized us of these battery in 2nd life application. Therefore, future scope of this study is to develop a rapid state of health measurement or estimation technique which can be effectively utilized to grade and group cells with similar SOH for the potential second life applications which can be utilized in energy storage applications or renewable energy storage applications and remote areas. With this method the battery's SOH can be assessed within a short period (<60 seconds).

References
[1] Li, X., Zhang, L., Liu, Y., Pan, A., Liao, Q., & Yang, X. (2020). A fast classification method of retired electric vehicle battery modules and their energy storage application in photovoltaic generation. International Journal of Energy Research, 44(3), 2337–2344. https://doi.org/10.1002/er.5083

[2] Liao, Q., Mu, M., Zhao, S., Zhang, L., Jiang, T., Ye, J., Shen, X., & Zhou, G. (2017). Performance assessment and classification of retired lithium-ion battery from electric vehicles for energy storage. International Journal of Hydrogen Energy, 42(30), 18817–18823. https://doi.org/10.1016/j.ijhydene.2017.06.043

[3] Xu, Z., Wang, J., Lund, P. D., & Zhang, Y. (2021). Estimation and prediction of state of health of electric vehicle batteries using discrete incremental capacity analysis based on real driving data. Energy, 225, 120160. https://doi.org/10.1016/j.energy.2021.120160

[4] Braco, E., San Martín, I., Berrueta, A., Sanchis, P., & Ursúa, A. (2020). Experimental assessment of cycling ageing of lithium-ion second-life batteries from electric vehicles. Journal of Energy Storage, 32(May), 101695. https://doi.org/10.1016/j.est.2020.101695

[5] Zheng, Y., Wang, J., Qin, C., Lu, L., Han, X., & Ouyang, M. (2019). A novel capacity estimation method based on charging curve sections for lithium-ion batteries in electric vehicles. Energy, 185, 361–371. https://doi.org/10.1016/j.energy.2019.07.059

[6] Xu, Z., Wang, J., Lund, P. D., Fan, Q., Dong, T., Liang, Y., & Hong, J. (2020). A novel clustering algorithm for grouping and cascade utilization of retired Li-ion batteries. Journal of Energy Storage, 29, 100303. https://doi.org/10.1016/j.est.2020.101303

[7] Lai, X., Qiao, D., Zheng, Y., & Yi, W. (2018). A novel screening method based on a partially discharging curve using a genetic algorithm and back-propagation model for the cascade utilization of retired lithium-ion batteries. Electronics, 7, 399. https://doi.org/10.3390/electronics7120399

[8] Lai, X., Deng, C., Li, J., Zhu, Z., Han, X., & Zheng, Y. (2021). Rapid Sorting and Regrouping of Retired Lithium-Ion Battery Modules for Echelon Utilization Based on Partial Charging Curves. IEEE Transactions on Vehicular Technology, 70(2), 1246–1254. https://doi.org/10.1109/TVT.2021.3055068

[9] Zhou, Z., Ran, A., Chen, S., Zhang, X., Wei, G., Li, B., Kang, F., Zhou, X., & Sun, H. (2020). A fast-screening framework for second-life batteries based on an improved bisecting K-means algorithm combined with fast pulse test. Journal of Energy Storage, 31, 101739. https://doi.org/10.1016/j.est.2020.101739

[10] Zhou, P., He, Z., Han, T., Li, X., Lai, X., Yan, L., & Lv, T. (2020). A rapid classification method of the retired LiCo$_x$Ni$_y$Mn$_{1-x-y}$O$_2$ batteries for electric vehicles. 6, 672–683.

[11] Lai, X., Qiao, D., Zheng, Y., Ouyang, M., Han, X., & Zhou, L. (2019). A rapid screening and regrouping approach based on neural networks for large-scale retired lithium-ion cells in second-use applications. Journal of Cleaner Production, 213, 776–791. https://doi.org/10.1016/j.jclepro.2018.12.210

[12] Zhang, Q., Li, X., Du, Z., & Liao, Q. (2021). Aging performance characterization and state-of-health assessment of retired lithium-ion battery modules. Journal of Energy Storage, 40, 102743. https://doi.org/10.1016/j.est.2021.102743

[13] Enache, B., Seritan, G., Grigorescu, D., Cepisca, C., Argatu, V., & Voicila, T. I. (2020). A Battery Screening System for Second Life LiFePO 4 Batteries. EPE, 298–301. https://doi.org/10.1109/EPE50722.2020.9305538

[14] Kamath, D., Arsenault, R., Kim, H. C., & Anctil, A. (2020). Economic and Environmental Feasibility of Second-Life Lithium-Ion Batteries as Fast-Charging Energy Storage. Environmental Science & Technology, 54, 6878−6887. https://doi.org/10.1021/acs.est.9b05883

[15] Li, J., Wang, Y., & Tan, X. (2017). Research on the Classification Method for the Secondary Uses of Retired Lithium-ion Traction Batteries. Energy Procedia, 105, 2843–2849. https://doi.org/10.1016/j.egypro.2017.03.625

[16] Luo, F., Huang, H., Ni, L., & Li, T. (2021). Rapid prediction of the state of health of retired power batteries based on electrochemical impedance spectroscopy. Journal of Energy Storage, 41, 102866. https://doi.org/10.1016/j.est.2021.102866

[17] Enache, B., Seritan, G., Cepisca, C., Grigorescu, S. Florin-Ciprian. A., Felilx-Constantin, A., Teodor, V., (2020). Comparative study of screening methods for second life lifepo4 batteries. Rev. Roum. Sci. Techn. -Électrotechn. et Énerg, 65, 71–74.

[18] Braco, E., Martin, I. S., Sanchis, P., & Ursúa, A. (2019), Characterization and capacity dispersion of lithium ion second-life batteries from electric vehicles. Proceedings - 2019 IEEE International Conference on Environment and Electrical Engineering and 2019 IEEE Industrial and Commercial Power Systems Europe, EEEIC/I and CPS Europe 2019. https://doi.org/10.1109/EEEIC.2019.8783547

[19] Patil, M. A. et al., (2015), A novel multistage Support Vector Machine based approach for Li ion battery remaining useful life estimation. Appl. Energy 159, 285–297. https://doi.org/10.1016/j.apenergy.2015.08.119

AI-Based PV Panels Inspection using an Advanced YOLO Algorithm

Agus HAERUMAN[1,2,a], Sami Ul HAQ[1,3,b], Mohamed MOHANDES[1,4,c *], Shafiqur REHMAN[1,5,d] and Sheikh Sharif Iqbal MITU[1,4,e]

[1]SDAIA- KFUPM Joint Research Center for Artificial Intelligence (JRC-AI) grant No. JRCAI-UCG-03

[2]Mechanical Engineering Department, King Fahd University of Petroleum and Minerals (KFUPM), Dhahran, Saudi Arabia

[3]Computer Engineering Department, King Fahd University of Petroleum and Minerals (KFUPM), Dhahran, Saudi Arabia

[4]Electrical Engineering Department, King Fahd University of Petroleum and Minerals (KFUPM), Dhahran, Saudi Arabia

[5]Research Institute, King Fahd University of Petroleum and Minerals (KFUPM), Dhahran, Saudi Arabia

[a]g202208480@kfupm.edu.sa, [b]g202315350@kfupm.edu.sa, [c]mohandes@kfupm.edu.sa, [d]srehman@kfupm.edu.sa, [e]sheikhsi@kfupm.edu.sa

Keywords: Solar Energy, PV Panel Thermal Inspection, Artificial Intelligence, Deep Learning, Object Detection, YOLOv7

Abstract. The rapid growth of solar photovoltaic (PV) systems as green energy sources has gained momentum in recent years. However, the anomalies of PV panel defects can reduce its efficiency and minimize energy harvesting from the plant. The manual inspection of PV panel defects throughout the plant is costly and time-consuming. Thus, implementing more intelligent ways to inspect solar panel defects will provide more benefits than traditional ones. This study presents an implementation of a deep learning model to detect solar panel defects using an advanced object detection algorithm called You Look Only Once, version 7 (YOLOv7). YOLO is a popular algorithm in computer vision for classification and localization. The dataset utilized in this study was sourced from ROBOFLOW, consisting of 1660 infrared images showcasing thermal defects in PV panels. The model was constructed to identify a broader range of images with heterogeneity, leveraging the aforementioned dataset. Following validation, the model demonstrates a mean Average Precision (mAP) of 85.9%. With this accuracy, the model is relevant for real-world applications. This assertion is affirmed by testing the model with additional data from separate video-capturing PV panels. The video was recorded using a drone equipped with a thermal camera.

Introduction

The increase in energy demand due to massive population growth and the requirement to minimize greenhouse gas emissions have motivated novel approaches to utilizing more clean and sustainable energy. Solar energy is one of the most abundant renewable energy sources. It has now become popular due to the increase in its efficiency and lower cost compared to the last decades. However, maintaining photovoltaic modules is essential to maximize energy harvesting and gain more efficiency.

Defects in PV modules, whether arising from installation or operational factors, can significantly reduce their power generation efficiency. Despite features such as frames made up of Aluminum or glass-lamination that protect panels from environmental factors like rain, wind, and snow, they may not be able to completely protect panels from mechanical stress during transport or in extreme weather conditions like hail [1]. In addition, manufacturing defects such as defective soldering or faulty wiring can also affect the efficiency of PV modules [2]. Thus, it is vital to employ timely and dependable inspection methods to evaluate and uphold the peak functioning of PV modules, ensuring the utmost effectiveness of solar PV plants.

Implementing more intelligent ways of detecting PV panel defects is one of the most important topics to be discussed. Some researchers have implemented an AI-based method for detecting PV panel defects. Akram et al., for instance, proposed isolated deep learning techniques and developed model transfer deep learning techniques for the detection of PV module defects [3]. Both methods require low computational costs and less time, so they are suited for hardware with less memory installed. Herraiz et al. presented a novel approach to identifying PV panel defects using convolutional neural networks (CNN). They combined thermography and telemetry data to monitor panel conditions [4]. Various alternative strategies employing deep learning have been introduced for the identification of flaws in solar-cell panels. These strategies encompass the application of transfer learning methods utilizing various architectures such as VGG16 [5], VGG19 [6], GoogLeNet [7], ResNet18 [8], Unet [9], FPN [10], LinkNet [11], and EfficientNet [12] to identify anomalies on solar-cell panels [13].

This study delves into the application of an advanced Object Detection Algorithm, specifically YOLOv7, in the thermal inspection of PV panels. The utilization of AI through YOLOv7 aims to revolutionize the detection and classification of potential issues, offering a faster and more precise alternative to conventional inspection techniques. The choice of YOLOv7 architecture for this study is motivated by its reputation for real-time object detection capabilities. YOLOv7 works by dividing the input image into grids, enabling simultaneous prediction of bounding boxes and class probabilities, thus streamlining the detection process. This approach aligns with the demands of thermal inspection for PV panels, where swift and accurate identification of anomalies is crucial for maintaining the efficiency and reliability of solar energy systems. Furthermore, YOLOv7 performs single forward-pass neural networks, thus making it faster and more efficient compared to the other object detection algorithms.

The remaining parts of this paper are organized as follows: Section II presents a comprehensive literature analysis, outlining existing approaches for PV panel inspection and emphasizing advances in AI applications for similar goals. Section III describes the materials and methods utilized in the implementation of the YOLOv7 architecture for thermal inspection, including the dataset and training procedure. Section IV includes the experiment data and analysis, evaluating the performance of the proposed AI-based thermal inspection approach. Section V finishes the work with a summary of major findings, consequences, and future research directions.

Literature Review
Recently, machine learning has grown in popularity as a method for studying PV panels. Various researchers used different ways to inspect and maintain the quality of PV panel modules. Visual inspection, current-voltage (I-V) curve analysis, infrared thermography, and Electroluminescence (EL) testing are among these methods. For instance, in [14], the author utilized a multi-scale CNN model in two modes: transfer learning-based (using two selected DNNs) and independent light-depth (CNN-ILD) CNN. The experimental data using the open ELPV dataset shows promising classifications for PV panel defects in EL images. However, the EL imaging system assesses the photovoltaic (PV) system in low-light conditions, concurrently administering a direct current (DC) to the PV. This approach enables the detection of minor defects, disconnected cell regions, shunts, and similar issues within the PV cells, making it well-suited for indoor inspections. Nevertheless, employing this method outdoors for extensive objects presents notable challenges [15].

Many researchers have also used other types of machine learning to classify solar panel flaws according to attributes extracted from EL images, including Support Vector Machines, Random Forests, and K-Nearest Neighbors [16], [17], and [18]. These methods require manual feature extraction, in which relevant features that capture the traits of various fault kinds are designed using domain expertise. However, these approaches' effectiveness usually depends on how well-engineered the features are, so they could not be as effective as deep learning approaches, in which the features can automatically learned. [19] and [20]. Similarly, several research studies have been done on deep learning techniques for detecting PV panel defects using EL and IR images. For example, in [15], the author proposed a remote sensing method using infrared radiation cameras installed on unmanned aerial vehicles (UAV) to capture images of solar panels and detect anomalies using CNN. Likewise, [21] proposed a deep-learning method to detect defective solar panels in EL images. They utilize two CNN architectures: a fine-tuned VGG16 model for classification and a lightweight CNN model created from scratch for baseline comparison. The proposed method achieved a 95.2% accuracy on the test dataset. However, these approaches have their limitations. The fine-tuned VGG16 model, while effective in classification, may face challenges in real-time performance, potentially hindering its practicality for dynamic PV panel inspections. On the other hand, the lightweight CNN model, though created for efficiency, might struggle to achieve the same level of accuracy as more established architectures, impacting its reliability in defect detection tasks. These drawbacks influenced our decision to explore alternative solutions better suited for the real-time and accuracy demands of our project.

While several research studies have delved into deep learning techniques for detecting defects in PV panels using EL and IR images, our focus on thermal IR image-based defect detection led us to explore alternatives better aligned with the real-time and accuracy demands of our project. Contrary to the drawbacks associated with the existing methods mentioned above, we opted for YOLOv7 for several reasons. Compared to [15], YOLOv7 provides better object recognition performance for outdoor PV panel inspection, with higher label assignment and bounding box localization. YOLOv7 is also based on neural network architectures that require no feature selection, as mentioned in [16], [17], and [18], which are based on Machine Learning models. The last thing to note here is that, compared to other deep learning models such as mentioned in [21], YOLOv7 requires much less computational cost, making it learn much faster with smaller datasets without pre-training needed. [22].

Methodology and Experiments
The dataset for this project was obtained from ROBOFLOW, an online resource for open-source datasets for computer vision. The dataset comprising thermal images of PV panels acquired by UAV was downloaded in YOLOv7 format. The dataset contains 5313 labeled images [23]. The label contains a bounding box information of the object, such as x-center, y-center, height, width, and the PV panel's status. This format is required for YOLOv7 to detect the object and do localization. Before training the model, the dataset was pre-processed. We chose only 1660 images for this study and removed those images that were taken from a short-distance shoot.

The dataset was then split into training, validation, and testing with a ratio of 70%, 20%, and 10% for training, validation, and testing, respectively. During the training process, we used only the training and validation dataset, while we reserved the testing dataset for testing unseen data.

Fig. 1a shows a sample PV panel thermal image with a single hot spot, while Fig. 1b shows the PV panel with a multi-hot spot. The brighter spots in the panels are the defective cells of the panel due to heat dissipation, and this phenomenon can decrease the efficiency of the panel. All the image sizes are 640x640 pixels.

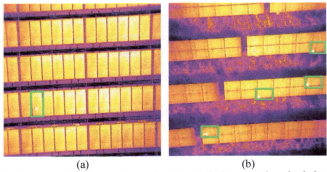

Fig. 1 Sample of images. (a) Image with single hotspot. (b) Image with multiple hotspots [23].

YOLOv7 is the most updated version of the YOLO family from the original authors of the YOLO architecture. This model outperformed all known predecessor object detectors in speed and accuracy, such as YOLOR, YOLOX, YOLOv5, Scaled-YOLOv4, and PPYOLOE. It has reduced a significant number of parameters and computational costs, leading to faster inference speeds and higher detection accuracy. The Basic YOLO Architecture is described in Fig. 2, and the YOLOv7 introduced some major changes in its new architecture, including the Extended Efficient Layer Aggregation Network (E-ELAN), compound model scaling, planned re-parameterized convolution, and coarse-to-fine lead guided assigner [24].

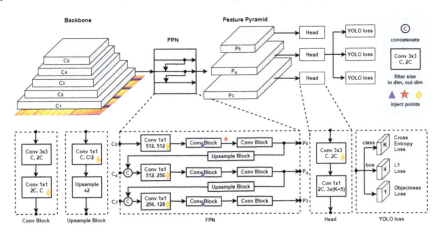

Fig. 2 Basic YOLO Architecture [25].

In this study, we built the model based on YOLOv7 using the mentioned dataset. We used the Nvidia GPU to accelerate the training process and a specific project directory in Google Drive mounted to the Google Colab. All the datasets were put in this directory. The original YOLOv7 repository and the pre-trained model were then acquired. With some adjustments to the dataset and label information in the COCO.yaml and YOLOv7.yaml files, the models were then trained. The parameters used during the experiment can be seen in Table 1. The Performance metrics we used to evaluate the accuracy of the model's performance are mAP, precision, and recall. Mean Average Precision (mAP) is commonly used to evaluate the model, with ranges of the evaluation from 0 to 1. On the other hand, we also use precision to indicate the accuracy of the detected objects, calculating how many detections were correct, while recall gives information about the ability of the model to identify all instances of objects in the images.

Table 1. Training parameters of the models.

Batch Number	Learning Rate	Momentum	Weight Decay	Number of epochs
8	0.01	0.937	0.0005	200

Result and Discussion
The objective was to evaluate the model's accuracy and investigate its performance in real-life applications. The model was built using heterogeneous images with different elevation drone cameras and more angles. When we test the model using testing data, it can be seen that the model can detect PV panel defects from various images. The data was split into 1170 training data points, 330 validation data points, and 160 testing data points. The batch size was 8, with a learning rate of 0.01, a momentum of 0.937, a weight decay of 0.0005, and a running of 200 epochs. After running the testing with the best weight, it yielded the mAP of 85.9% for threshold 0.5 IoU, the R-value of 83.2%, and the P-value of 75.9%, with 322 defect panels detected from 160 images, as described in Tabel 2. The sample of the testing image was used to see the performance of the model in detecting PV panel defects, as we described in Fig 3. It can be observed that the model performed well in detecting PV panel defects, with three bounding boxes detected with an IoU of 0.71, 0.79, and 0.80.

Table 2. Testing Result of the Model.

Images	Labels	Precision	Recall	mAP@0.5
160	322	0.759	0.832	0.859

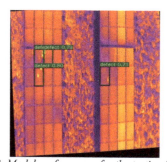

Fig. 3. Model performance for the testing image.

To see the performance of the models for a real-life application, the model was tested to run on a video file obtained from [26]. It was found that the Model performed well in the detection of PV panel defects, as can be seen in Fig. 4. This Model can also be applied for the online monitoring of PV panel defects when it is used to detect the panel using a drone-embedded camera during PV panel inspection.

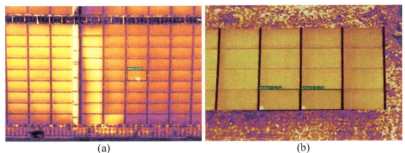

(a)   (b)
Fig. 4. Model performance on the testing videos.

Conclusion
Solar energy is one of the most popular renewable energy resources, and many countries are now utilizing photovoltaic (PV) panels to harvest energy from the sun. However, the occurrence of PV panel defects can reduce its efficiency and decrease power output. To address this challenge, a lot of research has been conducted to alternate manual PV panel inspection with the most effective and intelligent methods. This study presented an implementation of artificial intelligence, especially deep learning architecture for object detection, known as You Look Only Once, version 7 (YOLOv7). The object detection model was developed using a dataset comprising 1660 data points, achieving an mAP of 85.9%. The model can effectively learn a diverse set of images suitable for real-life applications. This capability was confirmed during testing on unseen images, where the model demonstrated good performance. Furthermore, the model underwent evaluation with a video file for real-life applications. To extend this study to future works, the model can be developed with more heterogeneous images with variations in elevations and angles to perform much better for online applications, especially when it comes to deploying the system in edge artificial intelligence environments.

Acknowledgment
The authors would like to acknowledge the support provided by Saudi Data & AI Authority (SDAIA) and King Fahd University of Petroleum & Minerals (KFUPM) under SDAIAKFUPM Joint Research Center for Artificial Intelligence (JRCAI) grant No. JRCAI-UCG-03.

References
[1] S. Djordjevic, D. Parlevliet, and P. Jennings. Detectable faults on recently installed solar modules in Western Australia. Renewable Energy, 67:215-221, 2014. https://doi.org/10.1016/j.renene.2013.11.036

[2] Ronnie O Serfa Juan and Jeha Kim. Photovoltaic cell defect detection model based on extracted electroluminescence images using SVM classifier. In 2020 International Conference on Artificial Intelligence in Information and Communication (ICAIIC), pages 578-582. IEEE, 2020. https://doi.org/10.1109/ICAIIC48513.2020.9065065

[3] M Waqar Akram, Guiqiang Li, Yi Jin, Xiao Chen, Changan Zhu, and Ashfaq Ahmad. Automatic detection of photovoltaic module defects in infrared images with isolated and develop-model transfer deep learning. Solar Energy, 198:175-186, 2020. https://doi.org/10.1016/j.solener.2020.01.055

[4] Alvaro Huerta Herraiz, Alberto Pliego Marugan, and Fausto Pedro Garcia Marquez. Photovoltaic plant condition monitoring using thermal images analysis by convolutional neural network-based structure. Renewable Energy, 153:334-348, 2020. https://doi.org/10.1016/j.renene.2020.01.148

[5] Karen Simonyan and Andrew Zisserman. Very deep convolutional networks for large-scale image recognition. arXiv preprint arXiv:1409.1556, 2014.

[6] Long Wen, X Li, Xinyu Li, and Liang Gao. A new transfer learning based on VGG-19 network for fault diagnosis. In 2019 IEEE 23rd international conference on computer supported cooperative work in design (CSCWD), pages 205-209. IEEE, 2019. https://doi.org/10.1109/CSCWD.2019.8791884

[7] Christian Szegedy, Wei Liu, Yangqing Jia, Pierre Sermanet, Scott Reed, Dragomir Anguelov, Dumitru Erhan, Vincent Vanhoucke, and Andrew Rabinovich. Going deeper with convolutions. In Proceedings of the IEEE conference on computer vision and pattern recognition, pages 1-9, 2015. https://doi.org/10.1109/CVPR.2015.7298594

[8] Kaiming He, Xiangyu Zhang, Shaoqing Ren, and Jian Sun. Deep residual learning for image recognition. In Proceedings of the IEEE conference on computer vision and pattern recognition, pages 770-778, 2016.

[9] Olaf Ronneberger, Philipp Fischer, and Thomas Brox. U-Net: Convolutional networks for biomedical image segmentation. In Medical Image Computing and Computer-Assisted Intervention–MICCAI 2015: 18th International Conference, Munich, Germany, October 5-9, 2015, Proceedings, Part III 18, pages 234-241. Springer, 2015. https://doi.org/10.1007/978-3-319-24574-4_28

[10] Tsung-Yi Lin, Piotr Dollar, Ross Girshick, Kaiming He, Bharath Hariharan, and Serge Belongie. Feature pyramid networks for object detection. In Proceedings of the IEEE conference on computer vision and pattern recognition, pages 2117-2125, 2017.

[11] Abhishek Chaurasia and Eugenio Culurciello. Linknet: Exploiting encoder representations for efficient semantic segmentation. In 2017 IEEE visual communications and image processing (VCIP), pages 1-4. IEEE, 2017. https://doi.org/10.1109/VCIP.2017.8305148

[12] Mingxing Tan and Quoc Le. Efficientnet: Rethinking model scaling for convolutional neural networks. In International conference on machine learning, pages 6105-6114. PMLR, 2019.

[13] Minhhuy Le, Dang Khoa Nguyen, Van-Duong Dao, Ngoc Hung Vu, Hong Ha Thi Vu, et al. Remote anomaly detection and classification of solar photovoltaic modules based on deep neural network. Sustainable Energy Technologies and Assessments, 48:101545, 2021. https://doi.org/10.1016/j.seta.2021.101545

[14] Hazem Munawer Al-Otum. Deep learning-based automated defect classification in electroluminescence images of solar panels. Advanced Engineering Informatics, 58:102147, 2023. https://doi.org/10.1016/j.aei.2023.102147

[15] Minhhuy Le, DucVu Le, and Hong Ha Thi Vu. Thermal inspection of photovoltaic modules with deep convolutional neural networks on edge devices in AUV. Measurement, 218:113135, 2023. https://doi.org/10.1016/j.measurement.2023.113135

[16] Mustafa Yusuf Demirci, Nurettin Besli, and Abdulkadir Gumuscu. Efficient deep feature extraction and classification for identifying defective photovoltaic module cells in electroluminescence images. Expert Systems with Applications, 175:114810, 2021. https://doi.org/10.1016/j.eswa.2021.114810

[17] Stefan Bordihn, Andreas Fladung, Jan Schlipf, and Marc Kontges. Machine learning based identification and classification of field-operation caused solar panel failures observed in electroluminescence images. IEEE Journal of Photovoltaics, 12(3):827-832, 2022. https://doi.org/10.1109/JPHOTOV.2022.3150725

[18] Harsh Rajesh Parikh, Yoann Buratti, Sergiu Spataru, Frederik Villebro, Gisele Alves Dos Reis Benatto, Peter B. Poulsen, Stefan Wendlandt, Tamas Kerekes, Dezso Sera, and Ziv Hameiri. Solar cell cracks and finger failure detection using statistical parameters of electroluminescence images and machine learning. Applied Sciences, 10(24), 2020. https://doi.org/10.3390/app10248834

[19] Ronnie O. Serfa Juan and Jeha Kim. Photovoltaic cell defect detection model based on extracted electroluminescence images using SVM classifier. In 2020 International Conference on Artificial Intelligence in Information and Communication (ICAIIC), pages 578-582, 2020. https://doi.org/10.1109/ICAIIC48513.2020.9065065

[20] Sergiu Deitsch, Vincent Christlein, Stephan Berger, Claudia BuerhopLutz, Andreas K. Maier, Florian Gallwitz, and Christian Riess. Automatic classification of defective photovoltaic module cells in electroluminescence images. ArXiv, abs/1807.02894, 2018.

[21] Abraham Kaligambe and Goro Fujita. A deep learning-based framework for automatic detection of defective solar photovoltaic cells in electroluminescence images using transfer learning. In 2023 4th International Conference on High Voltage Engineering and Power Systems (ICHVEPS), pages 81-85, 2023. https://doi.org/10.1109/ICHVEPS58902.2023.10257399

[22] Christine Dewi, Abbott Po Shun Chen, and Henoch Juli Christanto. Deep learning for highly accurate hand recognition based on yolov7 model. Big Data and Cognitive Computing, 7(1), 2023. https://doi.org/10.3390/bdcc7010053

[23] Solveview. Thermal defects dataset. https://universe.roboflow.com/solveview/thermal-defects , Aug 2023. visited on 2023-11-20.

[24] Chien-Yao Wang, Alexey Bochkovskiy, and Hong-Yuan Mark Liao. Yolov7: Trainable bag-of-freebies sets new state-of-the-art for real-time object detectors. In Proceedings of the IEEE/CVF Conference on Computer Vision and Pattern Recognition, pages 7464-7475, 2023.

[25] Xiang Long, Kaipeng Deng, Guanzhong Wang, Yang Zhang, Qingqing Dang, Yuan Gao, Hui Shen, Jianguo Ren, Shumin Han, Errui Ding, et al. Pp-yolo: An effective and efficient implementation of object detector. arXiv preprint arXiv:2007.12099, 2020.

[26] www.ThermalCamUSA.com. Infrared solar panel inspection by drone. https://www.youtube.com/watch?v=1E-_N-KtQTQ, Feb 2018. visited on 2023-11-20.

Impact of artificial intelligence (AI) in Martian architecture (exterior and interior)

Lindita Bande[1,a*], Jose Berengueres[1], Aysha Alsheraifi[1], Anwar Ahmad[1], Saud Alhammadi[1], Almaha Alneyadi[1], Amna Alkaabi[1], Maitha Altamimi[1], Yosan Asmelash[1]

[1] United Arab Emirates University, UAE

[a]lindita.bande@uaeu.ac.ae

Keywords: Martian Architecture, AI Tool, Critical Thinking, RHINO/Grasshopper

Abstract. Martian architecture has gained interest in the recent year. Several grand architectural studios have designed hypothetical buildings as part of a colony of the red planet. This study is a continuation of a previous research on mars Habitat. The use of AI to generate alternatives of design based on an initial idea gives insight of how technology can assist us in such major projects. The methodology followed in this study is as per the below steps: 1- General Description of the initial concept: Organic Architecture, 2- General Description of the initial concept: Minimal Architecture, 3- Use of AI in the selected projects, tool description, 4- Results: Outcomes of AI Applications. The aim of this study is to investigate the impact of the AI on Space Architecture, more specifically Martian Architecture. The initial step in the methodology is to design a colony that connects together but as also well distributed in the plan. The following step is using an AI tool to generate processed (rendered) images of the base image. These AI renders will then be analyzed and the final implication of the findings for the project will be described. The findings of this study can be relevant to relevant authorities in space exploration and space architecture with the help of AI tools.

Introduction
All manuscripts must be in English, also the table and figure text.

A new research outlines the design of Martian Habitat Units (MHUs) for extended human missions on Mars, prioritizing functionality, and aesthetics. Circular clusters of MHUs, each accommodating nine crew members, incorporate solar farms, nuclear fission, and wind turbines for energy. Lighting simulations demonstrate that a radial configuration maximizes natural light usage, meeting 36–44% of the lighting load. This information is crucial for planning energy-efficient systems on Mars. [1]

As per a recent article that discusses the design of a Mars research base for long-term habitation, incorporating art and architecture for a thriving lifestyle. Martian Habitat Units (HMUs) are designed with nuclear power, solar farms, and wind turbines, utilizing innovative features like Anti-Dust Settlement Membranes (ADSMs) for solar farm maintenance. The construction involves local grain 3D-printing, and the design prioritizes fail-safe procedures for crew safety. The technologies are based on current advancements, with potential reconsideration closer to the mission date. [2]

The text underscores escalating challenges in Mars missions due to increasing distance from Earth, jeopardizing ground support and crew capabilities. While fast-transit solutions may alleviate some hazards, they introduce novel risks and complexity. Extended exposure to microgravity raises concerns, requiring research into effective countermeasures. The communication delay and resupply constraints necessitate a shift towards autonomous human-system integration, urging urgent attention to develop and validate suitable architectures. [3]

The Mars Quantum Gravity Mission (MaQuIs) focuses on exploring Mars' gravitational field to study subsurface water occurrences and planetary dynamics. The paper outlines current knowledge, proposes satellite gravimetry using quantum technologies, and discusses scenarios for mission simulation. Authors highlight roles in assembling the consortium, estimating gravitational signals, and detailing inertial measurement systems for the mission. Future steps involve simulation scenarios, evaluating dependencies, and identifying limiting factors for the mission concept and technologies in the Martian environment. [4]

The passage highlights the application of machine learning in processing mass spectrometry data for future space missions. By using various artificial intelligence models, accurate results can be obtained quickly, benefiting in-flight processing. Root transformation and 2D spectrograms enhance accuracy, and pretrained convolutional neural networks (CNNs) perform exceptionally well. Generalization, model assembling, and proper training procedures are essential for small datasets. Increased data availability is crucial for further improvement, and machine learning analysis can be effectively run on the edge for future missions, particularly in analyzing sediments from Mars and other celestial bodies. [5]

The article introduces a GeoAI framework for accurate crater detection by integrating domain knowledge and scale-aware learning. Collaboration between computer scientists and Earth/space scientists enhances object detection models. The methodology aims to reduce the time for cataloging new craters, with future improvements planned for data completeness and efficiency optimization. The research envisions incorporating multi-source data to enhance detection accuracy, and the model and data will be open-sourced to encourage collaborative research in the field. [6]

A new research explores the application of AI methods for cyber risk analytics in extreme environments like outer space. It emphasizes adapting data strategies for collecting relevant cyber-risk data and leveraging IoT systems for diverse data streams. The review identifies potential impact assessment approaches, and the con-clusion introduces a quantitative version of the NIST 'traffic lights' system with multiple risk calculation metrics, enhancing cost and risk evaluation. The study pre-sents a mathematical formula for future cyber risk developments, focusing on coor-dination and reliability in AI/ML-based cyber protection for supply and control sys-tems while anonymizing risk data. [7]

Referring to an article that highlights the rapid growth in conversational studies, particularly in open-domain dialogue systems, with a thriving research community and increased industry efforts. The overview summarizes current progress and antic-ipates a promising future in the AI era, characterized by abundant data and powerful learning techniques. Despite facing challenges in improving conversational AI, the authors express optimism about advancing dialogue systems through dedicated ef-forts and resolving key issues. [8]

New research argues that despite a deep faith in human reason and experience, the main obstacle to a human mission to Mars is a technological barrier, compounded by increasing threats and risks. The author questions the urgent need for a long-term interplanetary program, citing challenges such as overpopulation, limited resources, and climate change. Additionally, the psychological challenges of living in a con-fined space, whether on Earth or in space, are highlighted, raising ethical questions about the quality of life in such conditions. The author suggests that mission plan-ners should prioritize the human factor and consider broader interconnected factors in their planning. [9]

As per a new article reviews the current applications of Artificial Intelligence (AI) in environmental disciplines, emphasizing its role in managing and analyzing large datasets related to demographics, traffic, and energy usage. The integration of big data and AI provides new opportunities for environmental tasks such as model-ing, monitoring, and research. AI tools, particularly machine learning algorithms, contribute to real-time monitoring of air and water quality, prediction of future trends, accurate detection of key fish species, and optimization of energy efficiency. Collaborations between ecology and data science are highlighted as crucial for effec-tive conservation efforts. [10]

Methodology
This study is a continuation of a previous research on mars Habitat. The use of AI to generate alternatives of design based on an initial idea gives insight of how technology can assist us in such major projects.
The methodology followed in this study is as per the below steps:
- General Description of the initial concept: Organic Architecture,
- General Description of the initial concept: Minimal Architecture,
- Use of AI in the selected projects, tool description.
- Results: Outcomes of AI Applications

General Description of the initial concept: Organic Architecture
The Martian Habitat based on Space Architecture concepts, designed in the likeness of flower petals, symbolizes the resilience of life in the Martian setting. It communicates a powerful message, affirming the existence and vitality of life in this challenging environment (figure 1-5).

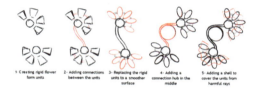

Figure 1. The design concept.

Figure 2. Function distribution in 3D.

Figure 3. Unit Development, Exterior-Interior.

Figure 4. Intererior development, concept.

Figure 5. 3D Model from Rhino/Grasshoper, view 1.

General Description of the initial concept: Minimal Architecture
The minimalist Architectural approach uses the dome as the main element of construction. Considering the high radiation in the red planet this shape seems to be the most functional one on an initial stage. Furthermore, much deeper investigation is needed to explore the nature of the materials on the planet, their resistance, endurance, lifecycle. For the current study this particular process is focused on the design of the unit that would create a colony of humans.

Use of AI in the selected projects, tool description.
The use of AI to generate renders based on figure 5 was quite interesting. By adding several keywords and uploading the image the tool generated renders that are shown in the results. In this trial 8 options were selected for the evaluation. There are several options how an image is processed, realistic, creative, more AI Render tools - https://mnml.ai/app/exterior-ai (table 1).

Table 1. Keywords used in the AI Tool.

Options	Keywords
Organic Architecture	
Option 1	Martian architecture, desert landscape, universal light, rocks
Option 2	Islamic architecture, mountain view, desert landscaping and bright light
Option 3	modern architecture, desert view, hard landscape and day light
Option 4	parametric architecture, desert view, soft landscape and evening light
Option 5	minimal architecture, grey field view, soft landscape and morning light
Option 6	classic architecture, desert view, hard landscape and morning light

Option 7	concrete facade, modern architecture, rock landscape and evening light
Option 8	concrete facade, modern architecture, desert view, hard landscape and evening light

Meanwhile the tool used for the Minimalistic Design is PromeAI (Free AI art generator).The use of AI in architecture helps the designer to visualize and get a creative idea easily and continuously, with just a simple image you can create different buildings using different architectural styles. Everyone can create shapes using any 3d software like Revit or SketchUp and insert it in any AI rendering tools like Prome AI or ReRender AI and write any keyword like modern or classic and decide if the AI tool should change the shape and decide if the AI tool should be creative completely or to keep the shape but just put some ideas to the shape. AI tool will generate images that would save the architect some time in rendering because the architect will have a reference images. Prome AI allow you to generate 200 images every day, but the architect or student that have a membership in the website the images that is generated using the tool will be unlimited. Prome AI generate 3 images so that you choose the best one. There are many tools in the Prome AI for example erase and replace tool which you can select what you want to change in the image. 2d images also can be used in AI tools to generate a creative piece of art. Choosing the mood and the place that you want the building to be in can be easilty created using AI for example of you want the building to be in mars AI can make it happen. For example, if you want the building to float in the sea AI will generate it to your imaginations. Some AI tool have some restrictions to what you can do with them (number of imaged to generate, the architecture style, the creativity, Etc.).

Results

The results of the 8 trials in organic architecture and minimalistic architecture are shown in table 2. Each option has a different outcome impacting the shape of the building, the openings, the relation to the outdoor. However, the shell structure remains in both scenarios.

Table 2. Ai Generated images from AI Tool

Ai Generated images from AI Tool	
Organic Architecture	Minimal Architecture
Option 1	
Option 2	

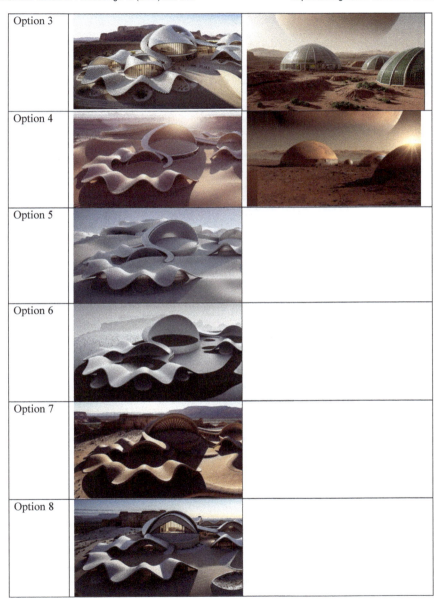

Discussion and Conclusions
Discussions
This study highlights the benefits of AI tools in architectural education, demonstrating how it can assist students, especially in designing sustainable houses in challenging environments like the Mars. The results are in agreement with similar studies elsewhere (Tholander et al. 2023; Ildirim et al. 2023). Said tools offer a variety of design options, which are instrumental in aiding critical assessment and decision-making. Regarding the implications for Martian architecture itself, the results underscore an awareness for the need for energy-efficient design while considering Mars' unique environmental challenges, such as radiation and dust storms. We note the ability of AI generated designs to incorporate and balance both functionalities, like energy systems and dust protection, and aesthetics. Furthermore, the broader implications of these findings extend beyond academic settings. They offer valuable insights for the space exploration industry, indicating that AI tools can significantly enhance the added value of the design process (figure 6).

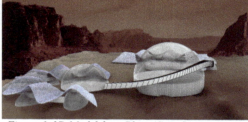

Figure 6. 3D Model from Rhino/Grasshoper, view 2.

Conclusions
The profound impact of Martian exploration efforts in driving progress is undeniably significant. Similar to the Apollo missions five decades ago, as noted by Comstock (2007), the research and development (R&D) outcomes from Mars missions are anticipated to enhance life quality on Earth as well. For instance, the development of computers is attributed to the Apollo space missions. Notably, the Apollo 11 moon landing in 1969 played a crucial role, perhaps unintentionally, in advancing computer technology. The Apollo Guidance Computer (AGC), created for the Apollo program to navigate and guide spacecraft, was among the first to incorporate integrated circuits (ICs). Other notable developments from the Apollo missions included advancements in satellite TV, water purification systems, and new insulating materials for spacecraft, now utilized in building construction (Denver et al., 1981). These findings hold particular relevance for space exploration authorities and architectural firms specializing in extreme environments, demonstrating, as corroborated by other sources, the potential of AI to enhance creative processes. (figure 7).

References
[1] Amini, K., Janabadi, E.D. and Fayaz, R. (2022) 'Lighting and illumination investigation of long-term residence on Mars for the case of a set of designed martian habitat units (mhus)', Acta Astronautica, 192, pp. 210–232. https://doi.org/10.1016/j.actaastro.2021.12.021

[2] Amini, K. et al. (2022) 'Design of a set of habitat units and the corresponding surrounding cluster for long-term scientific missions in the pre-terraforming era on Mars', Icarus, 385, p. 115119. https://doi.org/10.1016/j.icarus.2022.115119

[3] Valinia, A. et al. (2023) 'Risk trade-space analysis for Safe Human Expeditions to Mars', Acta Astronautica, 211, pp. 192–199. https://doi.org/10.1016/j.actaastro.2023.05.039

[4] Wörner, L. et al. (2023) 'Maquis—concept for a mars quantum gravity mission', Planetary and Space Science, 239, p. 105800. https://doi.org/10.1016/j.pss.2023.105800

[5] Nasios, I. (2024) 'Analyze mass spectrometry data with artificial intelligence to assist the understanding of past habitability of Mars and provide insights for future missions', Icarus, 408, p. 115824. https://doi.org/10.1016/j.icarus.2023.115824

[6] Hsu, C.-Y., Li, W. and Wang, S. (2021) 'Knowledge-driven geoai: Integrating spatial knowledge into multi-scale deep learning for Mars Crater Detection', Remote Sensing, 13(11), p. 2116. https://doi.org/10.3390/rs13112116

[7] Radanliev, P. et al. (2020) 'Design of a dynamic and self-adapting system, sup-ported with Artificial Intelligence, machine learning and real-time intelligence for predictive cyber risk analytics in extreme environments – cyber risk in the colonization of Mars', Safety in Extreme Environments, 2(3), pp. 219–230. https://doi.org/10.1007/s42797-021-00025-1

[8] Yan, R. and Wu, W. (2021) 'Empowering conversational AI is a trip to Mars: Progress and future of open domain human-computer dialogues', Proceedings of the AAAI Conference on Artificial Intelligence, 35(17), pp. 15078–15086. https://doi.org/10.1609/aaai.v35i17.17771

[9] Szocik, K. (2019) 'Should and could humans go to Mars? yes, but not now and not in the near future', Futures, 105, pp. 54–66. https://doi.org/10.1016/j.futures.2018.08.004

[10] Konya, A. and Nematzadeh, P. (2024) 'Recent applications of AI to Environmental Disciplines: A Review', Science of The Total Environment, 906, p. 167705. https://doi.org/10.1016/j.scitotenv.2023.167705

Sliding mode control for grid integration of point absorber type wave energy converter

Abdin Y. ELAMIN[1,a] and Addy WAHYUDIE[1,b*]

[1]Department of Electrical and Communication Engineering, United Arab Emirates University (UAEU), Al Ain, United Arab Emirates

[a]201890100@uaeu.ac.ae, [b]addy.w@uaeu.ac.ae

Keywords: Current Controllers, DC Bus Regulation, Grid Integration, Phase Locked Loop Point Absorber, Resistive Loading, Sliding Mode Control, Wave Energy

Abstract. This paper addresses the integration of a point absorber type wave energy converter into power grids, a process complicated by wave energy's intermittent and unpredictable nature. It proposes a sliding mode control strategy with an exponential reaching law in a voltage-oriented control architecture for the grid-side converter of the point absorber. The control objective is to maximize the transfer of generated power to the electrical grid, while concurrently stabilizing the DC bus voltage at a predetermined value and achieving a unity power factor. Results from simulations conducted within a MATLAB framework underscore the efficacy of the sliding mode control approach in sustaining specified DC bus voltage values, and in regulating the direct and quadrature currents by minimizing their respective tracking errors.

Introduction

Wave energy has a potential of 0.534 to 17.5 PWh/year and offers a high energy density with minimal environmental impact [1]. Its availability rate of up to 90% is significantly greater than the 20%-30% of solar and wind energies [2-3]. Wave energy converters (WECs), including various types such as point absorbers (PAs), are used to transform wave energy into electricity. Despite its benefits, wave energy's integration into power grids faces challenges due to its intermittent and unpredictable nature, causing fluctuations in power output that lack synchronization with key grid parameters, such as voltage, phase angle, and frequency. Asynchronization can significantly degrade grid stability and may introduce unwanted effects such as harmonics, frequency deviation and voltage collapse. The criticality of this issue escalates when the grid faces disturbances and unbalanced faults. The back to back converter topology is particularly effective for ensuring the poor power quality of the WEC doesn't propagate into the grid.

The role of the machine-side converter (MSC), i.e., AC/DC converter, is to maximize wave energy extraction, while the grid-side converter (GSC), i.e., DC/AC converter, is tasked with ensuring the synchronization of a constant voltage and frequency to the grid. In terms of GSC control strategies, Proportional-Integral (PI) controllers have been popularly suggested for regulating DC bus voltage and ensuring unity power factor by synchronizing current waveforms with the grid's frequency [4]. However, their performance is constrained by their reliance on linear models and sensitivity to parameter variations. Given the strongly coupled and nonlinear nature of the GSC system, a nonlinear control strategy could potentially enhance both static and dynamic performance. A study in [5] proposed a Lyapunov-based nonlinear controller for active power regulation and zero reactive power injection into the grid.

Sliding mode control (SMC) techniques have be used extensively as nonlinear controllers in GSCs for wind energy and photovoltaic applications [6], yet they haven't been implemented in WEC systems. Their robustness to parameter variations and disturbances [7], make them prime candidates in grid tied WEC applications. This paper proposes a SMC strategy for a GSC in a PA, utilizing a voltage-oriented control (VOC) architecture. The VOC incorporates two cascaded control loops, with the outer-loop focusing on DC bus voltage regulation. The inner-loop ensures that the GSC output currents accurately track the reference currents from the upper-loop and are synchronized with the grid frequency. A phase-locked-loop (PLL) is adopted for grid synchronization. The resistive loading control employed by the MSC for maximum wave energy absorption is not discussed in this paper, with a comprehensive explanation available in [8]. The paper is organized in the following manner: Initially, it provides a comprehensive model of the PA system from wave to grid. Subsequently, it delves into the design aspects of the GSC control strategy. This is followed by an analysis of simulation results. Finally, the paper ends with a conclusion.

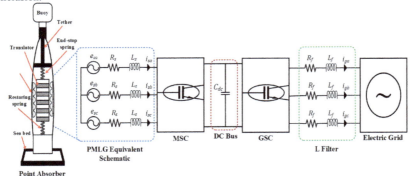

Figure 1: Point absorber system schematic.

System Model
A visual description of the PA system is shown in Fig.1. The PA consists of a floating cylindrical buoy, tethered to a three-phase permanent magnet linear generator (PMLG). A back-to-back converter topology, linking the MSC and GSC via a DC bus, serves as the intermediary between the PA and the electrical grid. Finally, an L filter is used to mitigate the harmonics distortion in the GSCs' output. This topology offers the ability to decouple the PA from the grid, enabling the system model to be split into two independent subsystems: the machine-side and the grid-side.

Machine-side Model

The machines-side model incorporates the wave-buoy hydrodynamic interaction and the electrical dynamics of the PMLG. By assuming only linear forces, the buoy heaving motion during wave interaction can be formulated using Newton's second law of motion given by

$$M\ddot{z}(t) = f_{ex}(t) - f_r(t) - f_b(t) - f_{rs}(t) + f_u(t). \tag{1}$$

where M being the buoy mass, and $\ddot{z}(t)$ is the heave acceleration. $f_{ex}(t)$ corresponds to the wave excitation force, $f_r(t)$ to the radiation force, $f_b(t)$ to the hydrostatic buoyancy force, $f_{rs}(t)$ to the restoring spring force, and $f_u(t)$ to the electromagnetic control force exerted by the PMLG.

The translator in the PMLG is mechanically linked to the buoy, thereby mirroring its heave motion. Consequently, this translator movement induces electromotive force (EMF) voltages in the stationary windings of the PMLG. The PMLG model in the abc frame is given as

$$\boldsymbol{e_s}(t) = R_s \boldsymbol{i_s}(t) + j\omega_e(t) L_s \boldsymbol{i_s}(t) + \boldsymbol{v_s}(t). \tag{2}$$

where $\boldsymbol{e}_s(t) = [e_{sa}(t) \ e_{sb}(t) \ e_{sc}(t)]^T$, $\boldsymbol{i}_s(t) = [i_{sa}(t) \ i_{sb}(t) \ i_{sc}(t)]^T$, R_s, L_s and $\boldsymbol{v}_s(t) = [v_{sa}(t) \ v_{sb}(t) \ v_{sc}(t)]^T$ are the three phase EMF voltage, stator currents, stator resistance, stator inductance and the three phase voltage terminals. The resistance and inductance represent the internal losses of the generator due to the copper windings and the magnetic field, respectively. By applying Parke transformation to Eq. 2 and including the MSC switching inputs, the machine-side model can be expressed in the synchronous reference (d-q) frame as

$$\frac{di_{sd}(t)}{dt} = \frac{-R_s i_{sd}(t)}{L_s} + \omega_e(t) i_{sq}(t) - \frac{u_{sd}(t)}{L_s}. \tag{3}$$

$$\frac{di_{sq}(t)}{dt} = \frac{-R_s i_{sq}(t)}{L_s} - \omega_e(t) i_{sd}(t) - \frac{\omega_e(t)}{L_s} \varphi_{pm} - \frac{u_{sq}(t)}{L_s}. \tag{4}$$

where $i_{sd}(t)$ and $i_{sq}(t)$ are the direct and quadrature components of the stator current, whereas $u_{sd}(t)$ and $u_{sq}(t)$ are the d-q components of MSC switching signals. The flux linkage is denoted by φ_{pm} and the angular frequency $\omega_e(t)$ is obtained as follows

$$\omega_e(t) = \frac{\pi}{\tau_p} \dot{z}(t). \tag{5}$$

with $\dot{z}(t)$ and τ_p being the heave velocity and the PMLG pole pitch, respectively. The captured wave power $P_m(t)$ and PA generated electrical power $P_e(t)$ are given by

$$P_m(t) = f_u(t) \dot{z}(t) = \frac{3\pi \varphi_{pm}}{2\tau_p} i_{sq}(t) \dot{z}(t). \tag{6}$$

$$P_e(t) = \frac{3}{2}\left(v_{sd}(t) i_{sd}(t) + v_{sq}(t) i_{sq}(t)\right) = v_{dc}(t) i_{dc}(t). \tag{7}$$

with $v_{dc}(t)$ and $i_{dc}(t)$ being DC bus voltage and DC current output of the MSC, respectively.

Grid-side Model

The grid-side model includes the GSC, L filter and electric grid. The grid is modeled as a voltage source in series with an impedance based on its Thevenin equivalent circuit. The impedance consists of a resistor R_g and inductance L_g to account for transmission line and power transformer effects. The grid-side model can be written using Kirchhoff's laws as follows

$$\boldsymbol{v}_g(t) = (R_f + R_g)\boldsymbol{i}_g(t) + (L_f + L_g)\frac{d\boldsymbol{i}_g(t)}{dt} + \boldsymbol{e}_g(t). \tag{8}$$

where $\boldsymbol{e}_g(t) = [e_{ga}(t) \ e_{gb}(t) \ e_{gc}(t)]^T$, $\boldsymbol{i}_g(t) = [i_{ga}(t) \ i_{gb}(t) \ i_{gc}(t)]^T$ and $\boldsymbol{v}_g(t) = [v_{ga}(t) \ v_{gb}(t) \ v_{gc}(t)]^T$ are the three phase grid voltages, grid currents and GSC voltage output, respectively. R_f represents a small resistor to account for losses in the filter. By applying Parke transformation to Eq. 8, the grid-side model in the (d-q) frame at grid voltage frequency ω_g is given by

$$\frac{di_{gd}(t)}{dt} = \frac{-(R_f + R_g) i_{gd}(t)}{(L_f + L_g)} - \frac{e_{gd}(t)}{(L_f + L_g)} + \frac{u_{gd}(t)}{(L_f + L_g)} + \omega_g(t) i_{gq}(t). \tag{9}$$

$$\frac{di_{gq}(t)}{dt} = \frac{-(R_f + R_g) i_{gq}(t)}{(L_f + L_g)} - \frac{e_{gq}(t)}{(L_f + L_g)} + \frac{u_{gq}(t)}{(L_f + L_g)} - \omega_g(t) i_{gd}(t). \tag{10}$$

here, $i_{gd}(t), i_{gq}(t), e_{gd}(t), e_{gq}(t), u_{gd}(t)$ and $u_{gd}(t)$ are the d-q components of the grid currents, grid voltages and GSC switching signals, respectively. The active and reactive power delivered to the grid are given respectively by

$$P_g(t) = \frac{3}{2}\big(e_{gd}(t)i_{gd}(t) + e_{gq}(t)i_{gq}(t)\big). \quad (11)$$

$$Q_g(t) = \frac{3}{2}\big(e_{gq}(t)i_{gd}(t) - e_{gd}(t)i_{gq}(t)\big). \quad (12)$$

Finally, the power exchange in the DC bus can expressed as follows

$$C_{dc} v_{dc}(t) \frac{dv_{dc}(t)}{dt} = P_e(t) - P_g(t). \quad (13)$$

where C_{dc} represents the DC bus capacitor.

Controller Design
The objective of the SMC is to govern the GSC operations to ensure the PA output power is grid compatible in terms of frequency synchronization, while regulating the DC bus voltage. Assuming no limitations on the grid power intake, the PA operates at peak generation capacity. A schematic of the SMC strategy in a VOC architecture is illustrated in Fig. 2.

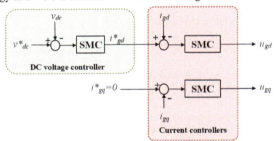

Figure 2: proposed GSC control strategy.

Outer Voltage Control Loop

As observed in Fig. 2, the voltage controller's output is the direct current reference $i_{gd}^*(t)$, which dictates the power level the GSC must either supply or absorb to keep the DC bus voltage at a fixed level. The quadrature current reference $i_{gq}^*(t)$ is set to zero to attain a unity power factor. Substituting Eq. 7 and Eq. 11 into Eq. 13, and simplifying that the quadrature grid voltage is zero $\big(e_{gq}(t) = 0\big)$, the DC voltage can be written as

$$\frac{dv_{dc}(t)}{dt} = \frac{1}{C_{dc}} i_{dc}(t) - \frac{3}{2C_{dc}v_{dc}(t)} e_{gd}(t)i_{gd}(t). \quad (14)$$

The SMC design requires formulating a sliding surface function $S_v(t)$ of the following form

$$S_v(t) = e_v(t) + \lambda_v \int e_v(t)\, dt. \quad (15)$$

here, $e_v(t)$ and λ_v are the voltage tracking error and convergence rate of $S_v(t)$. The tracking error $e_v(t)$ and its derivative are given by

$$e_v(t) = v_{dc}^*(t) - v_{dc}(t), \qquad \dot{e}_v(t) = \dot{v}_{dc}^*(t) - \dot{v}_{dc}(t). \quad (16)$$

where $v_{dc}^*(t)$ is the DC voltage reference. Based on Eq. 14, $\dot{e}_v(t)$ can be expressed as

$$\dot{e}_v(t) = \dot{v}_{dc}^*(t) - \frac{1}{C_{dc}} i_{dc}(t) + \frac{3}{2C_{dc}v_{dc}(t)} e_{gd}(t)i_{gd}(t). \quad (17)$$

By differentiating Eq. 15, yields the following

$$\dot{S}_v(t) = \dot{v}_{dc}^*(t) - \frac{1}{C_{dc}}i_{dc}(t) + \frac{3}{2C_{dc}v_{dc}(t)}e_{gd}(t)i_{gd}(t) + \lambda_v e_v(t). \tag{18}$$

Equating Eq. 18 to zero, and solving for $i_{gd}(t)$, yields the equivalent direct current reference $i_{gd,eq}^*(t)$ as follows

$$i_{gd,eq}(t) = \frac{2v_{dc}(t)}{3e_{gd}(t)}i_{dc}(t) - \frac{2C_{dc}v_{dc}(t)}{3e_{gd}(t)}\left(\dot{v}_{dc}^*(t) + \lambda_v e_v(t)\right). \tag{19}$$

Adding the exponential reaching law found in [7] to Eq. 19, gives the direct current reference $i_{gd}^*(t)$ as

$$i_{gd}^*(t) = \frac{2v_{dc}(t)}{3e_{gd}(t)}i_{dc}(t) - \frac{2C_{dc}v_{dc}(t)}{3e_{gd}(t)}\left(\dot{v}_{dc}^*(t) + \lambda_v e_v(t)\right) - \varepsilon_v \, sgn(S_v(t)) - k_v S_v(t). \tag{20}$$

with ε_v and k_v being positive tuning variables used to guarantee faster convergence speed, and the notation $sgn(S_v(t))$ represents the signum function of $S_v(t)$.

Inner Current Control Loop

Similar to the voltage controller, the current controllers require defining the d-q current tracking errors given by

$$e_{id}(t) = i_{gd}^*(t) - i_{gd}(t), \qquad\qquad e_{iq}(t) = i_{gq}^*(t) - i_{gq}(t). \tag{21}$$

The sliding surface functions can be defined as

$$S_d(t) = e_{id}(t) + \lambda_i \int e_{id}(t)\, dt, \qquad S_q(t) = e_{iq}(t) + \lambda_i \int e_{iq}(t)\, dt. \tag{22}$$

where i_{gd}^* and i_{gq}^* represent the d-q stator current references, whereas λ_i denote the convergence rate of the sliding functions $S_d(t)$ and $S_q(t)$. The derivatives of Eq. 22 using Eq. 9, Eq. 10 and Eq. 21 is given by

$$\dot{S}_d(t) = \dot{i}_{gd}^*(t) + \frac{(R_f+R_g)i_{gd}(t)}{(L_f+L_g)} + \frac{e_{gd}(t)}{(L_f+L_g)} - \frac{u_{gd}(t)}{(L_f+L_g)} - \omega_g(t)i_{gq}(t) + \lambda_i e_{id}(t). \tag{23}$$

$$\dot{S}_q(t) = \dot{i}_{gq}^*(t) + \frac{(R_f+R_g)i_{gq}(t)}{(L_f+L_g)} + \frac{e_{gq}(t)}{(L_f+L_g)} - \frac{u_{gq}(t)}{(L_f+L_g)} + \omega_g(t)i_{gd}(t) + \lambda_i e_{iq}(t). \tag{24}$$

Assuming $\dot{S}_d(t)$ and $\dot{S}_q(t)$ equal zero and incorporating the exponential reaching law, as delineated in [7], give the following GSC switching signals

$$u_{gd}(t) = (R_f + R_g)i_{gd}(t) + e_{gd}(t) - (L_f + L_g)\left(\dot{i}_{gd}^*(t) - \omega_g(t)i_{gq}(t) + \lambda_i e_{id}(t)\right) - \varepsilon_i \, sgn(S_d(t)) - k_i S_d(t). \tag{25}$$

$$u_{gq}(t) = (R_f + R_g)i_{gq}(t) + e_{gq}(t) - (L_f + L_g)\left(\dot{i}_{gq}^*(t) + \omega_g(t)i_{gq}(t) + \lambda_i e_{id}(t)\right) - \varepsilon_i \, sgn(S_q(t)) - k_i S_q(t). \tag{26}$$

here, ε_i and k_i are the tuning parameters for increasing the convergence rate.

Simulation Results

To assess the efficacy of the proposed SMC approach for GSC control, computational simulations were performed within a MATLAB framework. The voltage and current controllers operate at a sampling period of $1ms$ and their tuning parameters are set as $\lambda_v = 5$, $\lambda_i = 30$, $\varepsilon_v = \varepsilon_i = 3$ and $k_v = k_i = 1$. The system parameters used in this paper are depicted in Table 1.

Table 1: System parameters.

Subsystem	Specifications		
	Parameter	Symbol	Value
Machine-side	Total mass	m	3×10^4 kg
	PMLG flux linkage	φ_{pm}	19.8 Wb
	PMLG pole pitch	τ_p	0.045 m
	PMLG impedance	R_s, L_s	2 Ω, 32 mH
	DC bus capacitor	C_{dc}	2.7 mF
	Nominal DC bus voltage	v_{dc}	2000 V
Grid-side	L filter inductance	L_f	2 mH
	L filter resistance	R_f	40 mΩ
	Grid RMS voltage (line to line)	E_{gn}	563.4 V
	Grid frequency	f_g	50 Hz
	Grid impedance	R_g, L_g	20 mΩ, 0.1 mH

Fig. 3 demonstrates the efficacy of the voltage controller in the presence of polychromatic wave of significant height $4\ m$ and dominant frequency $0.65\ rad/s$. As evidenced in Fig. 3a, the DC voltage controller successfully sustained the bus voltage at its desired reference level of $2000\ V$. Voltage fluctuations were noted, particularly a $1.84\ V$ increase at $87.5\ s$, due to P_e exceeding P_g by $8.28\ kW$, as shown in Fig. 3b. Throughout the simulation from $60\ s$ to $140\ s$, the PA generated an average power output of $3.59\ kW$ and delivered an average power of $3.53\ kW$ to the electrical grid. This confirms that a majority of generated power is efficiently transferred to the grid with minor discrepancies due to switching losses.

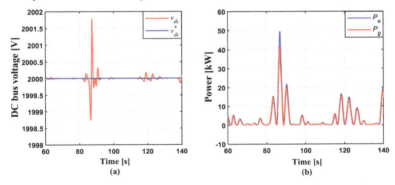

Figure 3: The voltage control loop performance: (a) DC voltage (b) power plots.

Fig. 4a and 4b show the performance of the low-level controllers for direct and quadrature currents, respectively. The direct current reference i_{gd}^* is determined by the DC voltage controller, which augments its value in response to a rise in DC voltage, thereby boosting the power injected into the electrical grid. The direct current reached a maximum peak of $86.7\ A$ in reaction to the highest DC bus voltage at $87.5\ s$. The quadrature current controller managed to regulate its respective current i_{gq} to zero to ensure a unity power factor. However, minor oscillations in both positive and negative directions were observed, attributable to the coupling effects between i_{gq}

and i_{gd}, especially at peak values of i_{gd}. Fig. 5 shows the power quality at the point of common coupling, with Fig. 5a displaying the three-phase grid injected currents and Fig. 5b confirming voltage and current of Phase A are synchronized for a unity power factor.

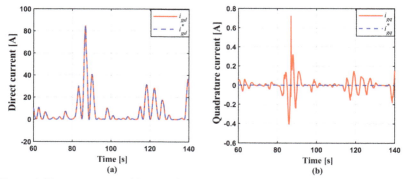

Figure 4: The current control loop performance: (a) direct current (b) quadrature current.

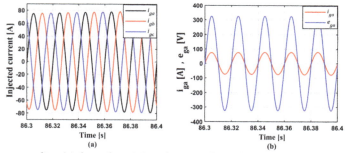

Figure 5: Power quality: (a) three phase injected current (b) grid voltage and current at phase A

Conclusion

This paper highlights the effectiveness of a sliding mode control strategy, incorporating an exponential reaching law within a voltage-oriented control architecture, for integrating a PA type WEC into the power grid. The study's primary goal was to enhance the transfer of generated power to the grid while maintaining the DC bus voltage at a set level and securing a unity power factor, addressing the inherent challenges posed by the variable nature of wave energy. Simulation outcomes, achieved through a MATLAB environment, validate the proposed control strategy's ability to maintain designated DC bus voltage levels and manage the direct and quadrature currents effectively by reducing tracking errors. This research contributes to the field by providing a robust control solution that can improve the reliability and efficiency of wave energy conversion systems when integrated into electrical grids.

References

[1] A. Wahyudie and M. Jama, "Perspective on damping control startegy for heaving wave energy converters," *IEEE Access*, vol. 5, 99. 22224-22233, 2017. https://doi.org/10.1109/ACCESS.2017.2757278

[2] O. Saeed *et al*, "Simple resonance circuit to improve power conversion in a two-sided planar permanent magnet linear generator for wave energy converters," *IEEE Access*, vol. 5, pp. 18654-18664, 2017. https://doi.org/10.1109/ACCESS.2017.2752466

[3] M. Jama *et al.*, "Self-tunable fuzzy logic controller for optimization of heaving wave energy converters," *in proc. of ICRERA 2012, pp. 6477273, 2012.* https://doi.org/10.1109/ICRERA.2012.6477273

[4] S. Rasool *et al*, "Coupled modeling and advanced control for smooth operation of a grid-connected linear electric generator based wave-to-wire system," *IEEE Transactions on Industry Applications*, vol. 56, no. 5, pp. 5575-5584, 2020. https://doi.org/10.1109/TIA.2020.3004759

[5] H. A. Said, D. García-Violini, and J. V. Ringwood, "Wave-to-grid (W2G) control of a wave energy converter," *Energy Conversion and Management: X,* vol. 14, p. 100190, 2022. https://doi.org/10.1016/j.ecmx.2022.100190

[6] K. Zeb *et al.*, "A comprehensive review on inverter topologies and control strategies for grid connected photovoltaic system," *Ren. and Sust. Energy Reviews,* vol. 94, pp. 1120-1141, 2018. https://doi.org/10.1016/j.rser.2018.06.053

[7] J. Liu, *Sliding mode control using MATLAB*. Academic Press, 2017. https://doi.org/10.1016/B978-0-12-802575-8.00005-9

[8] A. Y. Elamin *et al.*, "Real-time Model Predictive Control Framework for a Point Absorber Wave Energy Converter with Excitation Force Estimation and Prediction," *IEEE Access,* 2023. https://doi.org/10.1109/ACCESS.2023.3347731

Design and fabrication of low temperature flat plate collector for domestic water heating

Abdulkrim Almutairi[1,a*], Abdulrahman Almarul[1,b], Riyadh Alnafessah[1,c], Abbas Hassan Abbas Atya[1,d], Sheroz Khan[1,e], Noor Maricar[1,f]

[1]Department of Electrical Engineering, College of Engineering and Information Technology, Onaizah Colleges, 56447, Al Qassim, Saudi Arabia

[a]432110787@oc.edu.sa, [b]122382024@oc.edu.sa, [c]401100108@oc.edu.sa, [d]abbas.atya@oc.edu.sa, [e]cnar32.sheroz@gmail.com, [f] nmmaricar@yahoo.com

Keywords: Flat Plate Collector, Glazed, Unglazed, Daylight Time, Heat Transfer Rate, Mass Flow, Solar Irradiation, Specific Heat, Headers, Raisers, Thermal Conductivity, Apparent Solar Time, Absorber Plate

Abstract. Solar energy, specifically using flat plate collectors, shows very promising as a renewable energy source for meeting global energy demands. These collectors, commonly used to capture solar energy and convert it into heat for various purposes, offer advantages over conventional water heaters in terms of lifespan and energy consumption. This study focuses on understanding flat plate collectors with region specific parameters for domestic use in Onaizah, Saudi Arabia. It includes a detailed design process based on site-specific data and featured with settings adjustable for different seasons. The use of double glazing with a gap between glass panels provides heat insulation from the front. Initial experiments conducted involved temperature measurements of outlet water and calculation of the solar collector's efficiency, resulting in the highest temperature recorded at 62.3°C and an efficiency of 53.5%.

Introduction
Traditional water heaters rely on burning fossil fuels or using resistive heating elements, leading to carbon emissions or wastage of electrical energy. They struggle to meet hot water demands efficiently, resulting in unnecessary energy consumption. These systems also have maintenance issues and a shortened lifespan [1]. Flat plate collectors (FPCs) have undergone significant developments to improve performance. Researchers have worked on enhancing efficiency, minimizing thermal losses, advancing materials and coatings, and exploring innovative designs. The performance of FPCs depends on factors such as solar irradiance, temperature, design, and operating conditions [2, 5]. With history dating back to early 20th century, FPCs water heaters gained popularity in the 1940s and 1950s. Researchers then focused on selective coatings and improved glazing materials leading to efforts to improve insulation with materials of fiberglass, polyurethane foam, and mineral wool. Potential strategies and research directions included cost reduction, performance enhancement, and integration with energy storage systems [5, 9].

Advancements in materials science, manufacturing techniques, and system designs have continued to shape and optimize the FPCs technology, developing high-performance absorber coatings, exploring novel materials, and optimizing the design of flow channels. Innovations included integral collector-storage systems (ICS) and evaluated flat plate collectors (EFPCs), offering improved thermal performance and energy storage capabilities [12]. The orientation and tilt angle of FPCs play crucial role in their performance, typically aligned to the latitude of the location [13].

Calculations are used to determine the performance and heat output of FPCs, taking into account incident solar radiation, fluid flow rate, temperature differentials, and heat transfer coefficients.

Despite having advantages, FPCs have weather-coupled limitations of lower efficiency, having thus different types, including glazed, unglazed, hybrid, and high-temperature collectors [11,13].

FPCs are used in various applications such as residential water heating, space heating, commercial water heating, manufacturing processes, greenhouse heating, pool heating, and solar desalination. However, there are challenges that include improving thermal efficiency, minimizing heat losses, reducing costs, and developing efficient thermal energy storage systems. Supportive policies, financial incentives, and public awareness campaigns are important to promote the use of FPCs and maximize their potential usages. Continued research and collaboration among stakeholders are key to addressing these challenges and advancing FPC technology [10.13]

Problem:
The problem under focus inefficiency and high energy consumption associated with conventional water heaters used for residential and commercial purposes. Traditional water heaters often rely on fossil fuels or electric resistance heating elements. An ideal water heater should prioritize energy efficiency by utilizing renewable energy sources [13].

Objectives:
To determine the optimal design parameters for flat plate collectors in terms of the angle at which they are tilted to maximize solar energy absorption and overall system performance.

To assess the energy efficiency of flat plate collectors for domestic water heating by measuring heat absorption, transfer, and overall system efficiency under varying weather and operating conditions.

Significance of study
The significance of the study lies in addressing local needs by designing a flat plate collector tailored to the specific requirements and climatic conditions of the Qassim Region. Improving thermal performance using double glazing with a vacuum, reducing heat loss, and increasing overall system efficiency. Providing cost savings and affordability by reducing reliance on conventional energy sources and lowering utility bills. Supporting environmental sustainability by utilizing solar energy as a renewable and clean energy source.

Methodology
The design process involves considering various factors such as collector area, materials, fluid selection, and system configuration. The absorber plate area is a critical design parameter that determines the amount of solar energy that can be captured [12]. The solar radiation incident is measured throughout the day. For our application we need the measurements for a certain area in wintertime. Since, in Saudi Arabia the water heaters only needed in winter we need to take into consideration Daylight Time. Daylight Time in Saudi Arabia is 10 hours, this information can be used to calculate solar radiation incident on a specific area [6]. From the information about the location, we designed our collector for this application using equation (1).

$$A = \frac{Qu}{G\eta} \quad (1)$$

Where A is the collector area in square meters, Qu is the desired heat output in Watts, G is the solar radiation incident on the collector in watts per square meter, and η is the collector's thermal efficiency as in equation (2).

$$Qu = \dot{m}\, C_p(T_{out} - T_{in}) \quad (2)$$

Where $\dot{m}$ is the mass flow rate of the fluid in kilograms per second, C_p is the specific heat capacity of the fluid in joules per kilogram per °C for water, T_{out} is the outlet temperature of the fluid in degrees Celsius and T_{in} is the inlet temperature. The mass flow rate can be calculated from the capacity of the flat plate collector. If we assume that the capacity for our tank is 20 liters. To determine the required area for FPC, factors such as solar insolation data, daily energy demand, and assumed efficiency are considered. Design parameters of diameter, spacing, and number is crucial. The riser diameter is chosen based on cost, flow rate, and efficiency. A diameter of 9mm is selected in this case. The spacing between risers is decided on heat transfer, pressure drop, and cost. The number of risers is calculated based on the width of the absorber plate, riser diameter, and spacing. These parameters optimize performance, efficiency, and cost-effectiveness. Fig.1 shows a wooden case with insulation, absorber plate, pipes, and glazing.

Figure 1: The rendering of the flat plate collector configuration

Material Selection

The selection of materials is important when designing a flat plate solar collector. Factors like cost, availability, installation ease, and lifespan should be considered. Copper, aluminum, steel, and stainless steel are common materials for the absorber plate, each offering different thermal conductivity and corrosion resistance properties. Risers and headers are typically made of copper, aluminum, or stainless steel. Insulation materials like rock wool, cellulose, Styrofoam, polyurethane foam, and polyisocyanurate foam are commonly used. Glazing options include tempered glass, low-iron glass, and polycarbonate. The orientation of a solar collector should be such that it faces south for maximum solar radiation with tilt angle at around 26°±5° from the horizontal plane, with adjustable mechanisms. The apparent solar time can be determined by considering the time zone offset and longitude.

Results

The design of the solar water heater is based on the region average solar irradiation, which stands at 6000 kWh per day for Onaizah, and since the heater is used in the winter, which has ten hours of daylight, the average solar radiation per hour is 600 kW [14]. Assumptions are made for the inlet temperature (25°C), outlet temperature (70°C), and operation time (five hours). Assuming an efficiency of 50%, the area of the collector is calculated to be 0.6 square meters. The dimensions of the collector, including the number of risers, flow rate, and absorber plate area, are determined based on the desired specifications. The chosen dimensions for the design are 1m x 0.6m (length x width), with a header diameter of 22.225mm, riser diameter of 9mm, and a spacing of 90mm.

Copper is often chosen for risers and headers due to its high thermal conductivity and corrosion resistance. To enhance thermal absorbability, the plate and pipes are usually painted black. Insulation materials such as rock wool, cellulose, Styrofoam, polyurethane foam, and polyisocyanurate foam are considered, with the choice depending on factors such as thermal conductivity, moisture resistance, and cost. Rock wool is typically chosen when balancing these factors. For glazing material, options include tempered glass, low-iron glass, and polycarbonate.

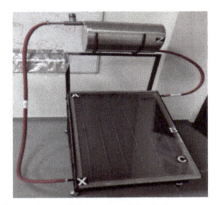

Figure 2: The fabrication of the design

Each offers advantages in terms of durability, transparency, and solar transmittance. The recommended spacing usually ranges from 10 to 25mm. In the described design, the spacing is 10mm and vacuumed to minimize heat loss from the glazing [13].

There are several different mechanisms that can be used to change the tilt angle of a flat plate solar collector. This design's mechanism is manual since it is the most reliable and can withstand the wight of the collector. The mechanism is two sets of steel tubes. One set is attached to the collector by a T-joint. And the other set is attached to a steal base holding the collector by a T-joint. As shown in Fig 3.

Figure 3: The mechanism of angle change.

The tank has many configurations based on the system design expected pressure and temperature. The tank is made from two sheets of galvanized steel and a layer of rock wool for insulation. The tank has an inlet at the bottom and an outlet at the top. From the picture below in Fig 4 it is demonstrated that the tank is connected to the collector though two hoses. The hoses are flexible to move with collector when changing the angle.

Figure 4: The tank for the

There are five sensors in the collector which are A, B, C, Z, and X as illustrated in Fig. 1. Sensors A, B, and C are temperature sensors on the plate, which provide the average temperature of the absorber plate. Sensors Z and X are the outlet and inlet sensors, respectively. All five monitors for the sensors are on the board on the side of the collector.

On Wednesday the twenty-first of February 2024 a preliminary test was performed on the collector. From this test the efficiency of the collector is calculated. The water outlet temperature increased to maximum value 62.3°C at around 4:00 PM and then decreased as shown in Fig. 5.
As shown in Fig. 6 the efficiency of the collector increased with time and reached its maximum value 53.5% around 4:00 PM and then decreased.

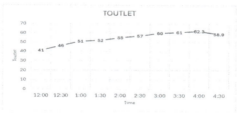

Figure 5: The temperature in the outlet vs

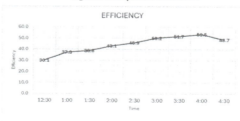

Figure 6: The Efficiency VS time chart.

Recommendations
The performance of the flat plate collector has been assessed after fabrication and testing, where its efficiency is a key factor. The actual collector efficiency may differ from the theoretical

efficiency depending on the design, mainly due to system losses. These losses can vary across regions and collectors but can be evaluated using specific criteria. These parameters must be tested in different conditions and in different seasons. The performance of a flat plate solar collector is evaluated based on its efficiency, which can vary from theoretical values due to system losses. These losses can be assessed using specific criteria such as the useful thermal load, thermal losses, and the heat removal factor. The useful thermal load refers to the amount of thermal energy provided by the collector to meet the desired demand, taking into account the heat energy transferred from the collector to the load. It depends on various factors including the collector's performance, heat transfer characteristics, specific requirements, and thermal losses within the system. The thermal losses are determined by calculating the overall heat loss coefficient, which considers the major areas of heat loss in the collector such as the top, bottom, and sides. The heat removal factor measures how effectively the collector transfers heat from the absorber plate to the heat transfer fluid, and it is calculated as the ratio of the actual heat transfer rate to the maximum possible rate under ideal conditions. This factor can be used to determine the absorbed useful energy by considering parameters such as the collector area, irradiance, absorptance and transmittance of the collector, overall heat loss coefficient, and the temperature difference between the fluid inlet and ambient temperature. These parameters should be tested under different conditions and seasons to accurately evaluate the performance of the flat plate collector.

Conclusion
In summary, the design and optimization of a flat panel solar collector involves various considerations. The collector is oriented south-ward to maximize solar radiation, with the tilt angle manually adjusted based on the latitude of Onizah, Saudi Arabia. Dimensions are determined based on the number of risers, water flow rate, and absorption plate area. Vertical orientation is preferred, with dimensions of 1m x 0.6m. Copper is chosen for risers and headers because of its thermal conductivity and resistance to corrosion. Black coating enhances thermal absorption ability. Rock wool insulation provides thermal insulation and moisture resistance. Tempered glass serves as glazing material to ensure durability and transparency. A 10mm vacuum gap between the glazing and absorber plate reduces heat loss. Manual adjustment includes two sets of steel tubes and a steel base to reliably change the tilt angle. Flexible hoses connect the collector to the tank, allowing for angle adjustments. Sensors monitor temperature variations. Initial tests show a maximum efficiency of 53.5%. Calculations evaluate thermal load and losses. This optimized collector effectively harnesses solar energy for domestic water-heating applications, providing a sustainable and cost-effective solution for the region's thermal energy requirements.

Acknowledgement
The authors express their gratitude to Department of Electrical Engineering, College of Engineering and Information Technology, Onaizah Colleges, Qassim, Saudi Arabia, for their laboratory/library support throughout the analytical derivation and design implementation.

References

[1] D. A. Bainbridge, The Integral Passive Solar Water Heater Book: Breadboxes, Batchers, and Other Types of Simple Solar Water Heaters. Davis, CA.: Passive Solar Institute, 1981.

[2] Y. Tripanagnostopoulos, M. Souliotis, and T. Nousia, "Solar collectors with colored absorbers," Solar Energy, vol. 68, no. 4, pp. 343–356, 2000. https://doi.org/10.1016/s0038-092x(00)00031-1

[3] K. Patel, "Comparative analysis of solar heaters and heat exchangers in residential water heating," International Journal of Science and Research Archive, vol. 9, no. 2, pp. 830–843, 2023. https://doi.org/10.30574/ijsra.2023.9.2.0689

[4] Shamsul Azha, Nurril Ikmal, Hilmi Hussin, Mohammad Shakir Nasif, and Tanweer Hussain. "Thermal performance enhancement in flat plate solar collector solar water heater: A review." *Processes* 8, no. 7 (2020): 756.

[5] El-Sebaee, I. M. "Develop of Local Flat Plate Solar Heater." *Journal of Soil Sciences and Agricultural Engineering* 12, no. 5 (2021): 385-390.

[6] L. C. Spencer, "A comprehensive review of small solar-powered heat engines: Part I. A history of solar-powered devices up to 1950," Solar Energy, vol. 43, no. 4, pp. 191–196, 1989. https://doi.org/10.1016/0038-092x(89)90019-4

[7] J. W. Twidell and A. D. Weir, Renewable Energy Resources. London: Taylor & Francis Ltd, 2015

[8] F. Struckmann, "Analysis of a Flat-plate Solar Collector," Academia, https://www.academia.edu/download/44502033/Fabio.pdf (accessed Sep. 19, 2023).

[9] H. Dagdougui, A. Ouammi, M. Robba, and R. Sacile, "Thermal analysis and performance optimization of a solar water heater flat plate collector: Application to Tétouan (Morocco)," Renewable and Sustainable Energy Reviews, vol. 15, no. 1, pp. 630–638, 2011. https://doi.org/10.1016/j.rser.2010.09.010

[10] L. C. Spencer, "A comprehensive review of small solar-powered heat engines: Part I. A history of solar-powered devices up to 1950," Solar Energy, vol. 43, no. 4, pp. 191–196, 1989. https://doi.org/10.1016/0038-092x(89)90019-4

[11] Putra, Andika, K. Arwizet, Yolli Fernanda, and Delima Yanti Sari. "Performance Analysis of Water Heating System by Using Double Glazed Flat Plate Solar Water Heater." *Teknomekanik* 4, no. 1 (2021): 1-7.

[12] S. Algarni, "Evaluation and optimization of the performance and efficiency of a hybrid flat plate solar collector integrated with phase change material and heat sink," Case Studies in Thermal Engineering, vol. 45, p. 102892, 2023. https://doi.org/10.1016/j.csite.2023.102892

[13] P. Khobragade, P. Ghutke, V. P. Kalbande and N. Purohit, "Advancement in Internet of Things (IoT) Based Solar Collector for Thermal Energy Storage System Devices: A Review," 2022 2nd International Conference on Power Electronics & IoT Applications in Renewable Energy and its Control (PARC), Mathura, India, 2022, pp. 1-5. https://doi.org/10.1109/PARC52418.2022.9726651

[14] "Solar Resource Maps of Saudi Arabia," solargis.com. https://solargis.com/maps-and-gis-data/download/saudi-arabia (accessed Oct. 25, 2023).

The utilization of IoT-based humidity monitoring method and convolutional neural networks for orchid seed germination

Muhammad Ridhan FIRDAUS[1,a], Hilal H. NUHA[1,b], Mohamed MOHANDES[2,c *], and Agus HAERUMAN[2,d]

[1]Telkom University, Bandung, Indonesia

[2]King Fahd University of Petroleum and Minerals (KFUPM), Dhahran, Saudi Arabia

(KFUPM), Dhahran, Saudi Arabia

[a]ridhanfirdaus@student.telkomuniversity.ac.id, [b]hilalnuha@ieee.org, [c]mohandes@kfupm.edu.sa, [d]g202208480@kfupm.edu.sa

Keywords: CNN, IOT, Orchid Seeding

Abstract. This research endeavors to advance orchid seed germination efficiency through the development of an Internet of Things (IoT)-based humidity monitoring system integrated with Convolutional Neural Networks (CNN). Recognizing the pivotal role of proper humidity in orchid seed sowing, the proposed system employs humidity sensors connected to an IoT platform for real-time data collection. The collected data undergoes analysis and prediction by CNN, elucidated through graphical representations such as histograms, line charts, and scatterplot charts. By synergizing IoT technology with artificial intelligence, this innovative system contributes positively to orchid seed sowing efficiency, empowering farmers and orchid cultivators to optimize plant growth conditions. Furthermore, the adaptability of this approach extends beyond orchids, making it applicable to various crop seeding applications through parameter modifications tailored to specific needs.

Introduction

Smart farming is taken from the word 'smart' in smart city, smart farming which was originally called 'precision farming' will become a mandatory agricultural concept in the future due to limited land. Smart farming utilizes technology such as big data, GPS, and the Internet of Things (IoT) to improve the quality and quantity of production in the agricultural industry. Things like this should help simplify and streamline all agricultural processes from production to marketing [1].

The Smart Farming concept began to be developed as an effort to increase the efficiency and productivity of the agricultural sector which is still dominated by traditional methods. Rapid technological developments enable farmers to utilize technological information and communication in agricultural land management, production and marketing of agricultural products. Smart Farming has many benefits, including increasing the efficiency of using resources such as air and energy, increasing the quality and quantity of harvests, reducing production costs, and minimizing negative impacts on the environment. Apart from that, Smart Farming can also help farmers integrate and control the agricultural environment in real-time, which can help in overcoming problems that arise quickly and in a timely manner [2][3].

Convolutional Neural Networks (CNN) is a method of smart farming. CNN itself is a type of neural network commonly used on image data. CNN can be used to detect and recognize objects in an image. CNN is a technique inspired by the way mammals — humans, produce visual perception.

In general, a Convolutional Neural Network (CNN) is not much different from a regular neural network (NN). NNs typically transform input by passing it through a series of hidden layers. Each layer consists of a collection of neurons, where each layer is fully connected to all the neurons in the previous layer. Finally, a fully connected layer (output layer) is used to represent the predictions. The CNN consists of neurons that have weights, biases and activation functions. Convolutional layers also consist of neurons arranged in such a way as to form a filter with length and height (pixels) and CNN utilizes the convolution process by moving a convolution kernel (filter) of a certain size to an image, the computer obtains new representative information from the results of multiplying parts of the image depending on the filter used [4][5]. CNN is used for sowing orchid seeds, because CNN can help with visual analysis, such as identifying problems, and providing information for the maintenance and development of orchid seeds.

The paper is organized into distinct sections to present the research on utilizing an Internet of Things (IoT)-based humidity monitoring system combined with Convolutional Neural Networks (CNN) for efficient orchid seed germination. The abstract succinctly introduces the research, highlighting the integration of humidity sensors with CNN analysis for optimal orchid seeding conditions. The introduction outlines the significance of smart farming, emphasizing the role of technology, particularly CNN, in agricultural advancements. The literature review delves into the importance of real-time humidity monitoring, the challenges in orchid cultivation, and the application of IoT and CNN in the context of smart farming. The methodology section details the problem identification process, the seven research stages, and the selection of evaluation metrics, showcasing the comprehensive approach taken. Device design elucidates the essential tools—NodeMCU, Water Pump DC 12V, and DHT11—employed in the research. Algorithm design provides a visual representation of the analysis method used in the Temporary Immersion System. The results section showcases temperature and humidity data, with a focus on accuracy testing using CNN and comparisons with other methods. The conclusion summarizes key findings, emphasizing the high accuracy of the IoT and CNN-based humidity monitoring methods and their potential applications in optimizing orchid seed sowing.

Literature Review
Moisture Monitoring. Monitoring soil moisture is very important in crop cultivation, especially at the time of planting. Proper soil moisture can increase the success of plant growth and reduce the risk of growth failure. The use of IoT sensor technology can monitor soil moisture in real-time and accurately [6][7].

Fig. 1 Humidity Monitoring [6].

Orchid Cultivation. Orchids are a type of ornamental plant that is very popular and has high economic value. However, orchid cultivation requires careful care and supervision, especially when sowing seeds. Real-time and accurate monitoring of humidity can help increase the success of sowing orchid seeds.

Internet of Things (IoT). IoT is a concept where electronic devices connected to the internet can communicate with each other and exchange data. In this case, the use of IoT sensor technology can monitor soil moisture in real-time and send data to a server that can be accessed from anywhere.

Convolutional Neural Network (CNN). CNN is one of the most popular types of Deep Learning algorithms in image processing. CNN can learn complex patterns in images and classify images into appropriate categories based on the learned features. In this case, CNN can be used to process humidity data from IoT sensors and classify humidity status to predict the success of orchid seed sowing [8][9].

Methodology

Identification of problems. The utilization of Root Mean Square Error (RMSE) is imperative in addressing the inherent challenges associated with orchid seed germination, particularly the unpredictability in factors influencing orchid growth outcomes. Orchid seeds pose unique difficulties in prediction due to various uncontrollable variables. Traditional prediction methods often fall short in accurately foreseeing the results of observed plant behavior, leading to suboptimal predictions. RMSE serves as a valuable metric, specifically gauging plant quality, enabling researchers to set high expectations for seeds predicted to be superior while discerning and discarding those anticipated to yield damaged or compromised outcomes. This strategic application of RMSE not only enhances predictive accuracy but also aids in making informed decisions regarding seed quality and subsequent plant development.

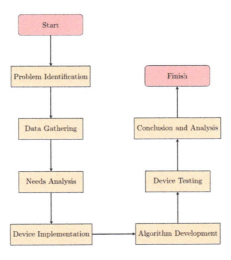

Fig. 2 Research Flow Diagram.

Research Stages. This research includes seven research stages which can be seen in Fig. 2. The beginning of this research is the problem identification process. After that, the data collection stage is carried out to process the information, then analyze the requirements needed to design the tool, including hardware, sensors and other components, with the aim of preventing errors when making the tool. After the tool has been successfully created, the research will proceed to the testing stage using the Convolutional Neural Network (CNN) method and making comparisons with other methods.

Method Selection. Mean Squared Error (MSE) is not a direct part of Convolutional Neural Networks (CNN), but MSE is one of the evaluation metrics commonly used in training and evaluating CNN models. Root Mean Square Error (RMSE) is a standard way to measure the error of a model in predicting quantitative data.

$$RMSE = \sqrt{\sum_{k=1}^{K} \frac{(\hat{y}_k - y_k)^2}{K}} \qquad (1)$$

MAE is one of the evaluation methods commonly used in data science. MAE calculates the average of the absolute differences between predicted and actual values.

$$MAE = \frac{1}{K}\sum_{k=1}^{K} |\hat{y}_k - y_k| \quad (2)$$

R^2 is basically used to see how adding independent variables helps explain the variance of the dependent variable.

$$R^2 = 1 - \frac{SSR}{SST} \quad (3)$$

where R^2 is the coefficient of determination, RSS is the sum of squares of residuals, and TSS is total sum of squares.

Device Design. Following items are some of the tools needed in designing:

NodeMCU [10] (Fig. 3a) is an open source IoT platform. Consists of hardware and software. The hardware is a development board integrated with an ESP8266 microcontroller and a USB to serial communication chip. The software is firmware that is compatible with the Arduino IDE.

Water Pump DC 12V (Fig. 3b). This type of pump is operated by immersing it in water and cannot work outside the water. The water pump in this design is designed to flow air from tank 1 to tank 2 periodically.

DHT11 [11] (Fig. 3c) is a sensor that can detect temperature and humidity around the area where the sensor is placed. This sensor consists of a thermistor which functions to measure temperature and a capacitive sensor which is used to measure humidity levels. Generally, this sensor is integrated in a module which is equipped with sensors and chips to convert analog signals into digital signals.

(a) (b) (c)
Fig. 3 Components (a) NodeMCU, (b) Water Pump, and (c) DHT11

Algorithm Design. In analyzing the impact of the time and frequency of watering in the Temporary Immersion System on the growth of orchid seeds and sowing orchid seeds, the algorithm method as depicted in Fig. 4 is used.

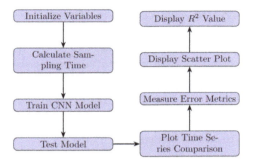

Fig. 4 Research Algorithm.

Based on the algorithm in Fig. 4, it can be explained that the code uses the predict() function to predict air humidity for each data in Xts. This function returns predicted values in vector form. The code uses the functions mse(), nmse(), rmse(), nrmse(), mae(), and mbe() to calculate various error measures between predictions and actual data. The first plot shows the actual and predicted humidity values for the first 500 minutes. The second plot shows the actual and predicted humidity values for the first 200 minutes. The R-squared value calculated with the code is 0.95, which means that the machine learning model can predict humidity with high accuracy.

Result and Discussion

Measurement results. During the period from 7 to 10 November 2023, measurements were carried out periodically every 10 seconds using the ESP32 device. A total of 4000 data were collected to record the temperature and humidity in vessel 1. The aim of this observation was to gain deeper insight into changes in temperature and humidity, especially in the context of vessel 1. The data processing was carried out using the Matlab application.

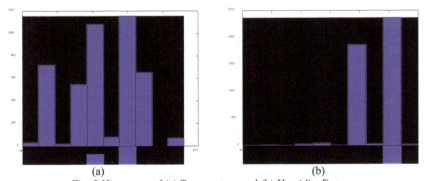

Fig. 5 Histogram of (a) Temperature and (b) Humidity Data.

Fig. 5(a) depicts a detailed graphical representation of the results of temperature measurements in vessel 1 during the observation period. The recorded temperature range was between 23° to 25.5° Celsius, providing a comprehensive picture of the temperature variations in the room. The peak frequency occurred at a temperature of 24.5° Celsius, with the amount of data reaching 1150, followed by a temperature of 24° Celsius which reached a frequency of 1100 data. Apart from that, a temperature of 23.3° Celsius was also recorded with a frequency of 750 data.

Fig. 5(b) depicts a detailed graphical representation of the measurement results of humidity in vessel 1. The data presented via a histogram highlights the humidity value range between 92% to 98%. From the analysis carried out, it can be seen that the dominant humidity in vessel 1 reached 97%, with a measurement frequency reaching 2400 data. Apart from that, humidity at the 96% level also shows significance, measured by a measurement frequency of 1800 data.

Accuracy Test Results Using CNN. In the framework of this research, accurate measurements were carried out using the CNN method, and the results were compared with several other methods. The goal of this comparison is to gain a thorough understanding of the performance of tree regression methods on accuracy measurements.

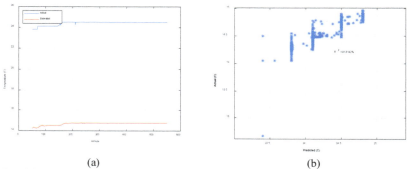

(a) (b)
Fig. 6 (a) Comparison of Estimated Temperature Prediction Results Using CNN and (b) its Scatter Plot

The graph presented in Figure 6(a) illustrates a discrepancy between actual and predicted temperatures at the 50th minute, with the actual temperature recorded at 23.8 Celsius and the predicted temperature at 14.2 Celsius. The actual temperature exhibited a gradual increase, in contrast to the forecasted temperature, which rose in an erratic manner. By the 250th minute, both temperatures had stabilized, with the actual temperature reaching 24.5 Celsius and the forecasted temperature leveling off at 14.7 Celsius. The graph features a red line representing the predicted temperature, which does not run parallel to the blue line that indicates the actual temperature, highlighting the inaccuracy of the temperature predictions. Furthermore, the temperature prediction scatter plot, as shown in Figure 6(b), reveals that the CNN method's tests display a general trend of rising actual temperature values in conjunction with increases in estimated temperatures, despite the fluctuations in this ascent.

In Figure 7(a), during the 50th minute, 9 different tests were conducted, revealing a discrepancy between the actual and the predicted humidities, despite their nearly identical graphical representations. The actual humidity was recorded at 97%, while the predicted humidity was slightly lower at 94.05%. Both humidities experienced a decline at the 237th minute, with the actual humidity dropping to 96% and the predicted humidity to 93.3%. By the 250th minute, humidities began to rise, eventually stabilizing at the original levels by the 270th minute. This pattern indicates a slight increase in the second humidity measurement before it too stabilized at the initial level, suggesting that the predictions do not perfectly align with the actual humidities. The Scatter Plot of the humidity prediction is shown in Fig. 7(b). The humidity accuracy test appears almost rectangular, with humidity values fluctuating between 93.3% and 94.2%, and reaching a peak at 94.6%. Concurrently, the CNN method achieved a high accuracy level, attaining 91.383%. Scatterplot analysis indicates a positive correlation between the actual and estimated humidity values.

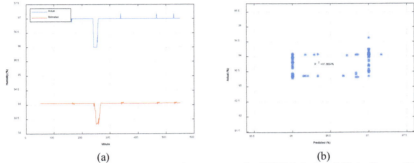

(a)　　　　　　　　　　　　　　　　(b)

Fig. 7 (a) Comparison of Humidity Prediction using the CNN Method and (b) its Scatter Plot.

Comparison of Results. Table I demonstrates that the training method outperforms the testing method in forecasting temperature and humidity levels. This superiority is evidenced by the higher values of R2, MSE, RMSE, MAE, MBE, NRMSE, and NMSE for the training method compared to those for the testing method. The overall analysis of the table within the figure suggests that the R2 method is more effective in predicting air temperature and humidity.

Table 1. Comparison of Results.

Method	Temperature			Humidity		
	Train	Validation	Test	Train	Validation	Test
R2	0.95	0.71	0.91	0.94	0.5	0.91
MSE	93.3	1.81	93.7	7.74	8.65	8.07
RMSE	9.66	13.4	9.68	2.78	93.01	284
MAE	9.65	13.4	9.68	2.77	93.01	28.3
MBE	-9.65	13.4	-9.68	-2.77	93.01	-28.3
NMSE	0.15	2.01	0.15	8.33	8.84	8.64
NRMSE	4.2	3.09	6.91	0.55	9.11	2.43

Conclusion

In conclusion, this study successfully demonstrates the efficacy of an Internet of Things (IoT) and Convolutional Neural Networks (CNN)-based air humidity monitoring approach for enhancing orchid seed germination. The research findings affirm that the integrated system, utilizing real-time data collection and CNN analysis, achieves a notable accuracy level, particularly highlighted by a 91.383% success rate in humidity prediction. The insights gained from this study offer valuable information for farmers, enabling them to make informed decisions about optimal watering schedules and nutrient supplementation, thereby increasing the likelihood of successful orchid seed sowing. Moreover, the adaptability of this method for broader agricultural applications underscores its potential as a transformative tool in smart farming practices. By promoting increased productivity, improved crop quality, and reduced risks of failure, the IoT and CNN-based humidity monitoring approach emerges as a promising avenue for sustainable agricultural practices.

References

[1] Adminlp2m, Application of Smart Farming 4.0 in Current Agricultural Technology, 5 November 2020. https://lp2m.uma.ac.id/2021/11/05/penerapan-smart-farming-4-0-dalam-teknologi-pertanian-masa-kini/

[2] Gifari Zakawali. Getting to Know Digital Farming, Here are the Benefits for Farmers, 21 November 2022. https://store.sirclo.com/blog/benefits-digital-farming/

[3] Advantages Of Smart Agriculture | Disadvantages Of Smart Agriculture. 28 January 2024. https://www.rfwireless-world.com/Terminology/Advantages-and-Disadvantages-of-SmartAgriculture-Farming.html

[4] Qolbiyatul Lina. Neural Networks Conclusion What Is It. 2 Jan 2019. https://medium.com/@16611110/apa-itu-convolutional-neural-network-836f70b193a4

[5] Trivusi. Understanding and How the Convolutional Neural Network (CNN) Algorithm Works. 28 July 2022.https://www.trivusi.web.id/2022/04/algoritma-cnn.html.

[6] Fig. Soil Monitoring with IoT – Smart Agriculture. February 24, 2021. https://www.manxtechgroup.com/soil-monitoring-with-iot-smart-agri.

[7] Irenasari, Almira Harwidya, and Soemarno Soemarno. " Soil Moisture Assessment Using Soil Moisture Index (SMI) Method at the Bangelan Coffee Plantation, Malang Regency, East Java." Jurnal Tanah dan Sumberdaya Lahan 9, no. 1 (2022): 1-12. https://doi.org/ 10.21776/ub.jtsl.2022.009.1.1

[8] Errissya Rasywir, Rudolf Sinaga, Yovi Pratama. Analysis and Implementation of Palm Disease Diagnosis using the Convolutional Neural Network (CNN) Method. September 2, 2020. https://doi.org/10.31294/p.v22i2.8907.

[9] Ibrahim, Shafaf, Noraini Hasan, Nurbaity Sabri, Khyrina Airin Fariza Abu Samah, and Muhamad Rahimi Rusland. "Palm leaf nutrient deficiency detection using convolutional neural network (CNN)." International Journal of Nonlinear Analysis and Applications 13, no. 1 (2022): 1949-1956.

[10] Global Studio Center. Arduino Series – 053: Soil Moisture Sensor Module. 12 April 2022. https://www.youtube.com/watch?v=0_dsTgPVado.

[11] Sintia, Wulantika, Dedy Hamdani, and Eko Risdianto. " Design of a Soil Moisture and Air Temperature Monitoring System Based on GSM SIM900A AND ARDUINO UNO." Jurnal Kumparan Fisika 1, no. 2 Agustus (2018): 60-65. https://doi.org/10.33369/jkf.1.2.60-65

Intelligent solutions for modern agriculture: Leveraging artificial intelligence in smart farming practices

Fatima Zahrae BERRAHAL[1,a*], Amine BERQIA[2,b]

[1]Smart Systems Laboratory, ENSIAS Mohammed V University in Rabat, Morocco

[a]fatimazahrae_berrahal@um5.ac.ma, [b]amine.berqia@ensias.um5.ac.ma

Keywords: Artificial Intelligence, Smart Agriculture, Machine Learning, Sustainable Agriculture, Precision Farming

Abstract. Rising global populations and climate change pose significant challenges to traditional farming methods. To address these issues, artificial intelligence (AI) is emerging as a transformative force in agriculture, often referred to as "Smart Agriculture" or "AI-powered Agriculture." This paper examines the multifaceted role of AI in revolutionizing farming processes. By leveraging AI technologies; farmers can enhance productivity, efficiency, and sustainability. This paper analyzes the diverse applications of AI in Agriculture, highlighting its potential to overcome critical farming challenges. It also explores the opportunities for AI-driven innovation in shaping the future of agriculture. With a specific focus on precision farming techniques, the paper investigates the implications and potential benefits of AI integration. This exploration sheds light on how AI can transform agricultural practices for a more sustainable future. This paper makes a comprehensive summary of the research on artificial intelligence technology in agriculture.

Introduction

Intelligent farming represents a cutting-edge approach to modern agriculture, leveraging advanced technologies such as Artificial Intelligence (AI) and Machine Learning (ML). This innovative farming paradigm marks a significant departure from traditional agricultural methods, harnessing the power of data-driven decision-making and precision techniques.

This paper delves into the convergence of smart agriculture and AI, revealing its potential to transform conventional agricultural practices and lay the groundwork for eco-conscious and more fruitful approaches.

The expansion of available land is vital to human survival and plays a crucial role in fostering economic growth and development [1]. Undoubtedly, this domain stands out as humanity's top priority, given that the overwhelming majority of our food supply relies on agriculture [2] Farming holds a vital position in the economy, acting as the bedrock of our monetary system. It serves as a fundamental pillar that supports the financial framework of our society.

With the growing unpredictability of environmental conditions, fulfilling the ever-increasing food requirements has become a progressively daunting task. Consequently, smart agriculture has emerged as a pivotal technology aimed at addressing these challenges [3].The agricultural industry is currently facing considerable obstacles due to various emerging factors, including global expansion, interests in food production, and the globalization of food markets, the unpredictability of food prices. Ensuring nutritional assurance has become a crucial goal, aiming to provide all individuals with consistent right to plentiful, nourishing sustenance. Therefore, the objective of infusing AI into cultivation techniques is to leverage AI to enhance and optimize assorted facets of farming and crop production practices.

Smart agriculture

Smart agriculture, as a comprehensive concept, incorporates a diverse array of technological innovations [4].The blending of contemporary information technology and conventional farming has led to Agriculture 4.0,known as smart agriculture. This era represents a seamless integration of advanced technologies into farming practices, offering intelligent solutions and automation to address various agricultural needs [5].The challenge posed by the spike in population and constrained grain production is sparking an increasing fascination with intelligent agriculture, leading to an upsurge in research in this area. The progress of agriculture relies on balancing productivity improvements with the constraints of the present era, and it is science and technology advancements that fuel the ongoing revolution in the agricultural sector [6]. The growing Tech-driven transition of farming procedures Is generating a need for cutting-edge technologies capable of facilitating the shift to Intelligent farming.The agriculture of the future is no longer solely about planting, fertilizers, and irrigation, but also about algorithms and Artificial Intelligence.

Artificial Intelligence(AI)

AI, also known as Artificial Intelligence, entails the programming of machines to simulate human intelligence and learning abilities. Since its inception, it has deeply influenced various scientific fields and everyday life for all individuals. Its fundamental components are increasingly acknowledged as vital in our society, and we frequently embrace the advantages it provides, often without even realizing it.

The IA can potentially elevate agricultural techniques. It can contribute to increased yields by helping them make informed choices about the most appropriate crop varieties, embracing enhanced soil and nutrient management methods, Efficiently controlling pests and diseases, estimating crop production, and predicting market prices. To attain these goals, AI employs sophisticated tools such as machine learning. These forefront tactics allow AI to confront the agricultural obstacles. Consequently, Growers have the chance to reap the advantages from Live surveillance of key agricultural variables For instance climate, heat and water utilization, Enabling them to make more informed choices [7].

The figure 1 illustrates the extensive application of artificial intelligence (AI) in agriculture, AI is revolutionizing how farmers make decisions and manage their operations. The figure likely depicts AI-driven solutions such as soil Managemen ,crop Disease Detection Through Machine Learning and Deep Learning, climate and Weather Forecasting ,Stronger and more resilient crop harvests,smart irrigation ,locating suitable areas for planting particular corps and Enhance Decision-Making for Sustainable Solutions .

To adequately feed the population, it has been calculated that global food production will need to increase by a margin of 60-110% [8].The realms of agricultural entrepreneurship, concentrate on enhancing their performance To fulfill the requirements of a burgeoning human civilization.

Fig.1 AI Harvest: Mapping the Pioneering Applications in Agriculture.

Unveiling the Promise of Artificial Intelligence in the Agrarian Sphere
Soil Management
Soil and crop management encounter various challenges, notably an escalating demand to sustain a burgeoning population and enhance food safety standards.

Precision farming employs advanced sensors and Forecasting analysis predictive analytics To capture instantaneous data related to soil information, Harvest readiness, Atmospheric condition, and Accessibility to improve crop production [9]. Precision spraying technology can lead to the use of unnecessary herbicide applications in redundant areas, potentially resulting in excessive herbicide consumption. However, this technology also offers the advantage of significantly, lowering the complete herbicide volume. When herbicides are used selectively in areas where weeds are present, it becomes feasible to lessen the environmental repercussions, cut costs, decrease crop damage to lessen the danger of chemical residue [10].The implementation of robots equipped with computer vision and artificial intelligence, which monitor and spray weeds, has the potential to reduce the current herbicide usage on crops by 80% .In precision fertilization,

A fertilizer dosage model is employed to calculate the necessary quantity of fertilizer, which is then administered through a variable-rate applicator [11].

Crop Disease Detection Through Machine Learning and Deep Learning
Plant diseases can significantly reduce crop yields, exerting direct repercussions on food production systems at both the national and global scales. The improper utilization of pesticides could lead to the development of long-term pathogen resistance, significantly reducing our ability to effectively combat them. This underscores the importance of precision farming as one of its key pillars [12].

Machine learning techniques have found significant uses in the agricultural industry. particularly in the analysis of soil fertility. The agricultural industry has consistently been a focal point for research in this domain [13].Automated techniques for plant disease detection are highly beneficial as they significantly reduce the extensive monitoring work required in large crop farms and enable the early detection of disease symptoms. Approximately 13% of global crop yield loss is attributed to plant diseases.The potential of machine learning in revolutionizing the way plant diseases are detected and managed became evident, leading to further exploration and development in this field [14].

Climate and Weather Forecasting
In the field of weather forecasting, machine learning has established it self as a powerful tool, revolutionizing how meteorologists analyze and predict weather patterns. Thanks to its ability to process vast amounts of atmospheric data, machine learning algorithms excel at capturing the intricacies of weather phenomena. AI techniques play a crucial role in analyzing vast quantities of unstructured and diverse data, enabling the identification and utilization of intricate relationships within this data, all without relying on explicit analytical methods. Embracing these AI techniques becomes essential to comprehend the ever-growing data influx and effectively meet the demanding requirements in Weather Forecasting [15].

Stronger and more resilient crop harvests
AI technology is being adopted by the agricultural industry to foster healthier crop production, managing pests and overseeing soil and growth conditions. With changing weather patterns and increasing pollution posing challenges for agriculturist to identify the ideal timing for sowing grains, AI, in conjunction with weather forecasts, can be used to analyze weather conditions. By utilizing AI and weather data, farmers can plan the type of crops to cultivate and identify the most favorable planting times, ultimately improving their crop selection and increasing potential profitability [16].

Crop and soil monitoring
Evaluating and closely observing the effects of agricultural systems on soil quality is crucial to establish optimal management methods and sustainable land utilization. This is essential for addressing climate change, preserving biodiversity, and ensuring food and energy security.Artificial Intelligence (AI) has diverse applications in the monitoring of crops and soil, Leveraging technologies such as drones.AI-based apps then analyze this data to identify optimal solutions[17,18]. These applications Improve the comprehension of the comprehensive condition and quality of the soil. AI can predict and identify potential pest attacks by analyzing satellite or drone images and tracking trends in pest activity. The system continually monitors incoming data to detect early signs of an impending infestation [19].

Smart irrigation
The adoption of innovations in irrigation systems is crucial for improving water-use efficiency and aligning with Sustainable Development Goals, contributing significantly to this effort.

Traditional irrigation methods frequently face inefficiencies and experience water wastage, resulting in significant economic and environmental difficulties[20].

Smart irrigation control systems have emerged as a promising solution to address these challenges. These systems leverage advanced technologies to optimize irrigation practices, resulting in decreased water consumption and improved crop productivity .Through the integration of sensors, actuators, and intelligent algorithms, smart irrigation control systems facilitate precise and timely water delivery, based on the specific needs of crops. The core of these systems is sensor-based irrigation scheduling, where soil moisture sensors, for example, offer real-time measurements of soil moisture content. As a result, farmers can make informed decisions about their crops' water requirements, leading to efficient water usage and optimal crop growth[21].

Locating suitable areas for planting particular corps
Artificial intelligence, leveraging drone-captured imagery, offers invaluable support to agricultural workers by aiding in the identification of optimal crop planting locations, considering geographical attributes, soil conditions, and pertinent variables. Utilizing supervised machine learning algorithms, AI facilitates the assessment of seed quality and recognizes pre-existing crops. Before the sowing process commences, AI scrutinizes seed images, juxtaposing them with visuals of healthy seeds to guarantee the most favorable planting conditions [22].

Enhance Decision-Making for Sustainable Solutions
The integration of AI technology within the agriculture sector is experiencing a notable surge, facilitating enhanced decision-making processes. A growing reservoir of data is being harnessed and processed to inform agricultural strategies. This advancement is fueled by a range of industry innovations, including the widespread deployment of sensors, expedited access to satellite imagery, lowered expenses associated with data loggers, increased utilization of drones, and improved accessibility to data repositories. Collectively, these developments are instrumental in refining irrigation methodologies and practices [23].

Discussion
The combination of smart agriculture and AI has revolutionized the traditional farming methods, offering farmers and stakeholders valuable tools and solutions to overcome challenges posed by the ever-increasing global population, climate change, and limited natural resources.

This paper provides a comprehensive discussion of the various applications and innovations in smart agriculture through the lens of artificial intelligence.

AI continues to make significant strides in revolutionizing the agricultural sector, enabling farmers to optimize their processes and increase productivity. Precision farming remains one of the primary applications of AI in agriculture. AI-powered sensors and drones collect vast amounts of data, including soil quality, weather patterns, and crop health. Machine learning algorithms then analyze this data to provide valuable insights for farmers, helping them make data-driven decisions about irrigation, fertilization, and pest control. This precise approach results in more efficient resource utilization and reduced environmental impact. AI-powered autonomous farm machinery is also gaining momentum. These machines can perform tasks such as planting, harvesting, and weeding with high precision and minimal human intervention. This technology significantly alleviates the labor burden on farmers and allows them to focus on more strategic aspects of their operations. Crop disease detection is another noteworthy application. AI algorithms trained on extensive datasets of diseased and healthy crops can quickly identify and diagnose diseases in plants, facilitating early intervention and preventing widespread outbreaks. However, despite the promising innovations, there are challenges and potential implications that need to be addressed. Data privacy and ownership remain A concern, with questions about who controls and accesses the sensitive information generated by AI technologies.

The digital divide in agriculture is another significant issue. Smaller farmers and those in remote areas might lack access to the necessary infrastructure and training to adopt AI technologies, creating a disparity between tech-savvy and traditional farming practices. Moreover, ethical considerations surround the use of AI in agriculture, particularly in relation to genetically modified crops and altering the natural course of farming practices.

In conclusion, in the present time, AI continues to offer immense potential to revolutionize smart agriculture. However, it is essential to address the challenges and implications thoughtfully and responsibly. Integrating AI technologies into agriculture in a sustainable, inclusive, and ethical manner is crucial to harnessing its benefits for the long-term development of the agricultural sector.

Feature direction and issues:
The future of agriculture gleams with the emergence of artificial intelligence (AI) as a powerful tool. There is also some feature direction and some issues for further exploration:

Feature direction:
- ✓ Develop AI algorithms for image recognition that can automatically detect pests and diseases in crops through drone or ground-based cameras, enabling early intervention.
- ✓ Train AI models on historical data to predict potential pest or disease outbreaks based on weather patterns and crop types, allowing for preventative measures.
- ✓ Train AI models to analyze soil sensor data and satellite imagery to identify nutrient deficiencies and create customized fertilizer plans for different zones within a field.
- ✓ Integrate AI with drone-based imaging or spectral sensors to identify nutrient stress symptoms in crops and deliver targeted fertilizer applications.
- ✓ Explore AI-powered robotic systems or smart applicators that can deliver precise fertilizer amounts based on real-time plant needs.
- ✓ Develop AI models to predict water needs based on weather forecasts, crop growth stages, and historical data. This allows for proactive irrigation adjustments.
- ✓ Refine AI-controlled robots for tasks like harvesting.

Issues for Further Exploration:
- ✓ Standardizing data formats and ensuring data quality across different farms and regions is crucial for accurate AI model training and application Developing robust security protocols to protect sensitive agricultural data from cyberattacks and ensuring farmer privacy is essential.
- ✓ Developing AI models that are interpretable and transparent to farmers allows for trust and informed decision-making.
- ✓ Accessibility and affordability: Bridge the digital divide by creating user-friendly interfaces and affordable AI technology for small and medium-scale farmers.
- ✓ Focus on human-centered AI design that empowers farmers and leverages their expertise alongside AI capabilities.
- ✓ Environmental impact assessment: Develop AI tools to optimize resource use and minimize environmental footprint associated with agricultural practices.

Conlusion
The growth of artificial intelligence (AI) is exponential and impacting numerous sectors, the swift advancement of information technology and data processing capabilities has given rise to a set of innovative tools commonly known as artificial intelligence (AI). The digital transformation occurring across all sectors is also making its way into agriculture. Farmers' living and working conditions are changing, thanks to new technologies.

As the world population continues to grow, agriculture plays a vital role in the future of our planet. Making it even more sustainable, efficient and productive is crucial, despite all structural and technological changes. One of the most powerful levers for action is the development of agriculture through artificial intelligence.

In conclusion, smart agriculture powered by AI holds immense potential for transforming the agricultural sector and addressing the challenges of food security and sustainability. However, to realize these benefits, it is crucial to address the challenges mentioned above and foster an enabling environment through collaboration between agricultural stakeholders, technology developers, and research institutions. By overcoming these obstacles, we can harness the power of AI to build a more efficient, resilient, and sustainable future for agriculture.

References

[1] Sarfraz .S, Ali .F, Hameed.A, Ahmad.Z, Riaz.K.:Sustainable agriculture through technological innovations.Sustainable agriculture in the era of the OMICs revolution ,2023, pp. 223-239. https://doi.org/10.1007/978-3-031-15568-0_10

[2] Sitharthan R, Rajesh M, Vimal S, Saravana Kumar E, Yuvaraj S, Abhishek Kumar, Jacob Raglend I, Vengatesan K.: A novel autonomous irrigation system for smart agriculture using AI and 6G enabled IoT network,Microprocessors and Microsystems,Volume 101, 2023,104905. https://doi.org/10.1016/j.micpro.2023.104905

[3] Ganesh Gopal Devarajan, Senthil Murugan Nagarajan, Ramana T.V., Vignesh T., Uttam Ghosh, Waleed Alnumay,DDNSAS: Deep reinforcement learning based deep Q-learning network for smart agriculture system,Sustainable Computing: Informatics and Systems,Volume 39, 2023,100890. https://doi.org/10.1016/j.suscom.2023.100890

[4] Stefano Cesco, Paolo Sambo, Maurizio Borin, Bruno Basso, Guido Orzes, Fabrizio Mazzetto, Smart agriculture and digital twins: Applications and challenges in a vision of sustainability, European Journal of Agronomy, Volume 146, 2023, 126809. https://doi.org/10.1016/j.eja.2023.126809

[5] Xing Yang, Lei Shu, Jianing Chen, Mohamed Amine Ferrag, Jun Wu, Edmond Nurellari and Kai Huang, "A Survey on Smart Agriculture: Development Modes, Technologies, and Security and Privacy Challenges," IEEE/CAA J. Autom. Sinica, vol. 8, no. 2, pp. 273-302, Feb. 2021. https://doi.org/10.1109/JAS.2020.1003536

[6] Y. Liu, X. Ma, L. Shu, G. P. Hancke, and A. M. Abu-Mahfouz, "From industry 4.0 to agriculture 4.0: current status, enabling technologies, and research challenges, " IEEE Trans. Ind. Informat., 2020. https://doi.org/10.1109/TII.2020.3003910

[7] S. Y. Liu, "Artificial Intelligence (AI) in Agriculture," in IT Professional, vol. 22, no. 3, pp. 14-15, 1 May-June 2020. https://doi.org/10.1109/MITP.2020.2986121

[8] J. Rockstrom, J. Williams, G. Daily et al., "Sustainable in- " tensification of agriculture for human prosperity and global sustainability," Ambio, vol. 46, no. 1, pp. 4–17, 2017. https://doi.org/10.1007/s13280-016-0793-6

[9] Ampatzidis Y, Partel V, Costa L (2020) Agroview: cloud-based application to process, analyze and visualize UAV-collected data for precision agriculture applications utilizing artificial intelligence. Comput Electron Agric 174:105457. https://doi.org/10.1016/j.compag.2020.105457

[10] Balafoutis A, Beck B, Fountas S, Vangeyte J, Wal TV, Soto I, Gómez-Barbero M, Barnes A, Eory V (2017) Precision agriculture technologies positively contributing to GHG emissions mitigation. Farm Prod Econ Sustain 9:1339. https://doi.org/10.3390/su9081339

[11] Elbeltagi A, Kushwaha NL, Srivastava A, Zoof AT (2022) Chapter 5: artificial intelligent-based water and soil management. Deep Learning for Sustainable Agriculture 2022:129–142. https://doi.org/10.1016/B978-0-323-85214-2.00008-2

[12] Kaur, S., Pandey, S. & Goel, S. Plants Disease Identification and Classification Through Leaf Images: A Survey. Arch Computat Methods Eng 26, 507–530 (2019). https://doi.org/10.1007/s11831-018-9255-6

[13] P. Tamsekar, N. Deshmukh, P. Bhalchandra, G. Kulkarni, K. Hambarde, S. Husen, Comparative analysis of supervised machine learning algorithms for GIS-based crop selection

prediction model, in: Computing and Network Sustainability, Springer, 2019, pp. 309–314. https://doi.org/10.1007/978-981-13-7150-9_33

[14] Witten, I.H., Holmes, G., McQueen, R.J., Smith, L.A., & Cunningham, S.J. (1993). Practical machine learning and its application to problems in agriculture.

[15] Dewitte, S.; Cornelis, J.P.; Müller, R.; Munteanu, A. Artificial Intelligence Revolutionises Weather Forecast, Climate Monitoring and Decadal Prediction. Remote Sens. 2021, 13, 3209. https://doi.org/10.3390/rs13163209

[16] Sharma S, Gahlawat VK, Rahul K, Mor RS, Malik M. Sustainable innovations in the food industry through artificial intelligence and big data analytics. Logistics. 2021; 5(4):66. https://doi.org/10.3390/logistics5040066

[17] Lowe M, Qin R, Mao X. A review on machine learning, artificial intelligence, and smart technology in water treatment and monitoring. Water. 2022;14(9):1384. https://doi.org/10.3390/w14091384

[18] Shelake S, Sutar S, Salunkher A, et al. Design and implementation of artificial intelligence powered agriculture multipurpose robot. International Journal of Research in Engineering, Science and Management. 2021;4(8):165–167.

[19] Marcu IM, Suciu G, Balaceanu CM, Banaru A. IoT-based system for smart agriculture. In: 2019 11th International Conference on Electronics, Computers And Artificial Intelligence (ECAI). IEEE; 2019, June:1–4. https://doi.org/10.1109/ECAI46879.2019.9041952

[20] C.A. Buckner, R.M. Lafrenie, J.A. D´enomm´ee, J.M. Caswell, D.A Want, Complementary and alternative medicine use in patients before and after a cancer diagnosis, Curr Oncol 25 (2018) e275-81. Available from, https://www.int echopen.com/chapters/83182. https://doi.org/10.3747/co.25.3884

[21] Y.D. Wu, Y.G. Chen, W.T. Wang, K.L. Zhang, L.P. Luo, Y.C. Cao, et al., Precision fertilizer and irrigation control system using open-source software and loose communication architecture, J. Irrig. Drain. Eng. 148 (2022) 1–9. https://doi.org/10.1061/(ASCE)IR.1943-4774.0001669

[22] Sane TU, Sane TU. Artificial intelligence and deep learning applications in crop harvesting robots-A survey. In: 2021 International Conference on Electrical, Communication, and Computer Engineering (ICECCE). IEEE; 2021, June:1–6. https://doi.org/10.1109/ICECCE52056.2021.9514232

[23] Banthia V, Chaudaki G. The study on use of artificial intelligence in agriculture. J. Adv. Res. Appl. Artif. Intell. Journal of Advanced Research in Applied Artificial Intelligence and Neural Network. 2022;5(2):18–22.

On the performance assessment of King Faisal University grid-connected solar PV facility

Mounir BOUZGUENDA

Department of Electrical Engineering, College of Engineering, King Faisal University,
Al- Hassa, PC 39182, Saudi Arabia

mbuzganda@kfu.edu.sa

Keywords: Energy Performance Index, Array Yield, Final Yield, Performance Ratio, Capacity Factor, System Efficiency, KSA 2030 Vision

Abstract. This paper evaluates the performance of the 9.8 kW grid-connected solar photovoltaic (PV) system mounted on the rooftop of the College of Engineering within the King Faisal University campus. The facility consists of four fields rated 2.45 kW each. Each field consists of ten 10 solar panels connected to a 3kW. The collected data include hourly and sub-hourly measurements of the radiation, ambient and module temperatures, AC and DC voltages, currents, and power outputs. Other measurements include cumulative produced energy and CO_2 savings. The assessed parameters of the PV facility include daily, and monthly energy, final yield, efficiency, and module performance ratio (PR). Aggregate performance analysis results show that the station operated as planned. The annual radiation measured in 2017 reached 2449 kWh/m² with a daily average of 6.710 kWh/m². The annual energy production reached 16.047 MWh with an average of 43.964 kWh/day. The system efficiency and capacity factor are 11.22% and 17.25%, respectively.

Introduction

Saudi Arabia's wealth in natural gas and oil has contributed to the country's rapid economic expansion over the last three decades. But the nation also produces the most carbon dioxide emissions, which suggests that it is becoming more and more dependent on gas and oil for electricity production. Saudi Arabia, having the biggest economy in the GCC, is heavily reliant on non-renewable energy sources, which makes the quest for alternate energy sources necessary. Saudi Arabia is viewed by experts as a potential hub for the production of photovoltaic solar energy. The European Commission Institute for Energy estimates that 0.3% of the light falling on the Middle East and Sahara deserts might supply Europe. Saudi Arabia, one of the largest developing countries, ranks eleventh in terms of electricity production globally. In particular, the oil and gas fuels dominate its electricity mix.

Meanwhile, evidence has shown that monitoring and performance analysis of both grid-connected and off-grid solar PV systems improve the performance of such systems. Several factors have a determinant impact on the performance of a PV system. Some of these factors have positive effects on performance while others have negative effects. Such factors and performance metrics are undertaken in this study.

The paper is organized as follows. Section 2 is devoted to Saudi Arabia's 2030 Vision and its renewable energy initiatives. The background, the motivation, and the objective of the present study are briefly introduced in this section. In Section 3, renewable energy prospects and operating large-scale facilities are discussed. In Section 4, solar PV performance metrics are discussed. Meanwhile, Section 5 and Section 6 are devoted to the analysis method and results discussion, respectively.

Renewable Energy Initiatives in KSA

Substantial evidence indicates that the shift to sustainable sources of energy will address environmental issues such as greenhouse gas emissions, air pollution, and climate change to a certain extent. This shift is occurring at a considerable rate in developing countries [1]. As a result of recognizing the need to diversify its energy sources, reduce its carbon footprint, and create a more sustainable and resilient energy sector, Saudi Arabia has launched ambitious renewable energy programs based on eight pillars.

- The Saudi Vision 2030 is a plan that aims to reduce the country's dependence on oil transform its economy and increase the contribution of renewable energies.
- The second pillar is the National Renewable Energy Program (NREP), initiated in 2017 and aims to add significant renewable energy capacity to the Saudi grid. Such projects include the Sakaka solar power plant, one of the country's first major solar projects.
- The King Salman Renewable Energy Initiative. It aims to install 41 GW of solar PV by 2032.
- The Renewable Energy Project Development Office (REPDO) was established to supervise and manage the development of renewable energy projects in the country. REPDO plays a crucial role in the acquisition and execution of renewable energy projects.
- The focus on solar and wind energy is due to KSA's abundant natural resources in these areas. Solar projects include photovoltaic and concentrated solar power systems, while wind projects utilize the country's coastal and inland wind resources.
- The Energy Efficiency and Conservation initiative to increase renewable energy capacity by focusing on improving energy efficiency in the construction, transport, and industry sectors.
- Diversification of energy sources: The government has been working to diversify its energy resources by integrating renewable energy into its energy mix and developing infrastructure for electricity storage and grid integration.

As a result of the framework of the National Renewable Energy Program, the number of renewable energy projects reached 13 projects with a total capacity of 4,470 MW of PV and 40 MW of wind. The expected annual energy produced amounts to 15,109 GWh with an estimated CO_2 emission reduction of 9,828,156 Ton/year by 2024. Table 1 lists some of the grid-connected PV systems in the country. With the increasing role of solar energy in meeting the kingdom's need for energy and the installation of many solar PV projects in the kingdom, it is necessary to assess the performance of existing and future off-grid and grid-connected solar PV systems in the kingdom. The proposed method has been applied in many countries and the study results will be compared to current studies in the field of solar PV system performance assessment.

Methodology

Research has been conducted recently to examine the effectiveness of grid-connected solar PV power systems in various settings. Reference yield, array yield, final array yield, capture losses, system losses, and performance ratio were among the performance metrics that were computed. Furthermore, it has been demonstrated that doing a performance analysis of the actual efficiency of photovoltaic installations is a difficult undertaking since it necessitates processing data from solar PV system operation monitoring. Procedures were therefore developed to assess solar power plant performance using data from actual system monitoring. Monitored data include radiation, ambient and module temperatures, AC voltage/current, DC voltage/current, and DC/AC power. Assessed parameters include daily, and monthly energy, final yield, efficiency, and module performance ratio (PR).

Table 1: Major solar PV projects in KSA [2]

Plant Name	Location	Size (MW)	Starting Year
Sakaka Solar PV Park	Al Jouf	405	2020
Haradh Solar PV Park	E. Province	30	2021
King Abdulaziz International Airport Solar PV Park	E. Province	10.5	2013
King Abdullah Petroleum Studies and Research Center Solar Park	E. Province	3	2013
King Abdullah Univ. of Science and Tech.-Solar Park	Jeddah	2	2010
Saudi Aramco Park Project	Khobar	10.5	2012
Al Fanar Jinko Solar PV Plant	Makkah	5.4	2018
Al Kharj Solar PV Park	Riyadh	15	2019
Matco Solar PV Plant	Riyadh	3.5	2019

Performance Assessment of PV Power Plant Performance Using Metric Indicators
IEC61724 standard mandates that the final yield (Y_f), array yield (Y_a), reference yield (Y_r), energy efficiency (η), and the total energy generated by the PV system E_{AC}, are used to evaluate the performance of a grid-connected PV installation.

Array yield.
The array yield (Y_a) is the ratio of the DC energy output delivered by the PV modules over a defined period divided by the PV-rated power and is given as [4]:

$$Y_a = \frac{E_{DC}(kWh)}{PV_{Rated}(kWp)} \tag{1}$$

The daily array yield ($Y_{a,d}$) and the monthly average array yield ($Y_{a,m}$) are given as [4]:

$$Y_{a,d} = \frac{E_{DC}(kWh/day)}{PV_{Rated}(kWp)} \tag{2}$$

$$Y_{a,m} = \frac{1}{N}\sum_{d=1}^{N} Y_{a,d} \tag{3}$$

E_{DC} is the DC energy output delivered by the PV modules (kWh) and N is the number of days in the month.

Final yield
The final yield is the total AC energy during a specific period divided by the rated power of the PV system. This metric is used to compare a given PV system with other existing PV systems. The daily array yield ($Y_{f,d}$) and the monthly average array yield ($Y_{f,m}$) are given as: [5]

$$Y_{f,d} = \frac{E_{AC}(kWh/day)}{PV_{Rated}(kWp)} \tag{4}$$

$$Y_{f,m} = \frac{1}{N}\sum_{d=1}^{N} Y_{f,d} \tag{5}$$

E_{AC} is the AC energy output delivered by the PV modules to the grid (kWh) *and N is the number of days in the month.*

$$E_{AC,m} = \frac{1}{N}\sum_{d=1}^{N} E_{AC,d} \tag{6}$$

With $E_{AC,m}$ being the monthly AC energy output and N the number of days in a month.

Reference yield
The reference yield is the ratio of the global solar radiation H_t (kWh/m^2) and the PV's reference irradiance. The reference yield is given as [6]:

$$Y_r = \frac{H_t \, (kWh/m^2)}{H_g} \qquad (7)$$

In this case, $H_g = 1 \text{ kW}/m^2$

Performance ratio
The performance ratio (PR) depends on the total losses by the PV system components (modules, inverters, trackers, and cables) and losses due to weather conditions such as ambient temperature, rain, shade, etc. The performance ratio (PR) is defined as the final yield divided by the reference yield and is expressed as:

$$PR = \frac{Y_f}{Y_r} \qquad (8)$$

Array capture losses.
The array capture losses (L_C) are due to the PV array losses and are expressed as:
$$L_C = Y_r - Y_a \qquad (9)$$

System losses
The system losses (L_S) are caused by inverter losses.
$$L_S = Y_a - Y_f \qquad (10)$$

Capacity factor
The annual capacity factor (CF) is:
$$CF = \frac{E_{AC}}{8760 * P_{PV,rated}} \qquad (11)$$

Where E_{AC} is the annual energy produced by the system and $P_{PV,\,rated}$ is the PV rated capacity.

System efficiencies
The system efficiencies consist of 3 components-namely the PV module efficiency, the inverter efficiency and the system efficiency. The PV module efficiency is expressed:

$$\eta_{PV} = \frac{100 * E_{AC}}{S * H_t} \, (\%) \qquad (12)$$

The monthly PV module efficiency ($\eta_{PV,m}$) expression is:

$$\eta_{PV,m} = \frac{\sum_{i=1}^{n} E_{DC}}{S * \sum_{i=1}^{N} H_t} \, (\%) \qquad (13)$$

Where E_{DC} is the total energy produced by solar PV modules, N is the number of days in a month and, and S is the total PV module surface (m²).
The inverter efficiency is given as

$$\eta_{PV} = \frac{100 * E_{AC}}{E_{DC}} \, (\%) \qquad (14)$$

The temperature loss coefficient (η_{tem}) is

$$\eta_{tem} = 1 + \alpha_P \, (T_C - T_{C,STC}) \qquad (15)$$

$$P_{PV} = Y_{PV} f_{PV} \left(\frac{G_T}{G_{C,STC}}\right) [1 - \alpha_P (T_C - T_{C,STC})] \qquad (16)$$

Where α_P indicates the power temperature coefficient (%/˚C), T_C is the PV cell temperature (˚C), and $T_{C,STC}$ is PV cell temperature at STC (˚C).

$$T_c = T_a + \frac{P}{800}(T_{NOCT} - 20) \qquad (17)$$

Where P is the power density at a specific time and T_{NOCT} is the normal operating cell temperature.

Description of KFU 9.8kW PV Facility
The 9.80 kW grid-connected PV facility has been operating since 2014. The facility consists of four arrays rated 2.45 kW each. Each array is made of ten 245Wp panels with two different solar cell specifications: efficiency and robustness in harsh weather conditions. Each field is connected to a 3-kW inverter with an efficiency of 97%. The characteristics of the solar panel technologies are listed in Table 2. The detailed performance analysis was done for Array 1.

Table 2. Characteristics of the Solar Modules

Parameter	SM-D245PC2 (Arrays 1@2)	SM-245PC8 (arrays 3&4)
Rated power (Pmax)	245W	245W
Voltage at Pmax (Vmp)	30.4V	30.4V
Current at Pmax (Imp)	8.06A	8.08A
Warranted minimum Pmax	245W	245 W
Short-circuit current (Isc)	8.39A	8.63A
Open-circuit voltage (Voc)	37.8V	37.4V
Module efficiency	14.91%	14.72%
Operating module temperature	-40°C to + 85°C	-40°C to + 85°C
Isc Temperature coefficient	0.061%/°C	0.052%/°C
Voc Temperature Coef.	-0.345%/°C	-0.312%/°C
Power Temperature Coef.	-0.458%/°C	-0.429%/°C
Inverter Efficiency	97%	97%

Performance Analysis
As discussed earlier, the solar PV facility consists of four arrays. Each array consists of 10 solar panels connected in series yielding a peak power of 2.45 kW. The first part of the performance analysis focuses on Array 1. Therefore, a detailed performance analysis is carried out. The second part of the analysis includes the system aggregate efficiency and capacity factor for all four arrays. Figure 2 displays (a) the daily AC energy produced in May 2017, (b) the weather statistics for June 13, 2017 and (c) that the cumulative energy produced by Array 2 reached 17.5 MWh in June.

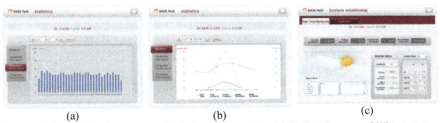

(a)   (b)   (c)

Figure 1: Array 2 Performance and Monitoring statistics (a) Daily AC energy (kWh) statistics May 2017.(b) Weather statistics, June 13, 2017. (c) Monitoring statistics, June 1, 2017.

Array 1 Performance Analysis
The nameplate efficiency of Array 1 is 14.91%. Tables 3 and 4 display the 2017 performance analysis results for Array 1. Based on these tables, the following observations are worth mentioning:
- The average plane of array radiation is 181 kWh/m². This is equivalent to 5.94 kWh/m²/day.
- Maximum power output ranges from 1.60 kW to 2.40 kW in March 2017.
- The highest monthly AC energy output of 400 kWh was produced in March and July.

- The annual AC energy output is 3.8 MWh. This is equivalent to an average of 10.42 kWh/day.
- The average efficiency exceeded 100%. This was because data are measured within one decimal point, which resulted in roundup and round-down errors. This has impacted the overall system efficiency and rendered higher than the solar conversion itself. In this case, the overall efficiency varied from 9.36% to 14.25%.
- The average performance varied from 55.8% to 84.9% with an overall average of 74.2%. The average monthly capacity factor varied from 13.97% to 21.59%.
- As seen in Tables 3 and 4 the obtained results of the capacity factors are aligned with the geographical location of the PV system and the harsh weather conditions. The system efficiency and capacity factors are compared with PV GIS based simulation results.
- The comparison results are shown in the last two rows of Table 4. The measured and simulated efficiency/capacity factor are slightly different with moderate deviation. However, they vary from one month to another.

Table 3: Array 1 Monthly Energy ambient and module temperatures

Month	Jan.	Feb.	Mar.	April	May	June	July	Aug.	Sept.	Oct.	Nov.	Dec.	Annual
POA (kWh/m^2)	160.6	130.8	162.3	204.7	209.4	188	209.9	211	211.4	214.1	169.8	188.8	181ᵧ
Pac_max (kW)	2.1	2.3	2.4	2.1	2.2	1.7	1.6	1.6	1.6	1.6	2	2.1	2.4*
POA_max (kWh/m^2)	1.000	1.041	1.041	1.041	1.041	1.008	1.004	1.041	1.041	1.041	1.041	1.041	1.041*
T_mod_max (°C)	47.7	53.4	60.4	69.1	67.5	66.1	70.1	71.5	70.4	63.8	55.9	52.2	71.5*
T_amb_max (°C)	31	30	37.8	48.1	48.8	50.1	53.3	53.2	50.8	44.3	37.6	33.8	53.23*
DC Energy (kWh)	302	275	322	381	351	352	336	331	312	293	246	350	321ᵧ
AC Energy (kWh)	300	200	400	300	400	300	400	300	300	300	300	300	317ᵧ

*Maximum value ᵧaverage value

Table 4: Array 1 efficiencies, performance ratio and capacity factor for 2017.

Month	Jan.	Feb.	Mar.	Apr.	May	June	July	Aug.	Sept.	Oct.	Nov.	Dec.	Annual
Eff_PV (%)	12.68	14.17	13.54	12.68	11.37	12.71	10.85	10.67	10.02	9.27	9.84	12.62	12.65
Eff_Inv (%)	100.3	100.6	100.3	100.5	100.9	100.9	101.0	100.8	100.9	101.0	100.7		99.5
Eff_sys (%)	12.72	14.25	13.59	12.75	11.47	12.82	10.96	10.76	10.12	9.36	9.94	12.71	12.44
Y$_{a,m}$ (kWh)	121.2	109.4	127.3	152.0		139	142.4	133.1	131.4	123.5	114.2	95.7	136.6	127.2
Y$_{f,m}$ (kWh)	122.1	111.1	131.4	155.5	143.1	143.7	137.1	135.3	127.4	119.4	100.6	143.0	129.0
Y_ref	160.6	130.8	162.3	204.7	209.4	188.0	209.9	211.0	211.4	214.1	169.8	188.8	181.0
P.R.(%)	76.1	84.9	81.0	76.0	68.3	76.4	65.3	64.1	60.3	55.8	59.2	75.7	74.2
C.F. (%)	16.42	16.53	17.66	21.59	19.23	19.95	18.43	18.7	17.7	16.05	13.97	19.22	18.61
Simulated Eff_PV (%)	13.6	13.2	12.7	12.4	11.9	11.6	11.6	11.8	12.0	12.4	13.1	13.5	12.4
Simulated C.F. (%)	18.6	18.2	20.5	18.7	19.8	19.7	20.1	20.9	21.2	21.5	18.2	19.0	19.7

Aggregate Performance Analysis

An aggregate performance analysis of the entire PV facility was conducted considering all four fields. The results are shown in Tables 5 and 6.

- Table 5 displays the monthly efficiency and the capacity factor for each of the four arrays. Array 4 has the highest annual values of 12.26% and 18.85%, respectively. Meanwhile, Array 2 has the lowest annual values of 9.67% and 14.87%, respectively.
- According to Table 6, the highest efficiency and the highest capacity factor were obtained in February and April, respectively. The annual efficiency and capacity factor were 11.22%

and 17.25%, respectively. Compared to reported capacity factors around the world shown in Figure 2, the recorded capacity factor at KFU is slightly below KSA figures and other countries.

Table 5. Aggregate Performance of the Entire Grid Connected PV Facility.

Month	Array 1 Eff_sys (%)	C.F. (%)	Array 2 Eff_sys (%)	C.F. (%)	Array 3 Eff_sys (%)	C.F. (%)	Array 4 Eff_sys (%)	C.F. (%)
Jan.	12.84	16.51	10.49	13.50	12.20	15.69	16.12	20.74
Feb.	14.40	16.70	11.99	13.91	13.82	16.03	17.91	20.77
Mar.	13.76	17.88	11.52	14.98	13.17	17.12	15.99	20.79
April	12.78	21.66	10.64	18.03	12.31	20.86	12.71	21.54
May	11.58	19.42	9.45	15.85	11.12	18.65	11.55	19.37
June	12.82	19.95	10.49	16.33	9.44	14.68	12.64	19.67
July	11.03	18.54	9.04	15.20	13.05	21.94	9.53	16.02
Aug.	10.84	18.32	8.80	14.87	10.39	17.56	6.69	11.30
Sept.	10.21	17.86	8.49	14.80	9.72	17.01	11.31	19.78
Oct.	9.44	16.18	7.68	13.17	8.93	15.31	13.47	23.10
Nov.	10.08	14.17	8.15	11.45	9.48	13.32	9.96	14.00
Dec.	12.81	19.37	10.74	16.24	12.26	18.54	12.75	19.27
Annual	11.74	18.06	9.67	14.87	11.21	17.24	12.26	18.85

Table 6. Aggregate Performance Analysis of the PV Facility.

Month	POA (kWh/m²)	Monthly Energy (kWh)	Daily Energy (kWh)	System efficiency η_{sys} (%)	Capacity Factor (%)
Jan.	160.6	1211.0	39.06	10.49	13.50
Feb.	130.8	1110.0	39.64	11.99	13.91
Mar.	162.3	1290.0	41.61	11.52	14.98
April	204.7	1448.0	48.27	10.64	18.03
May	209.4	1336.0	43.10	9.45	15.85
June	188.0	1246.0	41.53	10.49	16.33
July	209.9	1307.0	42.16	9.04	15.20
Aug.	211.0	1131.0	36.48	8.80	14.87
Sept.	211.4	1226.0	40.83	8.49	14.80
Oct.	214.1	1235.0	39.84	7.68	13.17
Nov.	169.8	934.0	31.13	8.15	11.45
Dec.	188.8	1338.3	43.17	10.74	16.24
Annual	188.4	1234.4	40.57	11.22	17.25

Summary

In this paper, a 9.8 kW grid-connected PV system in the College of Engineering Building at King Faisal University has been monitored since its commission. In particular, the 2017 performance was assessed on a daily, monthly, and annual basis. Array 1 performance analysis results show that this array's annual efficiency and capacity factor are 12.44% and 19.22%. respectively. The aggregate performance analysis results show that the station operated as planned. The overall efficiency and capacity factors are 11.22% and 17.25%. This is because Array 1 performed less than the other three arrays. The annual radiation measured in 2017 reached 2.449 kWh/m² with a daily average of 6.710 kWh/m². The annual energy production reached 16.047 MWh with a daily average of 43.964 kWh/day. Overall, the KFU facility capacity factor is slightly below KSA reported capacity factors. Further studies deem necessary to investigate the deviation of the capacity factor from KSA reported results.

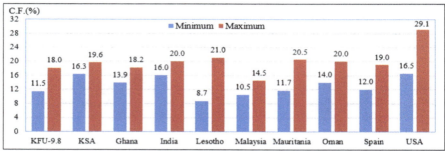

Figure 2: Capacity Factor for Selected Grid Connected PV Systems around the world [8-11].

References

[1] Maennel, Alexander, and Hyun-Goo Kim. "Comparison of greenhouse gas reduction potential through renewable energy transition in South Korea and Germany." Energies 11.1 (2018): 206. https://doi.org/10.3390/en11010206

[2] GlobalData's Saudi Arabia Solar PV Analysis: Market Outlook to 2035 report

[3] AlOtaibi, Z.S., Khonkar, H.I., AlAmoudi, A.O. et al., "Current status and future perspectives for localizing the solar photovoltaic industry in the Kingdom of Saudi Arabia", Energy Transit 4, 1-9 (2020). https://doi.org/10.1007/s41825-019-00020-y

[4] Vikrant Sharma, S.S. Chandel, "Energy Performance analysis of a 190 kWp grid interactive solar photovoltaic power plant in India", Energy, Volume 55, 15 June 2013, Pages 476-485? https://doi.org/10.1016/j.energy.2013.03.075

[5] Ayompe, et al., "Measured performance of a 1.72 kW rooftop grid connected photovoltaic system in Ireland", Energy Convers. Management 52, ??816-825. https://doi.org/10.1016/j.enconman.2010.08.007

[6] Emmanuel Kymakis et. Al., "Performance analysis of a grid connected photovoltaic park on the island of Crete", Energy Conversion and Management, Volume 50, Issue 3, March 2009, Pages 433-438. https://doi.org/10.1016/j.enconman.2008.12.009

[7] Ahmed Bilal et al. "Solar Energy Resource Analysis and Evaluation of Photovoltaic System Performance in Various Regions of Saudi Arabia", Sustainability 2018, 10, 1129. https://doi.org/10.3390/su10041129

[8] Nibras et al., "Performance Analyses of 15 kW Grid-Tied Photovoltaic Solar System Type under Baghdad city climate", Journal of Engineering 26(4):21-32, March 2020. https://doi.org/10.31026/j.eng.2020.04.02

[9] A.H. Al-Badi et al., "Economic perspective of PV electricity in Oman", Energy, Volume 36, Issue 1, January 2011, Pages 226-232. https://doi.org/10.1016/j.energy.2010.10.047

[10] Cristobal et al., "Self-consumption for energy communities in Spain: A regional analysis under the new legal framework", Energy Policy, March 2021, 150(4):112144. https://doi.org/10.1016/j.enpol.2021.112144

[11] Today in Energy, "Southwestern states have better solar resources and higher solar PV capacity, factors", https://www.eia.gov/todayinenergy/detail.php?id=39832

Solar energy powered smart water heating system

Ahmed ALTURKI[1,a], Yahya ALMAHZARI[1,b], Majid ALBLOOSHI[1,c], Faris ALGHAMDI[1,d], Ahmed A. HUSSAIN[1,e], Ala A. HUSSEIN[1,f], Jamal NAYFEH[1,g]

[1]Department of Electrical Engineering, Prince Mohammad Bin Fahd University, Khobar, Saudi Arabia

[a]201900814@pmu.edu.sa, [b]201701039@pmu.edu.sa, [c]201502824@pmu.edu.sa, [d]201700894@pmu.edu.sa, [e]ahussain1@pmu.edu.sa, [f]ahussein@pmu.edu.sa, [g]jnayfeh@pmu.edu.sa

Keywords: Photovoltaic (PV), Solar Energy, Temperature Control

Abstract. This paper proposes a smart system for controlling the temperature of a water heating system. The proposed system is powered by solar energy and it allows cooling and heating the water used in households throughout the year, particularly, when the outside temperature goes to extreme highs and lows. The system is equipped with sensors to measure the temperature of water inside the tank and show the range of water. A mobile application is developed to monitor and control the water temperature around the clock.

Introduction
Today, a country's Gross Domestic Product (GDP) is directly correlated with its energy consumption, which serves as a measure of that country's prosperity. As a result, the need for energy resources is growing every day. Energy resources come in many different forms, but they are primarily grouped into two categories: renewable energy sources (solar, air, and wind) and non-renewable energy sources (coal, petroleum).

Non-renewable energy resources speed up industrial expansion, but their supply is finite by nature. In order to meet the energy needs of the present and future generations, it is vital to find alternative energy sources due to the quick depletion of fossil fuel resources. Solar energy stands out among several options as having the best long-term prospects for supplying the world's rising energy needs. This resource's main shortcomings are its low intensity, sporadic nature, and nighttime non-availability.

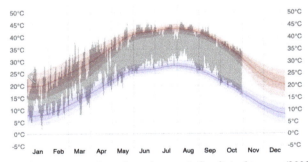

Figure 1. Temperatures throughout the year in Saudi Arabia year (2022).

Despite these drawbacks, solar energy still seems to hold the most promise of all the available renewable energy sources. Solar power is an abundant source of energy. Large amounts of renewable solar energy are produced by the sun and can be captured and used to generate heat and power. Water heating, air heating, building air conditioning, solar refrigeration, photovoltaic cells, greenhouses, photo-chemical power production, solar furnaces, and photo-biological co-versions are just a few of the many potentials uses for solar energy. The weather inside Saudi Arabia in the summer is excessively hot, which effects the water temperature inside the tanks, therefore it is difficult for people to take showers, use the toilet, do the dishes etc., therefore there is a high demand of water cooling systems, but a smart system does not exist, as a group we will design a smart system that solves this problem and make it easier for consumers to use them at home and make water more efficient which can be accessed at all times during the day.

The proposed system has several advantages over other similar systems proposed in [1] – [4]. The system in [1] is not solar powered, while the system in [2] does not have cooling feature, and it has no monitoring mechanism. The system in [3] allows heating and is solar powered, however, it has no cooling or monitoring features. In contrast, the proposed system allows both water cooling and heating. Also, it allows monitoring through a mobile application, and furthermore, it is solar powered. Details on the proposed system followed by technical specifications are demonstrated in the next section, followed by results, analysis and conclusion.

Proposed System
The block diagram of the proposed system is shown in Figure 2 below.

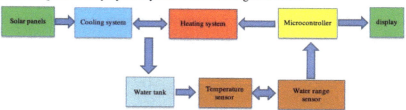

Figure 2. A block diagram of the proposed system.

The proposed system is divided into three subsystems as illustrated in Figure 3. Each of the subsystems are described in detail below.

A. Subsystem I: (Heating and Cooling System)
The majority of hot water systems include a central boiler where water is heated to a temperature between 60 and 83 °C (140 to 180 °F) before being routed through pipes to various rooms' coil units, such as radiators. Both pressure and gravity may circulate hot water, but forced circulation with a pump is more effective since it offers flexibility and control.

Hot-water systems circulate heated water using either a one-pipe or a two-pipe system. Compared to a two-pipe system, the one-pipe method utilizes fewer pipes; therefore, to accomplish the proper water temperature, we will be using the two-pipe system with cool water to achieve the water temperature that the user chose.

This system is also focused on finding the right size of cooling AC for water tanks; a cool water system should operate at a temperature between 5 to 40 °C (41 to 140°F), also finding a system that consumes the least amount of power that works on solar panels and combining this system to heating system; for finding the right water temperature that is needed. Major components used: Heating system, Cooling system, Water tank, Pipes

B. Subsystem II: (Photovoltaic System)
A photovoltaic (PV) system is made up of one or more solar panels, an inverter, and other mechanical and electrical components that harness solar energy to produce electricity. PV systems come in a wide range of sizes, from small rooftop or portable units to enormous utility-scale power plants. Although PV systems can function independently as off-grid PV systems, this article concentrates on grid-tied PV systems, which are PV systems that are linked to the utility grid.

The photovoltaic effect is the mechanism through which sunlight, composed of energy packets called photons, strikes a solar panel and generates an electric current. Each panel generates a very small quantity of electricity, but when connected to other panels, a solar array may generate much more energy.

To use all the systems on solar energy, we will need to calculate and test each system on how much power is being used; therefore, we will do the calculations and work out how many solar panels are needed and connect them to the system. Major components used: Solar panels, Inverter, Batteries, Charge controller.

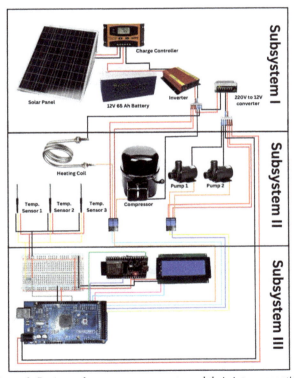

Figure 3. Project subsystems, components, and their interconnections.

C. Subsystem III: (Microcontroller and Communication System)
Information exchange between two points is described by the communication system. Communication is the process of sending and receiving information. The information transmitter,

the channel or medium of communication, and the information receiver are the three main components of communication.

To control a single device function, a microcontroller is integrated into the system. It accomplishes this by utilizing its core CPU to evaluate data that it receives from its I/O peripherals.

The microcontroller receives temporary data that is stored in its data memory, where the processor accesses it and employs program memory instructions to interpret and apply the incoming data. It then communicates and takes the necessary action via its I/O peripherals. Major components used: Sensors, Transmitter, Receiver, Arduino mega

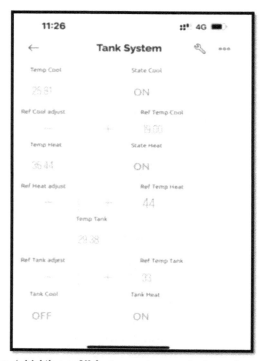

Figure 4. Mobile app UI for temperature monitoring and control.

Summary
This paper proposed a smart and efficient system for controlling and maintaining the temperature of water heating systems in residential units. It uses a renewable energy source to power the system making it environmentally friendly and sustainable. Since water-heating systems consume a lot of energy, the proposed system can help in significantly reducing the energy footprint and thus reduce the energy bills paid by the consumers.

References
[1] Water Tank System Using System Identification Method, International Journal of Engineering and Advanced Technology (IJEAT), Malaysia, 2013.

[2] Solar Water Heater Design, Al Akhawayn University, Morocco, 2019.

[3] Solar Water Heater Geyser, Undergraduate Senior Design Project, Prince Mohammed Bin Fahd University, Saudi Arabia, 2020.

[4] Powerful Tank Water Cooling Device, Global consumer bureau, Saudi Arabia, 2022.

[5] Choudhary, M., & Singh, R. (2018). Smart Water Heating System Using Internet of Things. In Proceedings of the International Conference on Information Systems Design and Intelligent Applications (pp. 573-582). Springer, Singapore.

[6] Li, X., Chen, X., & Li, M. (2018). Smart Water Heater Control System Based on Internet of Things. In 2018 International Conference on Computer, Information and Telecommunication Systems (CITS) (pp. 1-5). IEEE.

[7] Nikolic, I., & Nikolic, D. (2017). Smart Water Heating Systems—A Review. In 2017 5th Mediterranean Conference on Embedded Computing (MECO) (pp. 1-4). IEEE.

[8] Kumar, A., & Garg, R. (2019). Design and Implementation of a Smart Water Heating System Using IoT. In 2019 4th International Conference on Internet of Things: Smart Innovation and Usages (pp. 1-5). IEEE.

[9] Dervisoglu, E., & Karakose, M. (2020). Design and Implementation of a Smart Water Heating Control System Based on IoT. In 2020 5th International Conference on Computer Science and Engineering (UBMK) (pp. 1-5). IEEE.

[10] Chen, G., Zhang, X., & Liu, J. (2019). Design and Implementation of a Smart Water Heater Control System Based on IoT. IOP Conference Series: Earth and Environmental Science, 242(1), 012067.

Experimental investigation on the thermal and exergy efficiency for a 2.88 kW grid connected photovoltaic/thermal system

Mena Maurice FARAG[1,2,a*], Tareq SALAMEH[1,3,b], Abdul Kadir HAMID[1,2,c*], Mousa Hussein[4,d]

[1]Sustainable Energy and Power Systems Research Centre, Research Institute for Sciences and Engineering (RISE), University of Sharjah, Sharjah, United Arab Emirates

[2]Department of Electrical Engineering, College of Engineering, University of Sharjah, Sharjah, United Arab Emirates

[3]Department of Sustainable and Renewable Energy Engineering, College of Engineering, University of Sharjah, United Arab Emirates

[4]Department of Electrical and Communication Engineering, College of Engineering, United Arab Emirates University, Al Ain, UAE

[a]u20105427@sharjah.ac.ae, [b]tsalameh@sharjah.ac.ae, [c]akhamid@sharjah.ac.ae, [d]mihussein@uaeu.ac.ae

Keywords: Photovoltaic-Thermal, Thermal Efficiency, Electrical Efficiency, Exergy Efficiency

Abstract. Photovoltaic-thermal (PV/T) systems have been introduced recently for waste heat extraction, to improve electricity generation from photovoltaic (PV) systems and simultaneously utilize it for potential hot water for domestic or industrial use. This study investigated a 2.88 kW grid-connected PV/T system in the terrestrial weather conditions of Sharjah, UAE. The study was experimentally investigated during December when water as a working base fluid was evaluated for waste heat recovery. The electrical, thermal, and exergy efficiencies were examined for the given system, under five different hourly intervals across the experimental period. The results have shown a notable effect of the PV/T cooling method on the terrestrial weather conditions of the UAE. A peak total efficiency of 60% was observed, showing the effectiveness in improved thermal performance because of the active cooling procedure.

Introduction

Coal, oil, and natural gas are examples of fossil fuel resources that have actively supported the world's need to produce power [1]. Because of the world's population growth, it is predicted that this demand will increase by 48% in the next 20 years [2]. Global energy demand has accelerated the use of fossil fuels and their depletion, which has accelerated the rise in carbon emissions which has been a major factor in the phenomenon of global warming [3–5]. The rapid rise in global warming has resulted in significant changes to the climate that have been observed globally over the last ten years.

Solar technologies are now a vital source of electricity production for consumers, having advanced and grown tremendously over the last ten years [6]. To address the negative effects that fossil fuels have on pollution and climate change, photovoltaic plants have been installed and deployed at a rapid pace throughout the world [7]. Photovoltaic (PV) technology has become the dominant low-carbon technology in the world due to a noticeable drop in development costs. By the end of 2020, 627 GW of PV system installations are expected to have been installed globally. PV system installations have been deployed more often. It is anticipated that within five years, the ambitious deployment of PV system installations will rise globally by at least an average of 125 GW [8].

However, PV system technologies are highly dependent on environmental parameters particularly module temperature, which contributes to their degradation and longevity [9–11]. Photovoltaic-thermal (PV/T) systems have been introduced to tackle such a dilemma by introducing cooling methods to extract the dissipated heat generated from the PV cell's surface [12–14]. Various studies have demonstrated the use of PV/T systems using different cooling methodologies to discuss the thermal and electrical efficiencies of their demonstrated systems. A study was conducted by [15] demonstrating the performance of a PV/T system under open-loop and closed-loop connections. The study demonstrated the impact of open loop configuration in improving overall thermal efficiency. Another study demonstrated the use of nanofluids to observe the efficiencies of a PV/T system [16]. The study demonstrated that the exergy efficiency was 50% higher as compared to the non-cooling conditions. The demonstration of PV/T heat pipe for heat extraction was demonstrated in [17]. The useful exergy generated by the system were considered as objective functions, to assess the performance of the optimal systems. The optimal PVHT system demonstrated a 5.1% higher exergy as compared to other proposed designs. Moreover, a spray cooling-based system is proposed as a PV/T and heat recovery system for domestic applications, as reported in [18]. The exergy losses were computed based on simulation for four different seasons. A numerical and experimental investigation of a PV/T system was studied by [19]. The study utilized a mini-channel PV/T for waste heat recovery. The results demonstrated an electrical and thermal efficiency of 12% and 47%, respectively.

The literature has demonstrated that PV/T systems present large potential, particularly for harsh regions such as the United Arab Emirates [20]. The review didn't show a prior study discussing the electrical, thermal, and exergy efficiency of a PV/T system in the UAE. Therefore, this paper experimentally investigates the performance of a 2.88 kW PV/T system installed in the University of Sharjah main campus, in the terrestrial weather conditions of Sharjah, UAE. The system is exposed to the UAE weather conditions and examined using water as a cooling fluid.

Methodology
The University of Sharjah is located at the main campus in Sharjah, United Arab Emirates. The experimental setup was built on the rooftop of the central laboratories W12 building [21]. The site of the setup is located at a latitude coordinate of 25.34° N, whereas the longitude coordinate is 55.42° E [22]. The demonstrated PV/T system has of 2.88 kW capacity, where the system is connected to the local electrical grid for clean electricity supply [23–25]. Two separate PV modules of 320 W electrical rating are used during experimentation. The PV modules are raised above the ground with a tilt angle of 20°, to ensure the stable flow of fluid on the front surface of the PV module.

Experimental Procedure
The experimental measurements were conducted during the winter weather conditions, in December. The cooling methodology was conducted across five hourly intervals between 10 AM to 2 PM, for observation of thermal and electrical characteristics.

The electrical measurements are continuously measured through a Profitest PV analyzer, which provides an accurate capacitive load for DC power, voltage, and current measurements. Whereas K-type thermocouples are utilized to measure the thermal parameters such as inlet and outlet fluid temperature, PV module front and back surface temperature, and ambient temperature. A demonstration of the experimental procedure is presented in Fig. 1. Moreover, Table 1 briefly demonstrates the technical specifications utilized for the experimentation.

Table 1. Technical Specifications of 2.88 kW PV/T system experimental setup

Description	Specifications
PV Modules	Nine PV modules are connected in series. 320 W electrical rating per module. Tilt Angle 20°
Cooling Method	Fluid Type: Water. Inlet/Outlet Fluid Tank Capacity: 200 Gal. Inlet/Outlet Pump Capacity: 1 hp
Profitest PV Analyzer	P-V and I-V curve tracing
K-Type Thermocouples	Measurement range between -200° C to 1300 °C
HPS3008 Data logger	Temperature Data logging (8 Channels)
Irradiance Sensor	Solar Irradiance measurement (W/m²)
Anemometer	Wind speed measurement (m/s)

Fig. 1. Demonstration of Experimental Setup for the 2.88 kW PV/T system

Mathematical Model

The analysis results displayed in the subsequent sections are obtained by applying the following mathematical relations, equations, and formulas.

Electrical Performance

The solar-generated power (P_{in}) is a product of the solar irradiance (G) and the module area (A), which is expressed as follows:

$$P_{in} = G A \tag{1}$$

The maximum theoretical power (P_{theo}) can be computed based on the open circuit voltage (V_{oc}) and the short circuit current (I_{sc}), which is expressed as follows:

$$P_{theo} = I_{SC} V_{OC} \tag{2}$$

Similarly, the maximum output power (P_{max}) is computed based on the product of the maximum output voltage (V_{mp}) and maximum output current (I_{mp}), which is presented as follows:

$$P_{max} = I_{mp} V_{mp} \tag{3}$$

The electrical efficiency (η) is expressed as a ratio between P_{max} and P_{in}, represented as follows:

$$\eta_{Electrical} = P_{max}/P_{in} \tag{4}$$

Fill factor can be derived from the following equation:

$$FF = P_{max}/P_{theo} \tag{5}$$

Thermal Performance

Eq. 6 is used to determine the amount of solar heat Q_{solar} that the cooling water has captured. After differentiating it with respect to time, $m = \rho\dot{V}$, the mass flow rate, m, is computed using the mass formula. The volumetric flow rate is represented by $\dot{V}$, while the water density is denoted by ρ.

The specific heat capacity at constant pressure, $c_p = 4.18$ [J/kg. C], is used to analyze the heated property of water. By deducting the outflow temperature (T_{out}) and the input temperature (T_{in}), one may find the temperature difference (ΔT), which is then equal to $T_{out} - T_{in}$

$$Q_{solar} = m\, c_p\, \Delta T \tag{6}$$

Thermal Efficiency (η_t) can be computed through the following expression:

$$\eta_t = m\, c_p\, \Delta T\, /\, G\, A \tag{7}$$

The overall efficiency (U_F) by summing both electrical and thermal efficiencies can be described as the utilization factor as follows:

$$U_F = \eta_{Electrical} + \eta_{Thermal} \tag{8}$$

This overall efficiency represents the energy analysis that comprises of the electrical, thermal, and total efficiencies of the PV/T system.

Exergy Analysis
In this work, the second law of thermodynamics is utilized to perform exergy analysis. Calculating exergy is useful for energy systems to improve the sustainability and efficiency of the system, therefore utilizing the resources effectively and reducing the impact on the environment. The exergy analysis is typically used to mate the available energy. Therefore, the overall exergy balance for the given PV/T system is defined based on the following expression:

$$\sum \dot{E}x_{in} = \sum \dot{E}x_{out} + \sum \dot{E}x_{loss}$$
$$\Rightarrow \dot{E}x_{sun} + \dot{E}x_{mass,in} = \dot{E}x_{elc} + \dot{E}x_{mass,out} + \dot{E}x_{loss} \tag{9}$$

The in and out exergies are represented as Ex_{in} and Ex_{out}, respectively, the exergy destruction (Entropy) due to irreversibility is described as the exergy loss Ex_{loss}. Entropy is a measure of the disorder or randomness in an energy system due to irreversibility, such as heat transfer across finite temperature gradients, friction, and mixing; the relationship between exergy and entropy can be understood from the second law of thermodynamics. This irreversibility leads to the loss of available work (exergy) and increased entropy. This study used the Petela exergy conversion coefficient [26], which is described in the following expression:

$$\psi_s = 1 - \frac{4}{3}\left(\frac{T_{out}}{T_s}\right) + \frac{1}{3}\left(\frac{T_{out}}{T_s}\right)^4 \tag{10}$$

T_s represents the solar radiation temperature from the sun, which is previously reported to be equivalent to 5777 [K] as reported in [27,28]:

$$\dot{E}x_{in} = \dot{E}x_{sun} = \psi_s AG \tag{11}$$

Or $\dot{E}x_{sun} = G\left(1 - \frac{T_{amb}}{T_{sun}}\right)$

The thermal exergy $\dot{E}x_{th}$ typically represents the heat loss from the PV system exterior surfaces to the cooling fluid and ambient, which can be described as follows:

$$\dot{E}x_{mass,out} - \dot{E}x_{mass,in} = \dot{E}x_{th} = \dot{m}_{f,out}\left[(h_{f,out} - h_{f,in}) - T_{amb}(s_{f,out} - s_{f,in})\right]$$
$$h_{f,out} - h_{f,in} = C_{p,f}(T_{f,out} - T_{f,in})$$
$$s_{f,out} - s_{f,in} = C_{p,f}\ln\left(\frac{T_{f,out}}{T_{f,in}}\right) \tag{12}$$
$$\dot{E}x_{th} = \dot{m}_{f,out}\left[C_{p,f}(T_{f,out} - T_{f,in}) - T_{amb}C_{p,f}\ln\left(\frac{T_{f,out}}{T_{f,in}}\right)\right]$$

The exergy output for PV systems is typically equivalent to the total electrical energy $\dot{E}x_{ele}$, which can be described as follows:

$$\dot{Ex}_{ele} = V_m I_m \tag{13}$$

Therefore, the exergy losses of the PV system can be computed through the following expression:

$$\dot{Ex}_{loss} = \left(1 - \frac{T_{amb}}{T_{sun}}\right) G - \dot{E}_{ele} - \dot{m}_{f,out} \cdot C_{p,f} \left[(T_{f,out} - T_{f,in}) - T_{amb} \ln\left(\frac{T_{f,out}}{T_{f,in}}\right)\right] \tag{14}$$

Therefore, the computation of entropy generation by the PV system can be done as follows:

$$\dot{S}_{gen} = \frac{\dot{Ex}_{loss}}{T_{amb}} \tag{15}$$

As a result, the thermal and electrical exergy efficiencies can be computed based on the following equations:

$$\xi_{th} = \frac{\dot{Ex}_{th}}{\dot{Ex}_{sun}} \tag{16}$$

$$\xi_{ele} = \frac{\dot{Ex}_{ele}}{\dot{Ex}_{sun}} \tag{17}$$

Finally, the overall exergetic efficiency ξ of the PV/T system is computed based on a ratio of the output and input exergies of the system

$$\xi_{total} = \frac{\dot{Ex}_{ele} + \dot{Ex}_{th}}{\dot{Ex}_{sun}} \tag{18}$$

Results and Discussion

The closed-loop tests for front surfacing cooling are presented in Table 2. The experimental test was conducted on a fixed tilt angle of 20°. The experimental parameters are variable with time as demonstrated in Table 2, demonstrating the variability of different environmental and electrical parameters with time. Moreover, the weather profile is presented in Fig. 2(a), presenting a peak solar irradiance at noon.

The experimental measurements show the effectiveness of water as a working base fluid, by maintaining low front and back surface temperatures. This would reflect on enhancing the electrical efficiency of the PV/T system as illustrated in Fig. 2(b). The electrical efficiency (η) can be computed based on the numerical relations that were discussed in the previous section.

As demonstrated in Fig. 4, the PV/T system under cooling conditions would contribute significantly to maintaining high electrical efficiency, with a peak of 14.5%. Thereby, the effectiveness of water as a working base fluid can be attributed to the increase in electrical efficiency throughout the day, through the maintenance of the front and back surface temperatures. An inverse relationship between the module temperature and electrical efficiency can be observed, hence its reduction as the experimentation approaches noon time.

The observation of thermal performance is essential when discussing the performance of PV/T systems. The thermal study is conducted on the given PV/T system, as demonstrated in Fig. 3. The thermal investigation is computed based on the numerical equations presented in Eq. 6-7. Commenting on Fig. 3, an inversely proportional relationship between the thermal efficiency and the thermal energy losses can be observed. The highest losses are experienced during noon time due to the reduction of the cooling effect because of the increase in water temperature, thereby affecting the thermal and total efficiency of the system.

Table 2. Experimental measurements from 2.88 kW PV/T system

Time	G [W/m2]	T_A [°C]	T_{in} [°C]	T_{out} [°C]	T_{PVB} [°C]	T_{PVF} [°C]	I_{SC} [A]	V_{OC} [V]
10.00	709.81	23.01	20.32	21.84	25.75	22.52	6.48	45.02
11.00	800.02	21.07	20.61	21.57	26.58	23.03	7.27	44.79
12.00	820.50	22.89	24.80	25.78	30.60	26.17	7.53	44.51
13.00	765.24	26.05	25.16	26.28	32.87	28.00	6.94	44.15
14.00	639.94	26.83	26.17	27.38	33.46	29.54	5.24	43.87

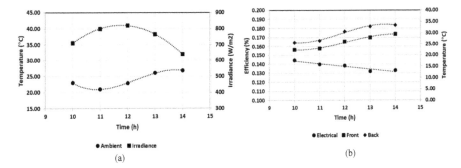

Fig. 2. Demonstration of (a) Weather profile for day of experimentation (b) Comparison of electrical efficiency with front and back module temperature

The exergy efficiency was computed based on the previously discussed numerical equations and illustrated in Fig. 4. A negative relation is presented between the exergy efficiency and exergy destruction, which are lost due to the irreversibility effect. Similarly, the highest exergy destruction is experienced during noon time, presenting the minimum exergy efficiency. In this notion, the thermal efficiency demonstrates the experimental test at the given weather conditions. Therefore, as the ΔT between inlet and outlet water temperature is significantly large, the higher the thermal efficiency. Additionally, ambient conditions are to be considered a critical factor for the thermal performance of any PV/T system.

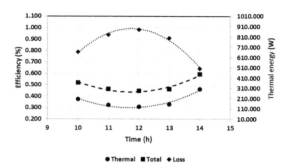

Fig. 3. Thermal performance across the period of experimentation

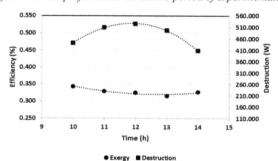

Fig. 4. Comparison of exergy efficiency with exergy destruction

Summary

This study discussed the electrical, thermal, and exergy performance of a 2.88 kW PV/T system in the terrestrial conditions of Sharjah, UAE. The experimental study was conducted during the winter conditions to assess the impact of water as a cooling fluid on both electrical and thermal efficiencies. The experimental measurements have concluded the inverse proportionality of the electrical efficiency with respect to the operating temperature, with a starting peak of 14.5%, which is close module efficiency described by the manufacturer. Moreover, a peak thermal efficiency of 48% is achieved, showing the beneficial use of PV/T systems in capturing waste heat for potential domestic applications. As a future work, the exergy analysis can be conducted for different seasons, to assess the significance of temperature difference in waste heat recovery, for total system efficiency enhancement.

References

[1] M.H. Mostafa, S.H.E. Abdel Aleem, S.G. Ali, A.Y. Abdelaziz, Energy-management solutions for microgrids, in: Distrib. Energy Resour. Microgrids, Elsevier, 2019: pp. 483–515. https://doi.org/10.1016/B978-0-12-817774-7.00020-X

[2] P. Moodley, C. Trois, Lignocellulosic biorefineries: the path forward, in: Sustain. Biofuels, Elsevier, 2021: pp. 21–42. https://doi.org/10.1016/B978-0-12-820297-5.00010-4

[3] M.M. Farag, F. Faraz Ahmad, A.K. Hamid, C. Ghenai, M. AlMallahi, M. Elgendi, Impact of Colored Filters on PV Modules Performance: An Experimental Investigation on Electrical and Spectral Characteristics, in: 50th Int. Conf. Comput. Ind. Eng., 2023: pp. 1692–1704.

[4] K.R. Abbasi, M. Shahbaz, J. Zhang, M. Irfan, R. Alvarado, Analyze the environmental sustainability factors of China: The role of fossil fuel energy and renewable energy, Renew. Energy. 187 (2022) 390–402. https://doi.org/10.1016/j.renene.2022.01.066

[5] M. Al-chaderchi, K. Sopain, M.A. Alghoul, T. Salameh, Experimental study of the effect of fully shading on the Solar PV module performance, E3S Web Conf. 23 (2017) 01001. https://doi.org/10.1051/e3sconf/20172301001

[6] M.M. Farag, R.C. Bansal, Solar energy development in the GCC region – a review on recent progress and opportunities, Int. J. Model. Simul. 43 (2023) 579–599. https://doi.org/10.1080/02286203.2022.2105785

[7] M.M. Farag, N. Patel, A.-K. Hamid, A.A. Adam, R.C. Bansal, M. Bettayeb, A. Mehiri, An Optimized Fractional Nonlinear Synergic Controller for Maximum Power Point Tracking of Photovoltaic Array Under Abrupt Irradiance Change, IEEE J. Photovoltaics. 13 (2023) 305–314. https://doi.org/10.1109/JPHOTOV.2023.3236808

[8] M.M. Farag, A.K. Hamid, T. Salameh, E.M. Abo-Zahhad, M. AlMallahi, M. Elgendi, ENVIRONMENTAL, ECONOMIC, AND DEGRADATION ASSESSMENT FOR A 2.88 KW GRID-CONNECTED PV SYSTEM UNDER SHARJAH WEATHER CONDITIONS, in: 50th Int. Conf. Comput. Ind. Eng., 2023: pp. 1722–1731.

[9] T. Salameh, A.K. Hamid, M.M. Farag, E.M. Abo-Zahhad, Energy and exergy assessment for a University of Sharjah's PV grid-connected system based on experimental for harsh terrestrial conditions, Energy Reports. 9 (2023) 345–353. https://doi.org/10.1016/j.egyr.2022.12.117

[10] T. Salamah, A. Ramahi, K. Alamara, A. Juaidi, R. Abdallah, M.A. Abdelkareem, E.-C. Amer, A.G. Olabi, Effect of dust and methods of cleaning on the performance of solar PV module for different climate regions: Comprehensive review, Sci. Total Environ. 827 (2022) 154050. https://doi.org/10.1016/j.scitotenv.2022.154050

[11] H. Rezk, I.Z. Mukhametzyanov, M.A. Abdelkareem, T. Salameh, E.T. Sayed, H.M. Maghrabie, A. Radwan, T. Wilberforce, K. Elsaid, A.G. Olabi, Multi-criteria decision making for different concentrated solar thermal power technologies, Sustain. Energy Technol. Assessments. 52 (2022) 102118. https://doi.org/10.1016/j.seta.2022.102118

[12] M.M. Farag, A.K. Hamid, Performance assessment of rooftop PV/T systems based on adaptive and smart cooling facility scheme - a case in hot climatic conditions of Sharjah, UAE, in: 3rd Int. Conf. Distrib. Sens. Intell. Syst. (ICDSIS 2022), Institution of Engineering and Technology, 2022: pp. 198–207. https://doi.org/10.1049/icp.2022.2448

[13] N.K. Almarzooqi, F.F. Ahmad, A.K. Hamid, C. Ghenai, M.M. Farag, T. Salameh, Experimental investigation of the effect of optical filters on the performance of the solar photovoltaic system, Energy Reports. 9 (2023) 336–344. https://doi.org/10.1016/j.egyr.2022.12.119

[14] T. Salameh, M. Tawalbeh, A. Juaidi, R. Abdallah, S. Issa, A.H. Alami, A novel numerical simulation model for the PVT water system in the GCC region, in: 2020 Adv. Sci. Eng. Technol. Int. Conf., IEEE, 2020: pp. 1–5. https://doi.org/10.1109/ASET48392.2020.9118264

[15] A. Alkhalidi, T. Salameh, A. Al Makky, Experimental investigation thermal and exergy efficiency of photovoltaic/thermal system, Renew. Energy. 222 (2024) 119897. https://doi.org/10.1016/j.renene.2023.119897

[16] S. Aberoumand, S. Ghamari, B. Shabani, Energy and exergy analysis of a photovoltaic thermal (PV/T) system using nanofluids: An experimental study, Sol. Energy. 165 (2018) 167–177. https://doi.org/10.1016/j.solener.2018.03.028

[17] A. Shahsavar, M. Arıcı, Energy and exergy analysis and optimization of a novel heating, cooling, and electricity generation system composed of PV/T-heat pipe system and thermal wheel, Renew. Energy. 203 (2023) 394–406. https://doi.org/10.1016/j.renene.2022.12.071

[18] H. Ma, Y. Xie, S. Wang, Y. Liu, R. Ding, Exergy analysis of a new spray cooling system-based PV/T and heat recovery with application in sow houses, Sol. Energy. 262 (2023) 111828. https://doi.org/10.1016/j.solener.2023.111828

[19] J. Zhou, X. Zhao, Y. Yuan, Y. Fan, J. Li, Mathematical and experimental evaluation of a mini-channel PV/T and thermal panel in summer mode, Sol. Energy. 224 (2021) 401–410. https://doi.org/10.1016/j.solener.2021.05.096

[20] A. Mokri, M. Aal Ali, M. Emziane, Solar energy in the United Arab Emirates: A review, Renew. Sustain. Energy Rev. 28 (2013) 340–375. https://doi.org/10.1016/j.rser.2013.07.038

[21] M.M. Farag, A.K. Hamid, Experimental Investigation on the Annual Performance of an Actively Monitored 2.88 kW Grid-Connected PV System in Sharjah, UAE, in: 2023 Adv. Sci. Eng. Technol. Int. Conf., IEEE, 2023: pp. 1–6. https://doi.org/10.1109/ASET56582.2023.10180880

[22] S.B. Bashir, M.M. Farag, A.K. Hamid, A.A. Adam, A.G. Abo-Khalil, R. Bansal, A Novel Hybrid CNN-XGBoost Model for Photovoltaic System Power Forecasting, in: 2024 6th Int. Youth Conf. Radio Electron. Electr. Power Eng., 2024. https://doi.org/10.1109/REEPE60449.2024.10479878

[23] M.M. Farag, F.F. Ahmad, A.K. Hamid, C. Ghenai, M. Bettayeb, Real-Time Monitoring and Performance Harvesting for Grid-Connected PV System - A Case in Sharjah, in: 2021 14th Int. Conf. Dev. ESystems Eng., IEEE, 2021: pp. 241–245. https://doi.org/10.1109/DeSE54285.2021.9719385

[24] M.M. Farag, F.F. Ahmad, A.K. Hamid, C. Ghenai, M. Bettayeb, M. Alchadirchy, Performance Assessment of a Hybrid PV/T system during Winter Season under Sharjah Climate, in: 2021 Int. Conf. Electr. Comput. Commun. Mechatronics Eng., IEEE, 2021: pp. 1–5. https://doi.org/10.1109/ICECCME52200.2021.9590896

[25] T. Salameh, A.K. Hamid, M.M. Farag, E.M. Abo-Zahhad, Experimental and numerical simulation of a 2.88 kW PV grid-connected system under the terrestrial conditions of Sharjah city, Energy Reports. 9 (2023) 320–327. https://doi.org/10.1016/j.egyr.2022.12.115

[26] R. Petela, Exergy of undiluted thermal radiation, Sol. Energy. 74 (2003) 469–488. https://doi.org/10.1016/S0038-092X(03)00226-3

[27] M. Romero, A. Steinfeld, Concentrating solar thermal power and thermochemical fuels, Energy Environ. Sci. 5 (2012) 9234. https://doi.org/10.1039/c2ee21275g

[28] E. Masana, C. Jordi, I. Ribas, Effective temperature scale and bolometric corrections from 2MASS photometry, Astron. Astrophys. 450 (2006) 735–746. https://doi.org/10.1051/0004-6361:20054021

Solar PV based charging station for electric vehicles (EV)

Raghad ALGHAMDI[1,a], Batool ALSUNBUL[1,b], Raghad ALMUTAIRI[1,c], Rania ALNASSAR[1,d], Maha ALHAJRI[1,e], Saifullah SHAFIQ[1,f], Samir EL-Nakla[1,g]*, Ahmed ABUL HUSSAIN[1,h], Jamal NAYFEH[1,j]

[1] Electrical Engineering Department, Prince Mohammad Bin Fahd University, P.O. Box 1664, Al Khobar 31952, Saudi Arabia

[a]201700819@pmu.edu.sa, [b]201700766@pmu.edu.sa, [c]201701747@pmu.edu.sa, [d]201701166@pmu.edu.sa, [e]201701472@pmu.edu.sa, [f]sshafiq@pmu.edu.sa, [g]snakla@pmu.edu.sa, [h]ahussain1@pmu.edu.sa, [j]jnayfeh@pmu.edu.sa

Keywords: Renewable Energy, Solar Energy, Electric Vehicle Charging, Batteries

Abstract. Electric Vehicles (EV) have been rapidly gaining attraction owing to their use of clean energy. The use of renewable energy sources, such as solar energy, is readily available to a wider community because of the falling costs of installing PV panels per watt. Saudi Arabia's Vision 2030, a sustainable vision for the future of Saudi Arabia focuses on new environmental and sustainable policies that are being developed to reduce carbon emissions, and to achieve that, there is a strong motivation to use clean energy and resources such as solar and wind energy. This paper presents the design of a stand-alone solar PV charging station for EV which includes additional features that allow the users to monitor the charging status of an EV via a smartphone application.

Introduction

Global warming is a fact, and the weather is becoming increasingly unpredictable as air is getting highly polluted. It is concerning that the world will address the climate situation too late, resulting in irreparable consequences. The dependence on non-renewable energy has been the major contributor of climate change since it results in the emission of greenhouse gasses [1,2].

Global electricity demand and cost is growing more rapidly as compared to renewables, driving a strong increase in the consumption of fossil fuels. According to EIA (Energy Information Administrations), the demand in energy keeps increasing higher than the global population, resulting in a rise in the average amount of power consumed for each person [3,4]. Another problem is the existing charging systems are overloading the grid, if several EV's are charged at the same time in the same place the power system may face excessive demands because the modern EV's consume more power equivalent to 10 homes [5,6].

The implementation of an off-grid power generation system can help resolve the problem of grid overloading as well as provide power in remote locations, where grid connectivity is an issue. Enhancing energy efficiency and lowering the grid energy demand are commonly regarded as the most valuable, quickest, least expensive, and safest way to fight climate change. EV charging reduces on-grid energy consumption by using solar panels and other renewable energy sources.

Shatnawi et al. presented a work of battery charging technologies and recent EV charging approaches and discussed the technical challenges in this field. The paper illustrated the importance of integrating renewable energy resources, particularly solar energy in the UAE and the Arab World, to provide clean and cost-effective public and private EV charging stations [7]. Another work conducted by Madhu et al. to design a smart charging station using Arduino and a range of sensors brings numerous advantages and functionalities. Through the integration of comprehensive sensors such as current, voltage, and temperature sensors, the charging station becomes capable of effectively monitoring and regulating the charging process, ensuring optimal battery conditions and charging safety [8].

A work done by Ballaji et al. investigated the simulation of an EV charging system. The system considered the EV battery as load and studied the effect of changing conditions in the environment due to varying irradiance throughout the day [9].

This paper presents the design of a stand-alone solar photovoltaic (PV) based charging station for an electric vehicle (EV) with multiple functions and fearures. It allows monitoring of the charging status through a mobile phone application with automated charging to prolong EV's battery life. It also monitors solar irradiance and related weather parameters such as ambient temperature and relative humidity that may affect the overall performance of the system. The proposed system has efficient maintenance and troubleshooting characteristics as different hardware modules of the system can be isolated for safety purposes, which can be handled by hardware means or through the mobile application.

The proposed system
The block diagram of the proposed system is shown in Fig. 1 which illustrates the overall design of the EV charging station. The proposed system consists of solar panels, a battery bank (battery storage system), an EV charger, and a DC/AC inverter.

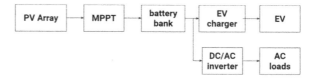

Figure 1: Overall Block Diagram of EV Charging Station System

The flow of the system shown in Fig. 1 above starts with the PV array connected to an MPPT charge controller that draws maximum power from the solar panels to be fed on to the battery bank. The battery bank acts as a solar energy storage system that provides a backup source of power on cloudy days or at night and to overcome the variation of power being produced by solar systems. It is necessary to include a battery bank especially when using a grid-independent (Off-grid) system. The system includes two types of loads, a DC load (EV battery) and an AC load (electrical appliances in the station). The EV charger consists of a DC-DC boost converter which operates as a step-up converter. Lastly, the system incorporates a DC/AC inverter that converts the DC output of the solar panels and battery bank to AC in order to power up appliances that are plugged in the station.

The EV charger is provisioned in the carport structure. The EV charger uses voltage and current sensors to measure the charging current and voltage in order to calculate the charging power. Additionally, the EV charger includes a battery monitoring system that displays to the user the charging status of the EV. The carport mounted solar panels provides an optimized fixed angle of 26 degrees for maximum sunlight collection.

System implementation
In order to design the system, it is essential to identify the loads first. The station is designed to charge an EV of 48V 7Ah and power up AC loads up to 150W minimum.

Storage batteries are needed because PV modules can generate power only when it is exposed to sunlight. They store the energy that is being generated by the PV modules during periods of high irradiance and make it available at night as well as during overcast periods. The batteries selected are two Lead-calcium batteries 12V 60Ah each, and they are connected in series as shown in Fig. 2. All the batteries used in a battery bank are the same type, same manufacturer, same age, and are maintained at equal temperature. Furthermore, the batteries have the same charge and discharge properties under these circumstances. If the above characteristics do not match, there is a high probability of huge energy loss within the battery bank.

Figure 2: Storage battery connection

MPPT charge controller is used to maximize the power transfer from the PV array to the battery bank. Next is a DC to AC inverter which is a 24V 3 kVA pure sine wave inverter as shown in Fig. 3.

Figure 3: (left) and MPPT

The EV charger is a DC to DC boost converter that will boost the voltage of the battery bank and match it to the voltage of the EV batteries and decrease the current to the rated charging current of the EV batteries. The design is developed using LM2588 5-A Flyback Regulator [10].

The method that was used for charging was the constant voltage (CV) charging method because it is the optimal method to get the most out of the batteries in terms of service life and capacity, as well as a reasonable recharge time and cost. A DC voltage between (50.1 V) and (50.4 V) is applied to the input terminals of the CV charge controller and the output terminals are connected to a (48V 7 Ah) lead acid battery to charge it.

To start wiring the components within the system, a wiring diagram and wire sizing are made in order to visualize the connection of the system components. The wiring diagram of the system is shown in Fig. 4.

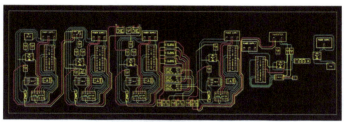

Figure 4: Wiring diagram

To display the charging status of the EV through a mobile application, a Raspberry Pi processor was used as a central server to record all the sensor readings and display the on the mobile platform to be accessed at any given time. The Raspberry Pi together with a home automation software as a server can also send data to the cloud which enables the user to view all sensor readings and to control the different parameters in the system even outside the local area network as shown in Fig. 5.

Figure 5: Mobile application user interface

Another set of parameters that was monitored in the system was the sensor outputs. The sensors include a light sensor, humidity and temperature sensor, pressure sensor, and an accelerometer sensor as shown in Fig. 6. The light sensor used is a BH1750 that measures the illuminance level of the light going to the solar panels. The humidity and temperature sensor used is DHT22 that will measure the humidity and temperature to be displayed to the user. In addition, a barometric pressure sensor BMP280 was used that detects the atmospheric pressure for weather forecasting. Lastly, an accelerometer & gyroscope (3-axis) sensor was used to measure the angle at which the solar panels will be adjusted to for maximum power.

Figure 6: Sensors Circuit

Finally, the structure of the charging station and electrical enclosure which is designed using Solid Works software and a ventilation system is added to maintain the ideal working temperature for the electronic components. Dimensions are 0.81 m in height, 1.76 m in width, and 0.3 m depth. All components were placed within a factory fabricated electrical enclosure as in Fig. 7. After that, the structure of the charging station was made whose dimensions are 2.47 m in height, 2 m in width, and 1.5 m depth, as shown in Fig. 8. The solar panels were placed at an angle of 26 degrees.

Figure 7: Electrical Enclosure

Figure 8: Charging Station & Carport

Testing and Results

The angle on which the PVs are mounted is a critical consideration on any solar power system installation. After positioning the panels at different angles during months of March and April, different output ratings were obtained as shown in Table 1. It was found that the optimum tilt angle for our location is 26 degrees to pull out every single watt hour out of the system.

Table 1: Test result of solar panels

Angle	Voltage (V)	Current (A)
22°	21.5	3.64
23°	21.5	3.67
24°	21.5	3.72
25°	21.5	3.77
26°	21.5	3.81

For the boost converter, all of the specifications stated previously have been met. The output voltage across the output capacitor is 54 V with no load and 50 V with load connected. To get the efficiency, the test was performed using 4 wire measurement method to get accurate data. All DC measurements were taken using a multimeter and electronic DC load tester as well as an infrared thermometer was used to get the temperature output of the circuit shown in Fig. 9. The below tables show the result of the boost converter when testing with a fan and without a fan.

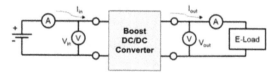

Figure 9: Testing converter

The results of the boost converter testing with a fan and without a fan were recorded as shown in Table 2 and Table 3 which show only the minimum and maximum readings recorded.

Table 2: Results of testing converter with fan

Vin(V)	In (A)	Pin (W)	Temp (C)	Vout (V)	Iout (A)	Iset (A)	Pout(W)	EFF %
24	0.37	8.88	25.4	50.4	0.15	0.15	7.56	85.13
24	1.45	34.8	24.4	50.4	0.65	0.65	32.76	94.13

Table 3: Results of testing converter without fan

Vin(V)	In (A)	Pin (W)	Temp (C)	Vout (V)	Iout (A)	Iset (A)	Pout(W)	EFF %
24	0.24	5.76	25.5	50.3	0.1	0.1	5.03	87.32
24	2.1	50.4	27.8	49.6	1	1	49.6	98.41

The converter with fan reached a maximum efficiency of 94.13%, and minimum of 85.13%. On the other hand, the converter without fan reached the maximum efficiency of 100.98%, and minimum of 100.87%. However, the converter without a fan is not efficient, it will stop working at some point as temperature increases the output voltage will decrease, and will not be able to give the desired output as well as the ICs will get damaged. The result shows the addition of a small fan played a very important role as the heat sink alone gets hot during operation and this can damage the ICs without a fan. The efficiency vs output current graph was generated as shown in Fig. 10 with average efficiency of converter of 92.47%.

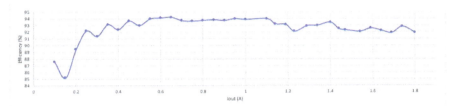

Figure 10: Efficiency Vs Output Current

The last part of the system is hardware and software integration. The circuits for all the parameters that should be monitored in our system were completed. Later, the sensors and communication modules were programmed to display the information to the user wirelessly.

Testing the battery status of the EV battery when charged, gave us accurate voltage readings for the battery used which was around 50.1 V as illustrated in Fig. 11 (a).

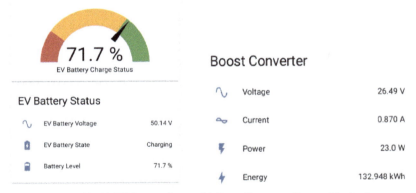

Figure 11: (a) EV Battery Status, (b) Boost Converter Energy Monitoring

The boost converter readings display the input coming from the battery bank entering the boost converter. Theoretically, the voltage of the boost converter should be 24 V. However, considering that 12 V batteries can go as high as 13.2 V when fully charged, when two 13.2 V batteries are connected in series as in our case it will give us a maximum output of 26.4 V, which can be observed in Fig. 11 (b) above.

Summary
The project involved building an off-grid solar based charging station for electric vehicles by fabricated the whole system design of the charging station as well as building the user interface of mobile application. Some important features that we included in our project is the monitoring of the charging status of the EV battery as well as the power consumption within the system. It also allows control of components and displays readings of the illuminance levels, temperature and humidity, etc.

For future work, the team will look to extend the current charging time of project which is around 3.7 hours to charge an EV to include fast charging methods for the battery to charge in a shorter period of time.

References

[1] Biya, T. S., & Sindhu, M. R., Design and power management of Solar Powered Electric Vehicle Charging Station with energy storage system. 2019, 3rd International Conference on Electronics, Communication and Aerospace Technology (ICECA). https://doi.org/10.1109/iceca.2019.8821896

[2] Information on https://climate.nasa.gov/evidence

[3] MOHAMED, K., etal, Opportunities for an off-grid solar PV assisted Electric Vehicle Charging Station. 2020, 11th International Renewable Energy Congress (IREC). https://doi.org/10.1109/irec48820.2020.9310376

[4] Information on www.who.int/news-room/air-quality-guidelines/ambient-(outdoor)-air-quality-and-health

[5] Information on https://www.eia.gov/outlooks/ieo/

[6] U.S. Department of Energy, Electric Vehicle Charging Infrastructure Analysis: A Case Study of California's Transportation Electrification. (2021)

[7] M. Shatnawi, K. B. Ari, K. Alshamsi, M. Alhammadi and O. Alamoodi, Solar EV Charging, 2021 6th International Conference on Renewable Energy: Generation and Applications (ICREGA), Al Ain, United Arab Emirates, 2021, pp. 178-183, https://doi.org/10.1109/ICREGA50506.2021.9388301.

[8] Madhu B. R, Anitha G. S, Mahantesha H. and Krishna S, "Smart Charging Station for Electric Vehicles Using Solar Power," 2023 7th International Conference on Computation System and Information Technology for Sustainable Solutions (CSITSS), Bangalore, India, 2023, pp. 1-5. https://doi.org/10.1109/CSITSS60515.2023.10334142

[9] A. Ballaji, R. Dash, V. Subburaj, K. J. Reddy, S. C. Swain and M. Bharat, "Design and analysis of EV Charging Station using PV Integrated Battery Sytsem," 2022 International Conference on Smart Generation Computing, Communication and Networking (SMART GENCON), Bangalore, India, 2022, pp. 1-6. https://doi.org/10.1109/SMARTGENCON56628.2022.10083931

[10] LM25885-A Flyback Regulator with Shutdown datasheet https://www.ti.com/lit/gpn/LM2588

Role of renewable energy in decarbonisation process: Case study in KSA

Samar DERNAYKA[1,a *], Saidur R. CHOWDHURY[1,b] and Mohammad Ali KHASAWNEH[1,c]

[1]Prince Mohammad Bin Fahd University, Civil Engineering Department, KSA

[a]sdernayka@pmu.edu.sa, [b]schowdhury1@pmu.edu.sa, [c]mkhasawneh@pmu.edu.sa

Keywords: Renewable, Wind Turbine, Photovoltaic, Solar, Carbon Emissions

Abstract. Currently, most countries are replacing the fossil fuel electricity generation with renewable technologies for their crucial role in mitigating the greenhouse gas emissions. This paper discusses the implementation of three power plants in Al Aziziya in the eastern province of KSA by deploying three different renewable technologies 1) Photovoltaic 2) Solar thermal power and 3) Wind turbine. Both the energy performance and rate of electricity exported to grid were predicted when the capacity varies from 1000 to 1000,000 KW. In addition, the role of the three different renewable technologies in the decarbonization process has been evaluated.

Introduction
Over the last century, the greenhouse gases including atmospheric carbon dioxide (CO_2) have potentially grown with the energy production, especially by burning fossil fuels to generate electricity. In fact, fossil fuel materials like oil and gas stand behind 75% of the global greenhouse gas emissions and almost 90% of carbon dioxide emissions [1]. The data analysis in [2] have shown that greenhouse gas emissions were raised by 4298.05 (MTCO2e) between 1990 and 2016 due to the tremendous electric energy consumption, which accounts of 40% of the total GHG emissions. The Kingdom faces relatively high-energy demands of energy in industrial as well as residential spots. According to The King Abdullah Petroleum Studies and Research Center (KAPSARC) Residential Energy Model (REEM), a residential Villa in Dhahran area consumes an average of 24,900 KWh annually, while apartments use around 17,200 KWh [3]. Further, a study conducted in 2012 shows that the annual average electricity consumption hits 176.5 kWh/m², exceeding international energy-efficiency benchmarks. This translates to around 21,180 kWh per year for a 120-m² house [4]. Internationally, a household consumes, on average 9,600 to 12,000 kWh annually. This translates to an average daily energy consumption of about 26 to 33 kWh, equivalent to 26,000 to 33,000 watt-hours [5].

Hence, promoting the decarbonisation concept becomes a necessity. This process is achieved either by reducing the energy consumption, or by applying efficient practice and technologies through Renewable energy [6, 7]. For a one percent shift away from the usage of oil, it is possible to reduce the carbon emissions by 1.288 [8]. With the growth of Energy demand, the decarbonisation process consists of reducing or cutting the greenhouse gas emissions, which can be achieved through zero-carbon renewable energy sources such as wind, Solar, Hydropower, Geothermal and Biomass. Simultaneously, the available carbon and methane in the atmosphere shall be continually captured, and stored to counter balance the released toxic gases. The implementation of renewable technologies is required to be in correlation with the actual needs and peak demands of power. Currently, Wind and Solar are two main clean energy resources that can be used to generate electricity with zero carbon emissions and consequently less greenhouse effect. The deployment of key technologies such as solar PV or Wind turbines is on track in many countries and it proved to present a beneficial role in the decarbonisation process. According to

[9], the slice of renewables in power production will rise from around one-quarter in 2015 to around 60% by 2030 and 85% by 2050 for energy sector decarbonisation. In Russia, the implementation of wind and solar have contributed to more than 5 GW since 2013 which is likely to exceed the targeted capacity of 5.9 GW in 2024. For Middle Eastern countries, Turkey is urging to increase the share of solar and wind as well to cover its demand of imported energy [10].

The Kingdom of Saudi Arabia has ongoing plans to halt the rise of greenhouse gas emissions, and contribute to decarbonisation in order to reach the zero-carbon emission target by 2060. The main milestone for this is certainly to generate electric power by shifting away from petroleum and fossil fuel towards Renewable energy sources. Yet, such projects are still challenging and they are affected by many factors such as the location of the facility, meteorological parameters, connections to grid and mostly the efficiency of the technology. The scope of this paper is to discuss the implementation of electric power plant based on three renewable technologies 1) Photovoltaic 2) Solar thermal power and 3) Wind turbine in Al Aziziyah, located in Dhahran, the eastern province of Saudi Arabia. The performance of the three different power plants is predicted by using Retscreen Expert, a clean energy management software that has been validated by a team of experts in the Canadian government [11] in addition to HOMER Energy software. The analysis of the results presents evaluation of the electric power exported to the grid as well as the role of the three renewable technologies in the decarbonization process.

Renewable Energy Technologies
Photovoltaic
A Solar PV system consists of panels that convert electromagnetic wave of sunlight into electricity using the photovoltaic (PV) effect. The PV modules are made up of multiple (series and parallel) interconnected solar cells, which are typically formed of semiconductor material such as silicon. When sunlight hits these cells, it excites electrons, generating an electric current. Connecting cells in series increases the voltage and connecting them in parallel increases the current of the panel [12, 13]. Solar PV capacity has experienced a growth more than any other source of electricity generation [7]. In general, the PV size is chosen based on the energy and power consumption. The process of calculation of solar photovoltaic modular system involves the following steps: 1) Determine electricity consumption by calculating the total amount of electricity appliances to be powered in kilowatt-hours (kWh). An electricity bill provides a preliminary idea of the average monthly or yearly consumption. (2) Assess the solar irradiation which depends on the geographic location, time of year, and weather conditions. (3) Account for system losses: These can typically range from 10% to 20% including shading, wiring losses, inverter efficiency, etc. (4) Calculate the required capacity as in Eq. 1 and Eq. 2 (5) Determine the number of panels based on Eq. 3 [14, 15]. The efficiency of a solar panel refers to the amount of sunlight it can convert into electricity. The average efficiency for most commercial solar panels is ranging between 15% and 20%. The typical power output of a standard solar panel is around 250 to 400 watts [12, 13 & 16]. Advanced PV modules can produce over 400 watts power with maximum efficiency that reaches 40%.

$$\text{Required Capacity (kW)} = \frac{\text{Electricity Consumption(kWh)}}{\text{Solar irradiation} * \text{System Losses}} \quad (1)$$

$$\text{Solar Irradiation per day} = \frac{P}{A} = \epsilon\, \sigma T^4 \quad (2)$$

$$\text{Number of panels} = \text{Required Capacity(kW)}/\text{Panel power Output(kW)} \quad (3)$$

Where, P and A represent solar power and Area; ϵ, σ and T indicate emissivity, Stefan-Boltzmann constant and surface temperature, respectively.

Solar Thermal Panel
A solar thermal panel absorbs the incoming solar radiation, converts it into heat [17, 18], and transfers this heat to a fluid (usually air, water, or oil) that flows and circulates (by mechanical component, such as a fan or pump) through the collector. A solar heating system utilizes solar heat (concentrated or not) without conversion in order to supply residential and commercial purposes [19]. The common types of solar panel or collectors are flat plate collector, integral collector-storage (ICS), double -glass solar collectors, compound parabolic collectors and evacuated tube collectors etc. [17, 18]. The concentration ratio, fluid flow velocity, and surface area would all have a significant influence on the collector's efficiency and heat removal factor. Compared to solar PV panels, solar thermal panels are more efficient, converting 70–90% of the energy input into heat, and they require less area. However, the capacity factor of the overall power utilization is in the range of 20 to 30% due to the intermittency of the naturel resources. Flat panel or collector, typically measuring one meter by two meters, is the most popular kind of solar thermal panel [20]. Figure 1 shows the system to convert sunlight to heat.

Figure 1 Converting sunlight to heat system

Again, concentrated solar power (CSP), a high temperature solar thermal system, uses groups of mirrors to concentrate solar energy at central collector. This produces a temperature high enough to generate steam, which then turns a turbine, driving a generator to produce electricity [19]. Electricity from a solar thermal system is an opto-caloric system that depends on solar radiation and its capture's capacity. Figure 2 shows the flow chart of electricity generation from solar radiation (by converting solar thermal radiation into electricity) with the support of a solar collector and other mechanical components.

Figure 2 Electricity generation from solar radiation using solar heat

High temperatures that are required to achieve the utmost efficiency can be obtained by increasing the energy flux density of the solar radiation incident on a collector. According to Lupu et al. [20] energy efficiency of a solar thermal collector is:

$$n_{en} = \frac{\dot{Q}_u}{GA_C} \qquad (4)$$

Q_u useful heat rate absorbed by the fluid; incident solar radiation, G (average 240W/m^2), A_c is area of the collector. C_p, m, and T are latent heat, mass of plate, and temperature, respectively. The useful heat rate absorbed by the fluid [20], Q_u, is

$$\dot{Q}_u = \dot{m} C_p (T_{fl,out} - T_{fl,in}) \qquad (5)$$

Wind turbine
Wind turbines harness the naturel wind energy to generate electricity. The blades start to rotate when they are stimulated by the wind's kinetic energy. The resulting mechanical energy created over a drive shaft is then converted to electric energy through a generator. A basic wind speed of 3 to 5m/s at 10m height is usually sufficient to have potential wind resource. The efficiency of a wind turbine is measured by calculating its capacity factor as per Eq. (6) [21].

$$Cp = \frac{\text{Actual Electricity Output (kWh)}}{\text{Maximum Electricity Output}} \qquad (6)$$

$$\text{Maximum Electricity Output (kWh)} = \text{Capacity(kW)} * \text{hours per year} \qquad (7)$$

The capacity factor indicates how fully the wind turbine capacity is used. It is affected by the availability of wind, the hub height, swept area of the unit and the size of its generator. The capacity factor ranges between 20 to 35% and it can go up to 52% for new giant devices.

Data Assessment
The potential electric power plants are located in Al-Aziziyah, in the vicinity of Dhahran, the eastern province of Saudi Arabia. The latitude and longitude for the facility location are 26.2 and 50.2 respectively. The climate zone is estimated to be extremely hot – Dry. The preliminary climate assessment is based on the meteorological data and NASA provided by RetScreen Expert Clean Energy Management Software platform. Fig.3 shows two meteorological parameters: the daily solar radiation and wind speed at 10m height above the ground.

The daily solar radiation ranges between 3.2 and 8 KWh/m². It is the global horizontal irradiance that presents the total solar radiation falling on a horizontal surface. The peak value of 7.8 KWh/m²/d attained in the month of June refers to 7.8 hours of sun received per day at 1 KWh/m² during this month. The extreme value of the annual average irradiance of 5.6 KWh/m²/d as per NASA as well as the high temperatures (above 30°c) make the area an excellent resource for solar applications. The annual wind speed variation shows an average of 4.4 m/s, which favors the installation of wind turbines. Besides, Dhahran area belongs to the moderate to good wind resource as per [22].

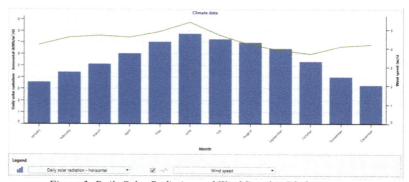

Figure 3- Daily Solar Radiation and Wind Speed in Dhahran area

RetScreen-Software based Simulation
The simulation of the three different renewable power plants has been conducted in RetScreen. Each RetScreen Technology model is developed within a workbook file that is been validated by

modelling experts and other simulation software. RetScreen evaluates the implementation of a renewable project based on several aspects: 1) Energy performance 2) Cost analysis 3) GHG emissions 4) Financial Summary and 5) Risk Analaysis. The input data is relatively small since RetScreen contains integrated products, cost and weather databases. The data that is required cover the location of the facility, climate data (hourly wind Speed for wind), selected technology (Manufacturer and model) and economic input data (Table 1). Other variables that are relevant to the energy pricing can also be filled by the user.

Table 1- Financial Indicators considered in RetScreen Simulation for the current simulation

Inflation rate: 2%	Debt Interest rate: 7%	Electricity Export rate: 0.10 $/kWh
Discount rate: 9%	Project Life: 20 years	Electricity Export Escalation rate: 2%

Results and Discussion
Renewable Energy Systems

Table 2 shows the technical specifications for the renewable technologies implemented in four potential power plants with different capacity each. Based on RetScreen predictions, the number of solar thermal and photovoltaic units is raised with the increase of the required capacity. For wind turbines, a larger electricity output is generated when the hub height increases and more turbines are added; which results in higher capacity factor. The gross energy production per turbine GE is different for each device model. For 1000,000 KW capacity, there is need for vast wind farm with 500 ENERCON 82 E2 2MW wind turbines at 138 m height. The solar thermal system presents the same performance for all capacities: 30% while photovoltaic cells have the least efficiency: the rate of miscellaneous losses considered is 15%, while the inverter capacity is 95%.

Table 2 – Technical Specifications for considered renewable based Power Plant

Capacity (KW)	Solar Thermal	Photovoltaic	Wind Turbine
1000	Abengoa Solar PS10 with parabolic mirrors Capacity factor: 30%	5000 units mono-Si Solar Collector: 3333 m² Efficiency: 17%	2 ENERCON 53 – 73m Capacity Factor: 25%
10,000	Abengoa Solar PS10 with parabolic mirrors Capacity factor: 30%	50,000 units mono – Si SP150 Capacity per unit: 150 W Solar Collector: 64103 m² Efficiency: 17.6%	5 ENERCON 82 E2 2MW – 78 m GE: 4771 MWh per turbine Capacity Factor: 24%
100,000	Abengoa Solar PS10 with parabolic mirrors Capacity factor: 30%	500,000 units mono – Si CS1H-320MS Capacity per unit: 320 W Solar Collector: 842105 m² Efficiency: 17.6%	44 ENERCON 82 E2 – 138 m GE: 5818 MWh per turbine Capacity Factor: 25%
1000,000	Abengoa Solar PS10 with parabolic mirrors Capacity factor: 30%	5,000,000 units mono – Si CS1H-320MS Capacity per unit: 320 W Solar Collector: 8421053 m² Efficiency: 17.6%	500 ENERCON 82 E2 2MW – 138 m GE: 5613 MWh per turbine Capacity Factor: 28%

Table 3 – Electricity outcome and CO_2 reduction for the three considered technologies

Designed Capacity (KW)	Electricity Exported to Grid (MWh)			GHG Emission Reduction (tCO2)		
	Solar Thermal	Photovoltaic	Wind Turbine	Solar Thermal	Photovoltaic	Wind Turbine
1000	2,628	1,543	1,968	1,243	730	1,650
10,000	26,280	11,576	20,675	13,582	5,475	9,778
100,000	26,280,000	246,950	221,668	124,291	116,795	104,837
1000,000	262,800,000	2,469,205	2,432,350	1,242,910	1,167,949	1,150,377

Table 3 shows the electricity exported to Grid as well as the rate of carbon reduction in tons of CO_2 relevant to each technology. The production of electricity from both photovoltaic and wind turbines is comparable specifically when the capacity exceeds 10,000 KW. The solar thermal system provides the highest rate of total electricity output at a specific designed capacity. It can play a pivotal role in the electricity generation in the area due to the abundance of sun and consequently the generated heat will be used in electricity generation. The electricity exported to grid will reach 262,800,000 MWh for a 1000,000 KW power plant capacity, which surpasses the output from Photovoltaic and wind. The amount of greenhouse gas displaced in tons of CO_2 is in correlation with the increase of the designed capacity. The implementation of photovoltaic results in the lowest carbon reduction rate in comparison with the solar and wind at 10 MW and smaller.

Role of Renewable energy technologies in the decarbonization process
Figure 4 shows the greenhouse gas emissions displacement with respect to the three different renewable technologies (solar thermal, Photovoltaic and Wind Turbine) in function of the required power plant capacity. At capacities equal to or lower than 10,000 KW, the solar thermal has the ability to displace the highest rate of carbon dioxide, followed by the wind turbine and then the photovoltaic system. For 100,000 and 1000, 000-KW power plants, the three technologies will result in comparable amount of greenhouse gas emissions. This is due to the expansion in the number of photovoltaic units and higher wind turbines that will capture more sun and wind and consequently release clean energy. The amount of displaced greenhouse gas emissions when setting up a 1,000,000 KW renewable power plant capacity attains 1,242,910 tCO₂, which accounts for 2,890,487 Barrels of crude oil not consumed.

The process of electricity generation by using renewable technologies instead of the conventional methods has a great impact on the decarbonization as it halts the release of toxic substances in the air. In fact, coal, natural gas or petroleum combustion are more carbon-intensive in electric power production in comparison with wind and solar as shown in Table 4. Besides, Coal combustion produces more greenhouse gases than the combustion of any other fossil fuels.

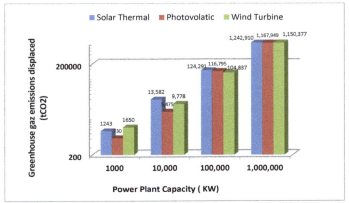

Figure 4 - Greenhouse emissions reduction with respect to renewable energy technologies (tCO₂)

Table 4 CO₂ emissions during electricity production for different energy sources (obtained from [23])

	Coal	Natural Gas	Fuel Oil	Wind	Solar	Petroleum
Electricity generation	0.88 kWh/Pound	0.13 kWh/cubic foot	12.90 kWh/gallon	2,4 kWh and 9,6 kWh/day	0.68 KWh/day	1.18 kWh/pound
CO₂ emissions	980 g CO₂/kWh	465 g CO₂/kWh	266.5 g CO₂/kWh	11 g CO₂/kWh	41 g CO₂/kWh	345.0 g CO₂/kWh

Reducing carbon emissions from the power industry is greatly impacted by moving electricity production from fossil fuel generation sources to renewable ones. Decarbonizing the grid is one potential option that calls for international cooperation and coordination between businesses and governments in order to reduce emissions as well as follow the Paris Agreement.

Conclusion

Renewable energy sources, particularly wind and solar, provide sustainable solutions to our power needs. They present a major role in the decarbonization process especially when used in the electricity generation process. A case study conducted in Al Aziziyah in the Kingdom of Saudi Arabia reveals the most beneficial impact on displacing the greenhouse gas emissions when implementing the Solar Thermal power plant. The area, prone to continuous extreme sunlight, and good wind potential offers an efficient and sustainable alternative for producing fossil fuel based electricity and facing the high demand with minimal environement effects. Finally, while renewable energy sources like wind and solar power significantly reduce greenhouse gas emissions, they are not entirely devoid of waste generation, particularly during the manufacturing and end-of-life stages.

References

[1] O. Edenhofer, R. P. Madruga, Y. Sokona, Renewable Energy Sources and Climate Change Mitigation, Cambridge University Press, New York, USA, 2012. ISBN 978-1-107-60710-1.

[2] R. Gh. Alajmi, Factors that impact greenhouse gas emissions in Saudi Arabia: Decomposition analysis using LMDI, Energy Policy 156 (2021) , 112454. https://doi.org/10.1016/j.enpol.2021.112454

[3] M. Aldubyan, M. Krarti,, & E. Williams, Residential Energy Model for Evaluating Energy Demand and Energy Efficiency Programs in Saudi Residential Buildings, (2020) KAPSARC, Riyadh, KSA. https://doi.org/10.30573/KS--2020-MP05

[4] F. Alrashed, & M. Asif, Trends in residential energy consumption in Saudi Arabia with particular reference to the Eastern Province. Journal of Sustainable Development of Energy, Water and Environment Systems, 2(4) (2014), 376-387. https://doi.org/10.13044/j.sdewes.2014.02.0030

[5] https://www.anker.com/blogs/home-power-backup/electricity-usage-how-much-energy-does-an-average-house-use

[6] M. Gul, Y. Kotak, & T. Muneer, Review on recent trend of solar photovoltaic technology. Energy Exploration & Exploitation, 34(4), (2016), 485-526. https://doi.org/10.1177/0144598716650552

[7] N. A. Ludin, N. I. Mustafa, M. M. Hanafiah, M. A. Ibrahim, M. A. Teridi, M. Sepeai & K. Sopian, Prospects of life cycle assessment of renewable energy from solar photovoltaic technologies: A review. Renewable and Sustainable Energy Reviews, 96 (2018) 11-28. https://doi.org/10.1016/j.rser.2018.07.048

[8] R. Gh. Alajmi, Carbon emissions and electricity generation modeling in Saudi Arabia. Environmental Science and Pollution Research (2022) 29:23169–23179. https://doi.org/10.1007/s11356-021-17354-0

[9] C. D. Gielena , F.Boshella , D. Sayginb , M. D. Bazilianc , N. Wagnera, R. Gorini, The role of renewable energy in the global energy transformation, Energy Strategy Reviews 24 (2019) 38 -50. https://doi.org/10.1016/j.esr.2019.01.006

[10] D. Saygin, M. Hoffman, P. Godron, How Turkey Can Ensure A Successful Energy Transition, Center for American Progress, Washington, DC, 2018.

[11] CLEAN ENERGY PROJECT ANALYSIS: RETSCREEN® ENGINEERING & CASES TEXTBOOK, Minister of Natural Resources Canada 2001-2005, ISBN: 0-662-39191-8 Catalogue no.: M39-112/2005E-PDF

[12] P. K. Nayak, S. Mahesh, H. J. Snaith, & D. Cahen, Photovoltaic solar cell technologies: analysing the state of the art. Nature Reviews Materials, *4*(4) (2019) 269-285. https://doi.org/10.1038/s41578-019-0097-0

[13] N. Rathore, N. L. Panwar, F. Yettou, & A. Gama, A comprehensive review of different types of solar photovoltaic cells and their applications. International Journal of Ambient Energy, *42*(10) (2021) 1200-1217. https://doi.org/10.1080/01430750.2019.1592774

[14] A. R. Jordehi, Parameter estimation of solar photovoltaic (PV) cells: A review. Renewable and Sustainable Energy Reviews, 61(2016) 354-371. https://doi.org/10.1016/j.rser.2016.03.049

[15] A. A. H. Hussein, S. Harb, N. Kutkut, J. Shen, & I. Batarseh, Design considerations for distributed micro-storage systems in residential applications. In Intelec *2010* (pp. 1-6). IEEE. https://doi.org/10.1109/INTLEC.2010.5525658

[16] G. K. Singh, Solar power generation by PV (photovoltaic) technology: A review. Energy 53 (2013) 1-13. https://doi.org/10.1016/j.energy.2013.02.057

[17] S. A Kalogirou, Solar thermal collectors and applications. Progress in Energy and Combustion Science 30 (2004) 231–295. https://doi.org/10.1016/j.pecs.2004.02.001

[18] L. Kumar, J. Ahmed, M.E. H. Assad, M. Hasanuzzaman. Prospects and Challenges of Solar Thermal for Process Heating: A Comprehensive Review. Energies 2022, 15, 8501. https://doi.org/10.3390/en15228501

[19] T. Merrigan. Solar Energy Technologies. National Renewable Energy Laboratory, U.S. Department of Energy. 2005. /https://www1.eere.energy.gov/solar/pdfs/solar_tim_merrigan.pdf

[20] A. G. Lupu, V. M. Homutescu, D. T. Balanescu, and A. Popescu, Efficiency of solar collectors – a review. IOP Conf. Series: Materials Science and Engineering 444 (2018) 082015. https://doi.org/10.1088/1757-899X/444/8/082015

[21] D. Song, Y. Yang, S. Zheng, W. Tang, J. Yang, M. Su, X. Yang, J. Y. Hoon, Capacity factor estimation of variable-speed wind turbines considering the coupled influence of the QNcurve and the air density, Energy, Vol. (183) (2019) pp 1049 – 1060. https://doi.org/10.1016/j.energy.2019.07.018

[22] Y. El Khchin, M. Sriti, Performance Evaluation of wind turbines for energy production in Morocco's coastal regions. Results in Engineering 10 (2021) 100215. https://doi.org/10.1016/j.rineng.2021.100215

[23] U.S. Energy Information Administration (EIA), Electricity generation from wind. 1000 Independence Ave., SW Washington, DC 20585 (2022) https://www.eia.gov/energyexplained/wind/electricity-generation-from-wind

An automated and cost-efficient method for photovoltaic dust cleaning based on biaxially oriented polyamide coating material

Said Halwani[1,2,a*], Mena Maurice Farag[1,2,b*], Abdul-Kadir Hamid[1,2,c*], Mousa Hussein[3,d]

[1]Sustainable Energy and Power Systems Research Centre, Research Institute for Sciences and Engineering (RISE), University of Sharjah, Sharjah, United Arab Emirates

[2]Department of Electrical Engineering, College of Engineering, University of Sharjah, Sharjah, United Arab Emirates

[3]Department of Electrical and Communication Engineering, College of Engineering, United Arab Emirates University, Al Ain, UAE

[a]Shalwani@sharjah.ac.ae, [b]U20105427@sharjah.ac.ae, [c]AKHAMID@sharjah.ac.ae, [d]mihussein@uaeu.ac.ae

Keywords: Dust Accumulation, Experimental Investigation, PV Systems, Cleaning Methods

Abstract. Photovoltaic (PV) systems have been at the forefront of renewable energy technologies. However, they are highly dependent on environmental parameters that affect their performance and longevity. Dust accumulation presents a critical factor in the performance of PV systems, leading to minimum system efficiency under largely dusty conditions. Several cleaning methodologies have been proposed in scientific literature to prevent dust accumulation at the forefront of PV modules. However, most cleaning methodologies are cost-consuming, time-consuming, complex in implementation, or require huge manpower to implement them. This paper proposes a cost-efficient and automated method for dust accumulation prevention and cleaning based on Biaxially Oriented Polyamide (BOPA) coating material. This transparent thin film is applied on the front surface and integrated based on an automated control scheme, for controlled rotation every 2 weeks, to prevent dust accumulation on the forefront of the PV surface. The performance of the BOPA coating film was experimentally assessed for 45 days, assessing the irradiance and electrical performance of the PV modules. The application of BOPA maintained a PV module electrical efficiency of 12.19%, while the dusty PV module electrical efficiency is reduced to 7.79% at high dust accumulation levels. Moreover, the BOPA material has demonstrated its ability in capturing solar irradiance, without losses for the visible light, hence maintaining an electrical current of 2.15 A, while the dusty PV module loses its electrical current by 40%, maintaining an electrical current of 1.28 A.

Introduction

Renewable energy resources (RES) have been at the forefront of electricity generation, due to their active contribution to clean energy generation and improved efficiency [1]. Renewable energy has been deployed due to its environmental and technical advantages in integration within the primary energy mix [2–4]. Solar energy has attracted attention in renewable energy technologies due to its abundance, reliability, and zero-cost availability [5,6]. Photovoltaic (PV) systems have been developed for the useful harvesting of solar energy, and its ability for clean energy generation and minimize the carbon footprint [7].

The global demand for solar power generation has been rapidly increasing, leading to the evolution of technology and being commercially available [8–10]. Therefore, it is essential to continuously develop the technology for efficient operation and generation [11]. Dust accumulation on the front surface of PV modules has been a commonly reported issue in the effectiveness and applicability of PV technologies [12]. Particularly, regions such as the Middle East, China, North Africa, India, and the United States have been developing PV plants with a capacity of Gigawatts. However, such regions are exposed to dusty conditions and harsh weather conditions, leading to severe degradation in PV system performance [13–15]. Dust accumulation is considered a critical factor in regulating the electrical efficiency of a PV system, specifically when large accumulation occurs.

The scientific literature focused on several methods to clean and sustain PV modules, preventing the severe impact of soiling [16–18]. The cleaning may occur naturally through rainfall or wind. Other cleaning methods have been introduced such as manual cleaning, dry cleaning based on a robotics system, electrostatic cleaning, and preventive cleaning based on coating materials.

As aforementioned, the cleaning techniques are of many types with each having several advantages and disadvantages. Natural cleaning techniques are relatively low cost due to the utilization of the natural gravity of water or wind. However, it's not physically possible to induce rainfall or wind in specific regions, particularly depending on the dust particle volume [19,20]. Additionally, manual cleaning benefits from minimum power consumption and allows for periodical cleaning. However, high labor costs are a drawback and other potential hazards may harm the system or laborers [21]. Furthermore, the installation of sprinklers is of good beneficial ability in cleaning PV systems under short periods with feasible installation [22], however, high water consumption and non-uniform cleaning are a potential drawback [23]. Recently, the development of robots for automated cleaning has allowed for the reduction of damage on PV module surfaces and lowering the of energy consumption, through dry cleaning. However, their operation and maintenance require significant investment, and their applicability is limited to small PV plants [24]. Emerging technologies such as electrostatic cleaning and surface coating materials are still developing and need to be studied for their effectiveness due to their cost and coating properties [25–28].

This paper proposes a cost-effective and simple methodology for dust cleaning based on the application of Biaxially Oriented Polyamide (BOPA) film. A thin layer of BOPA is assessed in outdoor experimental conditions in the terrestrial conditions of Sharjah, UAE. An automated and control scheme is proposed for rotation of the transparent layer every 2 weeks, to maintain PV modules in clean conditions for 45 days. The study assesses the impact of dust accumulation on solar irradiance and electrical performance, proposing a cost-efficient solution for dust cleaning and PV module maintenance.

Research Method
This research paper presents an experimental study conducted at the University of Sharjah during solar noon in November and December 2023. The experiment involved two identical photovoltaic (PV) panels, one serving as a reference and the other coated with a colorless and thin layer of Biaxially Oriented Polyamide (BOPA). The power output of both panels was measured before and after dust accumulation. The methodology included four distinct cases: Case one involved cleaning both PV panels to determine the power reduction caused by the BOPA coating. In cases two through four, varying degrees of dust accumulation were applied to the reference panel to observe its impact on power generation. The performance of photovoltaic (PV) panels is crucial for their practical application in solar energy systems. Dust accumulation on PV panels is a common issue, particularly in arid regions like the United Arab Emirates [29,30].

Several strategies have been proposed to mitigate the effects of dust on PV panel performance, including the application of coatings. This study focuses on investigating the effectiveness of a Biaxially Oriented Polyamide (BOPA) coating in enhancing the power output of PV panels under dusty conditions. By conducting a series of experiments with controlled dust levels, this research aims to provide empirical evidence supporting the efficacy of BOPA coatings in improving PV panel performance.

Biaxially Oriented Polyamide (BOPA) film offers a comprehensive spectrum of properties, making it a versatile choice for using it as a coat to the PV [31]. Its exceptional thermal stability ensures resilience to high temperatures, facilitating processes like heat sealing and sterilization. Coupled with impressive mechanical strength, BOPA provides robust protection during transportation and handling, ensuring the integrity of the coated PV [32]. Furthermore, its clarity and transparency options enhance product visibility and aesthetic appeal allowing the sunlight to reach the panel clearly, while its inherent chemical resistance safeguards against various substances, including oils and greases [33]. Embracing environmental sustainability, BOPA can be recycled and designed to reduce material usage, aligning with eco-conscious coating initiatives. This amalgamation of thermal resilience, strength, clarity, chemical resistance, and sustainability renders BOPA an indispensable solution across industries seeking performance and environmental responsibility in PV coating [34].

The following methodology was employed: Selection and Preparation of PV Panels: Two identical PV panels were chosen for the experiment. One panel was left uncoated (reference panel), while the other was coated with a colorless and thin layer of Biaxially Oriented Polyamide (BOPA). Both panels were cleaned thoroughly before the experiment to ensure baseline performance.

Experimental Setup

The PV panels were positioned outdoors during solar noon to receive maximum sunlight. The experimental setup included instruments for measuring the power output of the panels before and after dust accumulation. Fig. 1 demonstrates a diagram for the experiment setup.

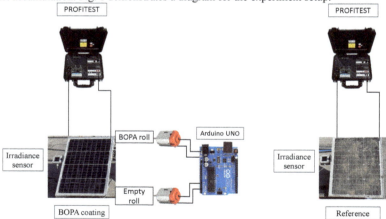

Fig. 1. Experimental setup with and without BOPA coating

The Profitest is used in this experiment to calculate the current, voltage, and power and it works as a load to the PV panel.

Dust Accumulation Procedure: Dust accumulated onto the panels for 45+ days for both panels. Four distinct cases were considered:

a. Case One (Clean Panels): Both PV panels were cleaned to determine the power reduction caused by the BOPA coating alone (day 1).

b. Case Two (Light Dust): After one week, the panels were subjected to light dust accumulation.

c. Case Three (Moderate Dust): The panels were exposed to a higher level of dust accumulation compared to case two after 3 weeks.

d. Case Four (Heavy Dust): The panels were heavily dusted after 45+ days.

The coating roll for the coated panel was connected to two small DC motors (6V) connected to Arduino Uno, this Arduino was programmed to power the motors for a short time allowing the coat to roll over the panel and have a new coat to the top of the panel, the number of rotation needed increases with time. The average number of rotations was measured (3.5 full rotations on average needed to cover the panel) so the motors were programmed to rotate 3.5 rotations every two weeks anticlockwise, one motor will rotate the BOPA roll and the other one will collect the dusty roll.

The coat was fully rotated and renewed every two weeks, keeping the panel clean and dust-free. The power output of both panels was measured before and after dust accumulation in each case. The data obtained were analyzed to assess the impact of the BOPA coating on PV panel performance under varying levels of dust accumulation. To calculate the cost a simple equation was used.

The number of rotations that the roll can cover the panel (R) is equal to the length of the roll (L_r) divided by the length of the panel (L_p), where the width of the roll and the panel are equal.

$$R = L_r/L_p \tag{1}$$

Then we divide R with 26.5 which is the number of weeks that the roll will rotate per year. Finally, the number of years is divided by the price of the BOPA roll, which gives 0.11 AED per rotation.

Experimental Results

The experimental results demonstrate that the BOPA-coated panel mostly produces more power than the uncoated reference panel, only after dust accumulation. These findings underscore the potential of BOPA coatings as a practical solution for optimizing PV panel performance in arid regions with high dust levels. Fig. 2 demonstrates the power produced in each case.

The results of the experiment indicate that the BOPA-coated panel outperformed the reference panel in all cases of dust accumulation. The reference power was more than the BOPA-coated only when both the panels were clean, because of the characteristics of the BOPA (thin and colorless) the reduction of the power was not high. The power output of the coated panel exhibited minimal reduction compared to the uncoated panel, demonstrating the effectiveness of the BOPA coating in mitigating the adverse effects of dust.

This study provides empirical evidence supporting the efficacy of BOPA coatings in enhancing the performance of photovoltaic (PV) panels under dusty conditions. The experimental results demonstrate that the BOPA-coated panel produces more power than the uncoated reference. The data in Table 1 shows that the BOPA coat is more effective when the dust accumulates, the reference power gets lower when the dust accumulation is high. The dust affected the irradiance, causing a reduction in the current and the power. In the fourth case, the efficiency was 4.4% improved with the coat, and 12.51 W increased in terms of power, even though it caused a 5°C increase in the overall temperature, and only 9 watts in 20s was the power needed for the two motors once every two weeks. Comparing other methods in the literature such as robotics systems and electrostatic cleaning consume more power for a longer time and require more maintenance.

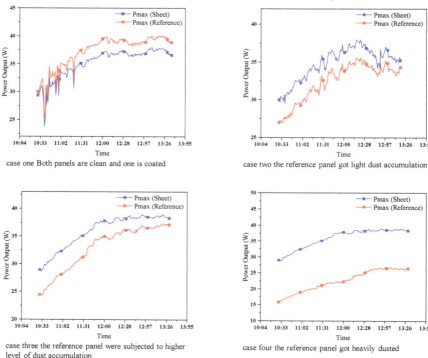

Fig. 2. Comparison of PV modules output power generation

Table 1. Summary of solar irradiance and electrical performance for coated and reference PV module

	Case one (ref)	Case one (coated)	Case two (ref)	Case two (coated)	Case three (ref)	Case three (coated)	Case four (ref)	Case four (coated)
Current (A)	2.19	2.17	1.84	2.11	1.93	2.23	1.28	2.15
Power (watt)	37.17	35.17	32.42	34.69	32.97	35.98	22.54	35.05
Irradiance (W/m2)	807.30	763.62	704.37	744.68	730.80	796.50	490.82	762.16
Efficiency (%)	12.93	12.24	11.53	12.00	10.95	12.35	7.79	12.19

Summary

This study has demonstrated the utilization of Biaxially Oriented Polyamide (BOPA) Coating material as a method for the prevention of dust accumulation. The proposed methodology introduced a thin film of BOPA material, optimizing the PV module performance in arid regions with high dust accumulation levels. An automated controlled scheme was integrated to allow rotation of film every 2 weeks, ensuring clean PV modules across an experimental period of 45 days. The findings suggest that BOPA coatings hold promise as a viable solution for improving the performance and longevity of PV modules under dusty conditions, through the improvement of solar irradiation exposure and electrical performance. The maximum electrical efficiency is maintained up to 12.19% with the BOPA material as compared to 7.79% under the highest dust accumulation. Additionally, solar irradiation is significantly reduced under high dust accumulation, leading to a severe reduction in electrical current, proving the viability of the proposed cleaning method. Further research may focus on optimizing the properties of BOPA coatings for maximum effectiveness and durability in real-world applications.

References
[1] M.M. Farag, N. Patel, A.-K. Hamid, A.A. Adam, R.C. Bansal, M. Bettayeb, A. Mehiri, An Optimized Fractional Nonlinear Synergic Controller for Maximum Power Point Tracking of Photovoltaic Array Under Abrupt Irradiance Change, IEEE J. Photovoltaics. 13 (2023) 305–314. https://doi.org/10.1109/JPHOTOV.2023.3236808

[2] Y.N. Chanchangi, A. Ghosh, S. Sundaram, T.K. Mallick, Dust and PV Performance in Nigeria: A review, Renew. Sustain. Energy Rev. 121 (2020) 109704. https://doi.org/10.1016/j.rser.2020.109704

[3] E.T. Sayed, T. Wilberforce, K. Elsaid, M.K.H. Rabaia, M.A. Abdelkareem, K.-J. Chae, A.G. Olabi, A critical review on environmental impacts of renewable energy systems and mitigation strategies: Wind, hydro, biomass and geothermal, Sci. Total Environ. 766 (2021) 144505. https://doi.org/10.1016/j.scitotenv.2020.144505

[4] S.B. Bashir, A.A.A. Ismail, A. Elnady, M.M. Farag, A.-K. Hamid, R.C. Bansal, A.G. Abo-Khalil, Modular Multilevel Converter-Based Microgrid: A Critical Review, IEEE Access. 11 (2023) 65569–65589. https://doi.org/10.1109/ACCESS.2023.3289829

[5] M.M. Farag, F.F. Ahmad, A.K. Hamid, C. Ghenai, M. Bettayeb, Real-Time Monitoring and Performance Harvesting for Grid-Connected PV System - A Case in Sharjah, in: 2021 14th Int. Conf. Dev. ESystems Eng., IEEE, 2021: pp. 241–245. https://doi.org/10.1109/DeSE54285.2021.9719385

[6] T. Salameh, A.K. Hamid, M.M. Farag, E.M. Abo-Zahhad, Energy and exergy assessment for a University of Sharjah's PV grid-connected system based on experimental for harsh terrestrial conditions, Energy Reports. 9 (2023) 345–353. https://doi.org/10.1016/j.egyr.2022.12.117

[7] T. Salameh, A.K. Hamid, M.M. Farag, E.M. Abo-Zahhad, Experimental and numerical simulation of a 2.88 kW PV grid-connected system under the terrestrial conditions of Sharjah city, Energy Reports. 9 (2023) 320–327. https://doi.org/10.1016/j.egyr.2022.12.115

[8] K. Obaideen, M. Nooman AlMallahi, A.H. Alami, M. Ramadan, M.A. Abdelkareem, N. Shehata, A.G. Olabi, On the contribution of solar energy to sustainable developments goals: Case study on Mohammed bin Rashid Al Maktoum Solar Park, Int. J. Thermofluids. 12 (2021) 100123. https://doi.org/10.1016/j.ijft.2021.100123

[9] M.M. Farag, R.C. Bansal, Solar energy development in the GCC region – a review on recent progress and opportunities, Int. J. Model. Simul. 43 (2023) 579–599. https://doi.org/10.1080/02286203.2022.2105785

[10] M.M. Farag, R.A. Alhamad, A.B. Nassif, Metaheuristic Algorithms in Optimal Power Flow Analysis: A Qualitative Systematic Review, Int. J. Artif. Intell. Tools. 32 (2023). https://doi.org/10.1142/S021821302350032X

[11] A.H. Alami, M.K.H. Rabaia, E.T. Sayed, M. Ramadan, M.A. Abdelkareem, S. Alasad, A.-G. Olabi, Management of potential challenges of PV technology proliferation, Sustain. Energy

Technol. Assessments. 51 (2022) 101942. https://doi.org/10.1016/j.seta.2021.101942

[12] M.M. Farag, F.F. Ahmad, A.K. Hamid, C. Ghenai, M. Bettayeb, M. Alchadirchy, Performance Assessment of a Hybrid PV/T system during Winter Season under Sharjah Climate, in: 2021 Int. Conf. Electr. Comput. Commun. Mechatronics Eng., IEEE, 2021: pp. 1–5. https://doi.org/10.1109/ICECCME52200.2021.9590896

[13] R. Shenouda, M.S. Abd-Elhady, H.A. Kandil, A review of dust accumulation on PV panels in the MENA and the Far East regions, J. Eng. Appl. Sci. 69 (2022) 8. https://doi.org/10.1186/s44147-021-00052-6

[14] M. Rashid, M. Yousif, Z. Rashid, A. Muhammad, M. Altaf, A. Mustafa, Effect of dust accumulation on the performance of photovoltaic modules for different climate regions, Heliyon. 9 (2023) e23069. https://doi.org/10.1016/j.heliyon.2023.e23069

[15] M.M. Farag, A.K. Hamid, T. Salameh, E.M. Abo-Zahhad, M. AlMallahi, M. Elgendi, ENVIRONMENTAL, ECONOMIC, AND DEGRADATION ASSESSMENT FOR A 2.88 KW GRID-CONNECTED PV SYSTEM UNDER SHARJAH WEATHER CONDITIONS, in: 50th Int. Conf. Comput. Ind. Eng., 2023: pp. 1722–1731.

[16] A. Younis, M. Onsa, A brief summary of cleaning operations and their effect on the photovoltaic performance in Africa and the Middle East, Energy Reports. 8 (2022) 2334–2347. https://doi.org/10.1016/j.egyr.2022.01.155

[17] T. Salamah, A. Ramahi, K. Alamara, A. Juaidi, R. Abdallah, M.A. Abdelkareem, E.-C. Amer, A.G. Olabi, Effect of dust and methods of cleaning on the performance of solar PV module for different climate regions: Comprehensive review, Sci. Total Environ. 827 (2022) 154050. https://doi.org/10.1016/j.scitotenv.2022.154050

[18] H. Aljaghoub, F. Abumadi, M.N. AlMallahi, K. Obaideen, A.H. Alami, Solar PV cleaning techniques contribute to Sustainable Development Goals (SDGs) using Multi-criteria decision-making (MCDM): Assessment and review, Int. J. Thermofluids. 16 (2022) 100233. https://doi.org/10.1016/j.ijft.2022.100233

[19] M.K. Smith, C.C. Wamser, K.E. James, S. Moody, D.J. Sailor, T.N. Rosenstiel, Effects of Natural and Manual Cleaning on Photovoltaic Output, J. Sol. Energy Eng. 135 (2013). https://doi.org/10.1115/1.4023927

[20] H.A. Kazem, M.T. Chaichan, A.H.A. Al-Waeli, K. Sopian, A review of dust accumulation and cleaning methods for solar photovoltaic systems, J. Clean. Prod. 276 (2020) 123187. https://doi.org/10.1016/j.jclepro.2020.123187

[21] M.Z. Al-Badra, M.S. Abd-Elhady, H.A. Kandil, A novel technique for cleaning PV panels using antistatic coating with a mechanical vibrator, Energy Reports. 6 (2020) 1633–1637. https://doi.org/10.1016/j.egyr.2020.06.020

[22] M.M. Farag, A.K. Hamid, Performance assessment of rooftop PV/T systems based on adaptive and smart cooling facility scheme - a case in hot climatic conditions of Sharjah, UAE, in: 3rd Int. Conf. Distrib. Sens. Intell. Syst. (ICDSIS 2022), Institution of Engineering and Technology, 2022: pp. 198–207. https://doi.org/10.1049/icp.2022.2448

[23] A. Al-Dousari, W. Al-Nassar, A. Al-Hemoud, A. Alsaleh, A. Ramadan, N. Al-Dousari, M. Ahmed, Solar and wind energy: Challenges and solutions in desert regions, Energy. 176 (2019) 184–194. https://doi.org/10.1016/j.energy.2019.03.180

[24] A. Amin, X. Wang, A. Alroichdi, A. Ibrahim, Designing and Manufacturing a Robot for Dry-Cleaning PV Solar Panels, Int. J. Energy Res. 2023 (2023) 1–15. https://doi.org/10.1155/2023/7231554.

[25] H. Kawamoto, Electrostatic cleaning equipment for dust removal from soiled solar panels, J.

Electrostat. 98 (2019) 11–16. https://doi.org/10.1016/j.elstat.2019.02.002

[26] N.K. Almarzooqi, F.F. Ahmad, A.K. Hamid, C. Ghenai, M.M. Farag, T. Salameh, Experimental investigation of the effect of optical filters on the performance of the solar photovoltaic system, Energy Reports. 9 (2023) 336–344. https://doi.org/10.1016/j.egyr.2022.12.119

[27] A. Syafiq, V. Balakrishnan, M.S. Ali, S.J. Dhoble, N.A. Rahim, A. Omar, A.H.A. Bakar, Application of transparent self-cleaning coating for photovoltaic panel: a review, Curr. Opin. Chem. Eng. 36 (2022) 100801. https://doi.org/10.1016/j.coche.2022.100801

[28] M.M. Farag, F. Faraz Ahmad, A.K. Hamid, C. Ghenai, M. AlMallahi, M. Elgendi, Impact of Colored Filters on PV Modules Performance: An Experimental Investigation on Electrical and Spectral Characteristics, in: 50th Int. Conf. Comput. Ind. Eng., 2023: pp. 1692–1704.

[29] M.M. Farag, A.K. Hamid, Experimental Investigation on the Annual Performance of an Actively Monitored 2.88 kW Grid-Connected PV System in Sharjah, UAE, in: 2023 Adv. Sci. Eng. Technol. Int. Conf., IEEE, 2023: pp. 1–6. https://doi.org/10.1109/ASET56582.2023.10180880

[30] M.A. M. Abdelsalam, F.F. Ahmad, A.-K. Hamid, C. Ghenai, O. Rejeb, M. Alchadirchy, W. Obaid, M. El Haj Assad, Experimental study of the impact of dust on azimuth tracking solar PV in Sharjah, Int. J. Electr. Comput. Eng. 11 (2021) 3671. https://doi.org/10.11591/ijece.v11i5.pp3671-3681

[31] L.-C. Liu, C.-C. Lai, M.-T. Lu, C.-H. Wu, C.-M. Chen, Manufacture of biaxially-oriented polyamide 6 (BOPA6) films with high transparencies, mechanical performances, thermal resistance, and gas blocking capabilities, Mater. Sci. Eng. B. 259 (2020) 114605. https://doi.org/10.1016/j.mseb.2020.114605

[32] X. Lin, Y. Wu, L. Cui, X. Chen, S. Fan, Y. Liu, X. Liu, W. Zheng, Crystal structure, morphology, and mechanical properties of biaxially oriented polyamide-6 films toughened with poly(ether block amide), Polym. Adv. Technol. 33 (2022) 137–145. https://doi.org/10.1002/pat.5497

[33] S. Rhee, J.L. White, Crystal structure, morphology, orientation, and mechanical properties of biaxially oriented polyamide 6 films, Polymer (Guildf). 43 (2002) 5903–5914. https://doi.org/10.1016/S0032-3861(02)00489-5

[34] L. Di Maio, P. Scarfato, L. Incarnato, D. Acierno, Biaxial orientation of polyamide films: processability and properties, Macromol. Symp. 180 (2002) 1–8. https://doi.org/10.1002/1521-3900(200203)180:1<1::AID-MASY1>3.0.CO;2-X

Vibration harvesting techniques for electrical power generation: A review

Omar D. MOHAMMED[1,a,*], Isha BUBSHAIT[1,b], Reyouf ALQAHTANI[1,c], Kawther ALMENAYAN[1,d], Semat ALZAHER[1,e], and Reem ALHUSSAIN[1,f]

[1] Mechanical Engineering Department - Prince Mohammad bin Fahd University PMU, Khobar, Saudi Arabia

[a]osily@pmu.edu.sa, [b]202000416@pmu.edu.sa, [c]202000155@pmu.edu.sa, [d]201902471@pmu.edu.sa, [e]201901224@pmu.edu.sa, [f]201902420@pmu.edu.sa

Keywords: Vibration, Piezoelectric, Mechanisms, Power Harvesting, Energy Generation, Battery Charging

Abstract. The ongoing demands of using mechanical motion and converting it to electrical power motivated designers to find new sustainable solutions for power generation. In the current article, different techniques of clean power generation are reviewed and discussed. The reviewed techniques use mechanical vibration to produce energy. The techniques using piezoelectric and mechanical design concepts are discussed and compared. The article sheds light on the importance of these techniques and concludes with the advantages and disadvantages of each applied technique.

Introduction
There are increasing demands to find new energy sources to reduce the use of fossil fuels. The new energy source has to be of better sustainability and environmental impact. The use of clean and renewable energy has increasing demands and thus puts a lot of pressure on engineers to find new ways to generate energy. Harvesting vibration energy has obtained a lot of researchers' attention, and that led to the development of new techniques or design concepts [1,2]. The use of mechanical motion in our daily activities and converting it to electrical power has been under investigation and application. This energy harvesting concept can be applied using two main techniques, first, using mechanisms or mechanical concepts, and second, using piezoelectric transducers.

The pendulum mechanism is discussed in [3] for energy harvesting. Different pendulum configurations for energy generation were applied, for example, the multi- or single-pendulum, and modulation-based pendulums. A few combinations of piezoelectric, electromagnetic and hybrid transducers were discussed. An electromagnetic vibration-based energy generator was proposed in [4]. The technique converted the linear vibration to a rotation. Linear motion can be converted to energy using a magnet, where the rotation in the magnetic flux can generate electricity.

Piezoelectric materials can be employed as devices to generate energy from mechanical motion, which is in the form of vibrations, into electrical energy that can power other gadgets. Power harvesting is the technique of collecting energy and turning it into useful power for a system. Portable systems that don't rely on conventional power sources, such as batteries, which have a finite lifespan, can be created by putting these power harvesting devices into use [1]. The vibration energy could come from many sources, such as industrial machinery, transit networks, or even everyday activities like walking or keyboard typing. By capturing and converting these vibrations into useful electricity, we can tap into a nearly unlimited and clean energy resource. Researchers and developers aimed to turn wave energy into power. Power harvesting using piezoelectric materials has been studied for a variety of possible applications. Different power transducer techniques were applied to convert the mechanical motion to electrical energy [5]. Piezoelectric transducers, which are small devices, can be used for energy conversion. The collected energy is usually vibrations which should be converted to electricity to be used in other daily used devices. The mechanical vibration energy is usually wasted as dissipated heat [6] but by applying vibration energy harvesting it can be converted to useful energy. However, conventional vibration energy harvesters can experience the issue of operating close to the resonance frequency and this can affect how efficient the operating range is in generating electrical energy. The overall power that can be used is small. The amount of the resulting power depends on the harvester design and also the amplitude and frequency of the vibration. Therefore, if the collected vibration energy is recorded properly the harvester must be designed and fabricated accurately for the best power results [7].

Researchers and developers aimed to turn wave energy into power. The use of piezoelectric transducers has been investigated for vibration energy harvesting in different applications. Several researchers examined the use of energy dissipated in human activities to generate energy that can power small electronic devices. Proposing Polyvinylidene fluoride films as a material for an implanted physiological power source. The prototype was shown to produce a peak voltage of 18V, corresponding to a power of around 17mW. Many researchers have also researched how to capture energy from mechanical structures. Using a piezoelectric vibrator and a steel ball to convert mechanical energy to electrical energy. Their research evaluated the amount of energy released when a steel ball struck a thin piezoelectric plate. A cantilever beam model was used in [8] for harvesting energy with piezoelectric transducers. The presented model was designed to accurately produce energy by adding the damping effect of power harvesting. A comparison approach to compare different designs and methods for vibration energy harvesting generators was presented and discussed in [9].

The unpredictability of ambient vibrations, the requirement to align the resonance frequency of the harvester with the prevailing vibration frequency, and the effective conversion and storage of the captured energy are some of the difficulties involved in vibration harvesting. Vibration harvesting has a lot of potential for powering low-power electronics, wireless sensor networks, and Internet of Things devices in applications that call for constant or long-term power sources, despite these obstacles. The goal of ongoing research and development in this area is to enhance vibration harvesting systems' scalability, dependability, and efficiency for a variety of real-world uses. The majority of research focuses on a specific technology, making it challenging to compare vastly different vibration-based energy harvesting designs and methodologies [2].

In this research paper, the process of comparing the selected harvesting energy techniques occurs. In our case, piezoelectric technique, and mechanisms technique.

Piezoelectric technique for vibration harvesting energy
Many scholars have looked into the possibility of harvesting energy from mechanical structures' ambient vibrations. The concept of a generator is examined by using a steel ball and a piezoelectric vibrator to convert mechanical energy to electrical energy. Their research measured the energy released upon the collision of a steel ball on a thin piezoelectric plate [10]. A power harvesting system model was developed and included a cantilever beam with piezoelectric patches fastened to it. Although the model's construction allowed for any combination of boundary circumstances and the piezoelectric material's position, it was tested on a cantilever beam that was suffering a base excursion from the clamped state. It was discovered that the model was accurate in estimating the energy produced and that it was also useful in illustrating how power harvesting damps energy [11]. The research that is presented in this article will concentrate on the piezoelectric techniques for vibration harvesting energy as well as utilizing the harvested energy. Fig.1 (a) demonstrates that it is based on an actual tile with a 150 × 150 mm2 area. Fig.1 (b) demonstrates that the piezoelectric tile is made up of four supporting springs, a bottom plate, a middle plate where the piezoelectric modules are installed, and an upper plate that must be physically walked on. The

upper plate is linked to the piezo-installed layer in the middle plate. The thickness of the top and bottom plates is 10 mm apiece, and the length of the four springs is 40 mm. Fig.1 (c) provides a thorough illustration of the centre section's cross-section, showing the locations of the modules. The piezoelectric material, measuring 47 × 32 × 0.2 mm3, is positioned on a stainless steel plate substrate, measuring 62 × 37 × 0.2 mm3. PZT-PZNM, produced by TIOCEAN (Korea), is a thick film piezoelectric material. Fig.2 shows the experimental setup of the piezoelectric technique. The system efficiency for converting the mechanical motion to electrical power was obtained in [12]. FEA calculations were made and the results show that a bigger conversion efficiency value could be obtained with a thinner and shorter beam of a higher resonance frequency.

Different vibration harvesting energy concepts were applied for generating energy from machines, heat exchangers, compressors and motors by piezoelectric and energy harvesting devices. It was noticed that piezoelectric harvesting of energy was practically good for energy generation from vibration motion but it generated low power amount [13]. Piezoelectric energy harvesting with magnetic coupling promises a more meaningful solution to narrow bandwidth and low energy efficiency [14]. Theoretically, the cantilever beam's resonance frequency depends on its effective stiffness, effective mass, and tip mass [2]. Calculations are demonstrated regarding the beam, both effective mass and stiffness.

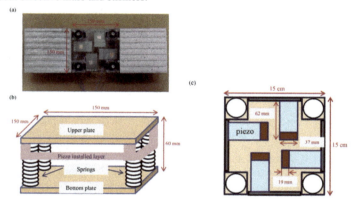

Figure 1: Piezoelectric tiles. (a) piezoelectric tile with a real tile. (b) schematic of the piezoelectric tile. (c) Piezo layer [1]

It had been selected the tip mass whose resonance frequency was closest to the tile's vibration frequency based on the experiment that was detailed [1]. Another research established a general theory containing many specifications where two piezoelectric generators were given the "effectiveness" design concept, and theoretical power outputs were computed. For one design, the power predicted by the effectiveness hypothesis was around 30% higher than the measured power output, and for the other design, it was 10% higher [10]. Three types of generators had been utilised and stated in detail initially and most importantly the piezoelectric generators, the electromagnetic generators, and the electrostatic generators. Familiarity with all types of generators is vital for predicting better alternatives depending on the required outcomes [15].

(a) (b)

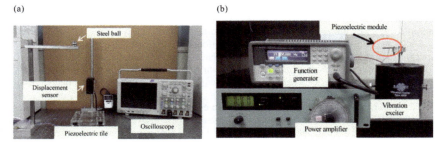

Figure 2: Piezoelectric tile system (a) vibration frequency measurement, (b) resonance frequency measurement (made by TIOCEAN CO.) [1].

While electromagnetic motors and piezoelectric materials have both been applied to vibration energy harvesting, their characteristics differ. Electromagnetic motors are velocity-induced transducers, whereas piezoelectric materials are force- or stress-induced transducers [16]. The active range of amplitude and frequency was studied in [17]. Successful lab and field tools were discussed, and the resulting energy level was compared. Keeping in mind the most recent developments in broadband energy harvesting methods, such as nonlinear methods, multimodal methods, and resonance tuning methods [18]. The application, limitations and advantages of several energy generation techniques were reviewed in [19]. Fig.3 shows a schematic diagram for two types of piezoelectric. Piezoelectric transducers were used in [20] to generate energy from a tile structure where it can produce energy when a person steps on it. A current of 140µA and power 2.8 µW could be generated. A small amount of power can be generated using piezoelectric transducers.

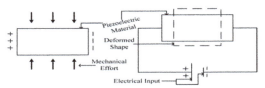

Figure 3: Piezoelectric EH types: (a) direct piezoelectric effect and (b) reverse piezoelectric effect [19].

Mechanisms technique for vibration-harvesting energy
Converting mechanical motion to electricity generation can be done by an electromagnet or a permanent magnet. Because it doesn't require power input, the permanent magnet is a better option for low-power devices than the electromagnet. Ferromagnetic or ferrimagnetic material is present in these permanent magnets. Despite producing a strong electric field, ferromagnetic materials are frequently employed because their increased electrical resistance reduces the influence of eddy currents.

A multi-degree freedom system can be introduced into an excitation structure to increase the bandwidth of the EH. Different subsystems were combined in the EH design to provide different modes. It was discovered that each resonance's average power generation by the EH differs significantly [19]. Harvesting energy response is discussed in [21], where a harvester of high sensitivity is used for generating energy. A harvesting system, including power electronics for managing the power, an electromagnetic converter and different mechanical parts was used. The system of high sensitivity can produce useful energy using vibration shocks. Energy harvesting techniques using bi-stable systems were reviewed in [22]. Electro-mechanical systems were presented to show the practical benefits of these techniques. The different bi-stable harvesting systems use magnetic repulsion, magnetic attraction and mechanical load to induce bi-stability. Energy generation using vibration harvesting that can be applied to self-powered micro and wireless systems was reviewed in [15]. Maximum harvesting energy is analyzed and optimum results were found in single- and multi-mass systems. The system sensitivity is obtained from the implemented simulation.

Vibration harvesting energy and conversion to electricity was applied in several mechanisms, using the electromagnetic principle [4, 23-28]. As seen in Fig.4, an energy generator using a speed bump that harvests vibrational motion is presented in [23]. The model uses a mechanism of rack and pinion and clutches. The mechanism is tested in the bump impact case and power generation is examined.

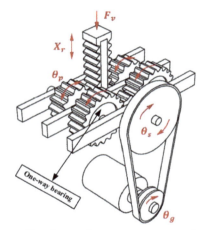

Figure 4: A speed breaker mechanism using one-way bearing [23]

Energy generation was investigated using mechanical concepts. A few mechanisms using spring were reviewed and discussed in [24]. The presented design configurations which are applied and tested work for power generation by using mechanical motion. A speed breaker design was presented and used in [25,26] for power generation. As seen in Fig.5, the mechanism can convert the linear motion of the rack to the rotary motion of the pinion and the rotational motion is converted to electrical energy using the magnetic field concept. As stated in [26], the used mechanism could produce 1.16 V, which is much higher the produced value in case of the piezoelectric. A compact energy harvesting system is presented in [27], which is an efficient, durable, and also feasible device. Modelling and simulation were used to validate the design model.

Figure 5: A speed breaker mechanism assembly [25]

Conclusions
Piezoelectric devices, which are small and light transducers, can be used to transform mechanical motion, often vibrations, into electrical energy that can power other devices used in our daily life. There are multiple energy harvesting techniques for converting vibration energy into usable electricity using mechanical design concepts. Power harvesting devices, using both piezoelectric and mechanical concepts, can be applied for vibration energy harvesting without relying on traditional energy sources, which have limited lifespans. Vibration energy is a free and sustainable source of energy that should get more attention to improve the applied techniques for increasing power generation, especially in the piezoelectric techniques. Mechanical concepts generate much higher energy than piezoelectric devices, however, they involve mainly springs, racks and gears which can result in system complexity, heavy weight, and energy losses. Moreover, the applied mechanisms can be a source of noise if the concept is applied to harvest energy from human walking. On the other hand, piezoelectric transducers, which are small and light, can produce energy with a small amplitude of vibration, but it is a much smaller amount of energy.

References
[1] S. J. Hwang, H. J. Jung, J. H. Kim, J. H. Ahn, D. Song, Y. Song, ... & T. H. Sung, Designing and manufacturing a piezoelectric tile for harvesting energy from footsteps. Current Applied Physics, 15(6) (2015), 669-674. https://doi.org/10.1016/j.cap.2015.02.009

[2] V. R. Challa, M. G. Prasad, Y. Shi, & Fisher, A vibration energy harvesting device with bidirectional resonance frequency tunability. Smart Materials and Structures, 17(1) (2008), 015035. 10.1088/0964-1726/17/01/015035

[3] T. Wang. Pendulum-based vibration energy harvesting: Mechanisms, transducer integration, and applications. Energy Conversion and Management, (2023) 276, 15, 116469. https://doi.org/10.1016/j.enconman.2022.116469

[4] Y. Wang, P. Wang, S. Li, M. Gao, H. Ouyang, Q. He, P. Wang. An electromagnetic vibration energy harvester using a magnet-array-based vibration-to-rotation conversion mechanism. Energy Conversion and Management, (2022) 253, 115146. https://doi.org/10.1016/j.enconman.2021.115146

[5] X. Tang, & L. Zuo . Enhanced vibration energy harvesting using dual-mass systems. Journal of sound and vibration, 330(21) (2011), 5199-5209. https://doi.org/10.1016/j.jsv.2011.05.019

[6] A. O. Odetoyan, & A. N. Ede. Energy harvesting from vibration of structures-a brief review. In IOP Conference Series: Materials Science and Engineering (Vol. 1107, No. 1, p. 012192) (2021, April). IOP Publishing. 10.1088/1757-899X/1107/1/012192

[7] T. Yildirim, M. H. Ghayesh, Li, W., &G. Alici, A review on performance enhancement techniques for ambient vibration energy harvesters. Renewable and Sustainable Energy Reviews, (2017). 71, 435-449. https://doi.org/10.1016/j.rser.2016.12.073

[8] H. A. Sodano, D. J. Inman, & G. Park, Generation and storage of electricity from power harvesting devices. Journal of intelligent material systems and structures, 16(1) (2005), 67-75. https://doi.org/10.1177/1045389X05047210

[9] S. Roundy, On the effectiveness of vibration-based energy harvesting. Journal of intelligent material systems and structures, 16(10) (2005), 809-823. https://doi.org/10.1177/1045389X05054042

[10] M. Umeda, K. Nakamura, & S. Ueha, Analysis of the transformation of mechanical impact energy to electric energy using piezoelectric vibrator. Japanese Journal of Applied Physics, 35(5S) (1996), 3267. 10.1143/JJAP.35.3267

[11] H. A. Sodano, G. Park, & D. J. Inman, Estimation of electric charge output for piezoelectric energy harvesting. Strain, 40(2) (2004), 49-58. https://doi.org/10.1111/j.1475-1305.2004.00120.x

[12] D. Koyama, & K. N& Nakamura. Electric power generation using vibration of a polyurea piezoelectric thin film. Applied Acoustics, 71(5) (2010), 439-445. https://doi.org/10.1016/j.apacoust.2009.11.009

[13] V. Jamadar, P. Pingle, & S. Kanase, Possibility of harvesting Vibration energy from power producing devices: A review. In 2016 International Conference on Automatic Control and Dynamic Optimization Techniques (ICACDOT) (2016, September) (pp. 496-503). 10.1109/ICACDOT.2016.7877635

[14] J. Jiang, S. Liu, L. Feng, & D. Zhao, A review of piezoelectric vibration energy harvesting with magnetic coupling based on different structural characteristics. Micromachines, 12(4) (2021), 436. https://doi.org/10.3390/mi12040436

[15] S. P. Beeby, Tudor, M. J., & White, N. M. Energy harvesting vibration sources for microsystems applications. Measurement science and technology, (2006).17(12), R175. 10.1088/0957-0233/17/12/R01

[16] V. L.Kalyani, V. L., Piaus, A., & Vyas, P. Harvesting electrical energy via vibration energy and its applications. Journal of Management Engineering and Information Technology, (2015)2(4), 2394 - 8124.

[17] A. Hosseinkhani, D. Younesian, P. Eghbali, A. Moayedizadeh, & A. Fassih, Sound and vibration energy harvesting for railway applications: A review on linear and nonlinear techniques. Energy Reports, 7 (2021), 852-874. https://doi.org/10.1016/j.egyr.2021.01.087

[18] L. Tang, Y. Yang, & C. K. Soh, Broadband vibration energy harvesting techniques. Advances in energy harvesting methods, (2013), 17-61. 10.1007/978-1-4614-5705-3_2

[19] R.K. Mohanty, A., Parida, S., Behera, & T. Roy. Vibration energy harvesting: A review. Journal of Advanced Dielectrics, (2019), 9(04), 1930001. https://doi.org/10.1142/S2010135X19300019

[20] K. Al-Hamoudi, J. Almudhaki, A. Khali, A. Alqahtani. Footstep Piezo Generator. Senior Design Project, Prince Mohammad bin Fahd University. 2019.

[21] Z. Hadas, V. Vetiska, V. Singule, O. Andrs, J. Kovar, and J. Vetiska. Energy harvesting from mechanical shocks using a sensitive vibration energy harvester. International Journal of Advanced Robotic Systems, (2012), 9(5), p.225. https://doi.org/10.5772/5394

[22] R. L.Harne,& K. W. Wang, A review of the recent research on vibration energy harvesting via bistable systems. Smart materials and structures, (2013), 22(2), 023001. 10.1088/0964-1726/22/2/023001

[23] A. Azam, A. Ahmed, N. Hayat, S. Ali, A. Khan, G. Murtaza, T. Aslam. Design, fabrication, modelling and analyses of a movable speed bump-based mechanical energy harvester (MEH) for application on road. Energy, (2021), 214, 118894. https://doi.org/10.1016/j.energy.2020.118894

[24] A.M. Patel, A. N. Patel. A Review on Different Mechanisms Used in Power Generation Through Speed Bumps. International Journal for Scientific Research & Development, (2017), 5(10) 2321-0613.

[25] J. Dara, C. Odazie, P. Okolie, A. Azaka. Design and construction of a double actuated mechanical speed breaker electricity generator. Heliyon, (2020), 6, e04802. https://doi.org/10.1016/j.heliyon.2020.e04802

[26] A. Al-Khaldi, A. Al-Jumaah, A. Al-Mubarak. Footstep Power Generator. Senior Design Project, Prince Mohammad bin Fahd University. 2021.

[27] N. Raj, A. Dasgotra, S. Mondal, S. Kumar, R. Patel. Deflection-Based Energy Harvesting Speed Breaker and It's Mechatronic Application. International Journal of Sustainable Engineering, (2021), 14 (5) 1033–1042. https://doi.org/10.1080/19397038.2020.1862352

[28] X. Zhang, G. Li, W. Wang, S. Su. Study on the energy conversion mechanism and working characteristics of a new energy harvester with magnetic liquid. Sensors and Actuators A: Physical, (2023), 359, 114409. https://doi.org/10.1016/j.sna.2023.114409

Multiport universal solar power bank

Abdullah ALTELMESSANI[1,a], Abdulqader ALJABER[1,b], Turki ALTUWAIRQI[1,c], Ala A. HUSSEIN[1,d], Jamal NAYFEH[1,e]

[1]Department of Electrical Engineering, Prince Mohammad Bin Fahd University, Khobar, Saudi Arabia

[a]201701254@pmu.edu.sa, [b]201902284@pmu.edu.sa, [c]201900905@pmu.edu.sa, [d]ahussein@pmu.edu.sa, [e]jnayfeh@pmu.edu.sa

Keywords: IEEE Smart Village (ISV), Battery Management System (BMS), Maximum Power Point Tracking (MPPT), Perturb and Observe (P&O)

Abstract. In an era of ubiquitous electricity dependence, the need for a reliable and portable power source is increasingly vital. This proposal advocates for the development of a solar-powered portable device capable of harnessing solar energy to charge itself and subsequently supply power to various electronic devices. The envisioned solution aims to address critical scenarios such as emergencies, where access to electricity is crucial, as well as recreational activities like camping, especially in remote locations. The proposed portable power bank will be equipped with solar panels for efficient energy absorption, ensuring self-sufficiency and sustainability. The device will feature both DC and AC outputs, catering to a wide range of electronic gadgets, thereby enhancing its versatility. In the event of a disaster, this innovation could prove invaluable by providing a reliable source of electricity when traditional power infrastructure is compromised. Moreover, in recreational settings, such as desert camping in locations like Saudi Arabia, users can harness the power of the sun to charge their devices in the open, offering convenience and environmental friendliness. This solar-powered portable energy hub embodies a step towards fostering energy independence and resilience in the face of unpredictable circumstances, catering to both emergency preparedness and everyday scenarios.

Introduction
To design Multiport Universal Solar Power Bank that can use the energy of the sun with the help of a photovoltaic system such as a solar panel. The system will have some subsystems that will monitor the power level of the battery.

- Project [1] Comparison

Our solar power bank project contrasts with Project [3], which is tailored for military use. Our design boasts a cooling system and wireless monitoring, enhancing its suitability for a range of civilian applications, including emergency and outdoor use. However, it may not match the extreme condition optimization of military-spec devices. Conversely, Project [3] excels in robustness and energy density for military environments but lacks the adaptability and user-friendly features of our project, such as wireless monitoring for everyday civilian use.

- Project [2] Comparison

We see significant differences in focus and application. While their project revolves around a specific application of a mobile charger using recycled materials, our project encompasses a broader spectrum of functionalities, including a cooling system and wireless monitoring, aiming to provide versatile and accessible electricity for a variety of scenarios. Our project emphasizes adaptability, environmental sustainability, and user-friendliness, demonstrating a comprehensive approach to portable solar energy solutions.

- Project [3] Comparison

Comparing our project with theirs, there are notable distinctions. Our project caters to a wider civilian audience, providing a versatile solar power bank with innovative features like a cooling system and wireless battery monitoring. This contrasts with the other project's military-focused design, which prioritizes high energy density and robustness for field operations. While their charger is specialized for military needs, ours is tailored for diverse environments and user convenience, highlighting our commitment to adaptable and user-friendly solar energy solutions.

- Project [4] Comparison

The SunBlazer IV focuses on larger-scale, community-based applications with a strong emphasis on scalability and modularity. It aims to empower entire communities by providing a robust, adaptable solar energy system. In contrast, our project centers on individual use with a portable solar power bank, enhanced with features like a cooling system and wireless monitoring. Our project is designed for personal convenience, adaptability, and a wide range of scenarios, whereas the SunBlazer IV caters to broader community needs and sustainable development goals.

Traditional energy sources, primarily fossil fuels, contribute significantly to carbon emissions, a leading cause of climate change. Your project, by using solar energy, helps reduce this carbon footprint. Also, fossil fuels are non-renewable and their extraction can be environmentally damaging. Solar energy, being renewable, offers a sustainable alternative. Fossil fuel combustion releases pollutants that harm air quality and public health. Solar power generation, in contrast, produces no air pollutants, making it a cleaner option.

The contributions of this research are as follows:
- **Easier access to electricity in remote areas.**
- **Use renewable resources to reduce the number of non-renewable resources used.**
- **To be used in case of a power outage.**
- **Create a solar power bank that is compatible with a wide range of electronic devices.**
- **Strive for a compact and lightweight design, making the power bank easy to carry and suitable for on-the-go activities, such as camping, or emergency situations.**
- **Implement safety features to protect both the power bank and connected devices from overcharging, overheating, and other potential risks, ensuring user safety and the longevity of the device.**

System Specifications
This project will involve the design and testing of different subsystems. Each subsystem will follow a specific job.
- Convert DC from solar panel to DC into the battery using (DC Regulator).
- Supply different voltages to different devices.
- Measures battery level.
- Can act as an AC power source with a max output of 400[W].
- The Capacity will be around 20000[mAH].
- The model will be 50[cm^2] with a height of 20[cm].

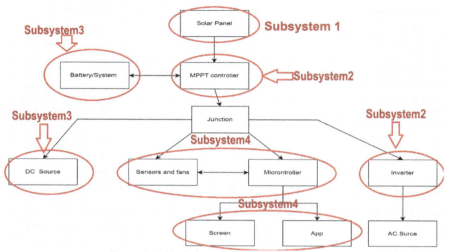

Figure 1: A block diagram of the proposed system.

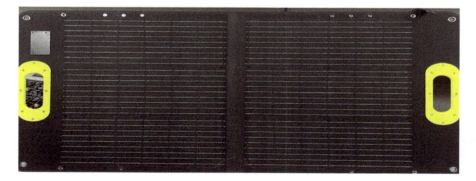

Figure 2: Solar pannel used for this project.

Figure 2: Inverter and controller used for this project.

The summary of the paper on the development of a Multiport Universal Solar Power Bank highlights the project's successful integration of innovative design, renewable energy technology, and user-centric features to address the growing need for portable, reliable, and sustainable power sources. By harnessing solar energy through efficient photovoltaic systems and incorporating versatile functionality, the project not only provides a practical solution for emergency and recreational power needs but also contributes to environmental sustainability by reducing reliance on non-renewable energy sources.

The project's emphasis on compatibility with a wide range of electronic devices, coupled with its compact and lightweight design, ensures that it meets the demands of both emergency preparedness and outdoor activities. The inclusion of safety features and adherence to engineering standards further enhances the reliability and user safety of the solar power bank.

Moreover, by focusing on sustainability through the use of renewable resources and eco-friendly materials, the project aligns with global efforts to mitigate environmental impact and promote energy independence. The consideration of economic, social, and ethical factors in the design and manufacturing processes demonstrates a comprehensive approach to addressing the multifaceted challenges of modern energy needs.

Summary and Conclusion

In summary, the Multiport Universal Solar Power Bank project represents a significant step forward in the development of portable solar energy solutions. Its ability to provide a reliable, safe, and environmentally friendly power source in a variety of scenarios underscores the potential of renewable energy technologies to improve quality of life and foster sustainable development. As the project moves forward, it holds the promise of broadening access to energy, enhancing emergency preparedness, and contributing to a more sustainable and resilient energy future. The proposed system can be used during outdoor activities, emergency cases, and backup power source for small appliances.

References

[1] Muhseen, Z., et al., 2020. Portable smart solar panel for consumer electronics. In: 2020 International Conference on Smart Technologies in Computing, Electrical and Electronics (ICSTCEE), Bengaluru, India, pp.494-499. https://doi.org/10.1109/ICSTCEE49637.2020.9277123.

[2] Venkataraman, K., Selvan, E.V., Gandhi, R.A., Chakravarthi, M.C.A., Varunanand, V. and Vasanthakumar, B., 2022. A mobile charger through a solar panel fabricated from silicon scrap.

In: 2022 International Conference on Power, Energy, Control and Transmission Systems (ICPECTS), Chennai, India, pp.1-4. https://doi.org/10.1109/ICPECTS56089.2022.10047178.

[3] Sitbon, M., Gadelovits, S. and Kuperman, A., 2014. Multi-output portable solar charger for Li-Ion batteries. In: 7th IET International Conference on Power Electronics, Machines and Drives (PEMD 2014), Manchester, UK, pp.1-7. https://doi.org/10.1049/cp.2014.0430.

[4] Larsen, R.S. and Estes, D., 2019. IEEE Smart Village launches SunBlazer IV and Smart Portable Battery Kits: Empowering remote communities. IEEE Systems, Man, and Cybernetics Magazine, 5(3), pp.49-51. https://doi.org/10.1109/MSMC.2019.2916247.

[5] Rodriguez, C. and Lee, S., 2023. Advances in lithium-ion battery technology for portable solar devices. Advanced Energy Materials, 11(4), 2003035.

[6] Johnson, K. and Patel, R., 2022. The role of renewable energy technologies in disaster risk reduction. International Journal of Disaster Risk Reduction, 49, 101925.

[7] Tran, E. and Al Fayed, M., 2023. Impact of solar power banks on outdoor recreation and emergency preparedness. Energy Policy, 129, pp.110-121.

[8] Doe, J. and Smith, J., 2023. On the efficiency and integration of photovoltaic systems in portable devices. Renewable Energy Focus, 30(2), pp.142-158.

[9] Patel, A. and Chung, L.W., 2022. Sustainable materials for next-generation energy storage devices. Journal of Power Sources, 450, 227690.

[10] Institute of Electrical and Electronics Engineers, 2024. IEEE standard for safety and performance of portable solar power systems. IEEE 1625-2024.

Electricity sector reforms in Saudi Arabia and their impact on demand growth and development of renewable energy

Samir El-NAKLA[1,a*], Chedly Belhadj YAHYA[2,b], Jamal NAYFEH[1,c]

[1]Department of Electrical Engineering, Prince Mohammad Bin Fahd University, AlKhobar, Saudi Arabia

[2]Electrical and Computer Engineering Technology, Chattahoochee Technical College, Marietta, GA, USA

[a]snakla@pmu.edu.sa, [b]chedly.yahya@chattahoocheetech.edu, [c]jnayfeh@pmu.edu.sa

Keywords: Electricity Reforms, Demand Growth, Renewable Energy, KSA

Abstract. The Kingdom of Saudi Arabia (KSA) has experienced widespread development over the last four decades. This has resulted in tremendous increase in electricity demand. This paper reviews the status of the KSA's electricity consumption and demand to date; discusses the current and future challenges facing the Saudi electricity sector; and the Saudi government's initiatives taken to address these challenges. The study shows that KSA government started to apply the long term strategy, Vison 2030 to reduce energy consumption and high energy demand in the country. The paper evaluates the outcomes of applying the tariff reform programs that introduced in 2016 and 2018 respectively and its impact on different consuming sectors especially the residential and also the renewable energy program.

Introduction

The Kingdom of Saudi Arabia (KSA) is located in the Middle East and spread over about 2.15 million km^2, which constitutes around 80 percent of the Arabian Peninsula [1]. The country has a very harsh environment where the temperature varies from as high as 50°C in the shade in mid-summer to 0°C or even lower in winter. The high variation in temperature produces strong variance in electricity demand over the year resulting mainly from high demand of electricity for air conditioning during the hot weather season.

In the last four decades, KSA has witnessed massive economic development coupled with high population growth and urbanization; driven by crude oil revenues [2,3]. The population of the kingdom has grown at an annual average rate of 3% over the last 40 years. Overall, the total population has dramatically increased from 7 million in 1975 to about 34 million in mid of 2019. Furthermore, urbanization in KSA increased from 48% in the early 1970s to around 80% in the year 2000 and projected to reach 88% in 2025 [4]. High economic development as well as population and urban growth have resulted in exponential growth of the country's electricity demand [5].

In 2015, the electricity peak demand reached as high as 62.3 Gigawatt (GW). The current demand is typically met through conventional crude, heavy oil, and gas powered plants spread across the country [6]. In addition to ever-increasing electricity demand, the electricity sector in KSA is struggling with many other issues that may jeopardize the sector sustainability including: aging power plants, suboptimal electricity tariff, inefficient legal and institutional framework, and low public awareness of electricity conservation.

Moving towards a more sustainable model, the KSA government established King Abdullah City of Atomic and Renewable Energy (KACARE) with the aim to utilize the indigenous renewable energy resources through science, research and industry [6]. The ambition of KACARE program is to generate 72 GW energy from renewable energy sources such as solar, wind, nuclear and waste-to-energy (WTE) by 2032 [7]. The performance of KACARE and the Saudi Government plan for electricity sector development have never been scientifically studied and evaluated. Additionally, there are very limited studies that document the status of electricity system in the Kingdom and little have been done to evaluate the system performance.

The Kingdom of Saudi Arabia's government has applied a new strategy for the future called Vision 2030, which was announced on April 2016, that emphasizes the challenges to meeting the future's growing requirements. The Council of Saudi Economic and Development Affairs introduced historic vision 2030 which aimed toward a number of targets and reform strategies for the Kingdom's future. Electricity demand is dramatically increasing in Saudi Arabia, with increasing peak load over the past two decades [8].

The Kingdom's energy plan is to eliminate oil overuse, to reduce subsidies by increasing the energy tariff in several stages over next ten years, and to conduct a large-scale energy reform to re-evaluate all the energy resources. The Kingdom government began its plan to reform energy prices in 2016 and 2018 which led to reduce the energy consumption [9]. The new Electricity Tariffs rate was the real challenge for the residential sector, especially for the consumer who used more than 6,000 (kWh/month) [10]. The residential sector is the main electricity consumer, constituting at least 50 percent of total consumption. Increased electricity tariffs did not affect the government and industry sectors as affected residential sector.

In order to control electricity consumption and energy management, the Kingdom government reconstructed and rehabilitated several organizations such as the Water & Electricity Regulatory Authority (WERA) which is the regulator body for electricity and water in the country and Saudi Energy Efficiency Center (SEEC) which an organization dealing with customer's awareness and educating the public.

The Saudi Energy Efficiency Center (SEEC) conducted a broad awareness campaign to educate the public about consumption reduction methods by issuing leaflets, preparing workshops that explain how to select the appropriate electrical equipment, and insisting that sales outlets adhere to the required specifications for consumption reduction [8].

In addition to the previous effective reforms implemented by energy programs, the Kingdom government launched the National Renewable Energy Program (NREP) as an essential long-term program in Future Vision 2030 which aims to substantially increase renewable energy's share in the total energy mix with carbon emission reduction [11].

The NREP during the National Transformation Program (NTP) towards which Vision 2030 will increase renewables' share in the energy mix from zero to four percent [11]. In addition, the Saudi government issued permission to households to install solar power systems. Although there are still several obstacles facing energy consumption reduction, the Kingdom's ambitious reform programs are progressing in the right direction.

The paper summarizing and presenting the data of power consumption, peak load, per capita and also looked at the consumption per different sectors within the last ten years. As tariff reforms introduced in the years 2016 and 2018, the paper looked at those increases in tariff on different consuming sector especially the residential and evaluated the impact on the government plan and vision and initiatives of renewable energy.

Challenges and Reforms
The high growth in electricity demand coupled with falling oil prices in 2014 made it very hard for the government to meet the demand without increasing help from private independent power producers (IPPs). These conditions pushed for serious reforms to tackle the demand and allowing for increased share for private sector towards a more stable and sustainable supply. Figures 1 and 2 show the growth of total electricity sold and the peak demand, the data is obtained from WERA (Water and Electricity Regulatory Authority) [5]. Both Figs. show a high growth mode up to 2015,

then from 2016 to 2021 a reduced growth mode as a result of the undertaken government reforms especially the increased tariffs as will be detailed later.

In 2010 the National Energy Efficiency Program (NEEP) was converted to the Saudi Energy Efficiency Center (SEEC) tasked with improving efficiency in buildings, transport and appliances. In coordination with MEWA and SASO (Saudi for building insulation and air conditioning Efficiency) [8]. The SASO 2662/2012 air conditioning Energy Efficiency Ratio (EER) have been Standards Organization), new standards were introduced implemented in stages beginning 2013 and by 2015 all appliances not meeting the EER requirement were taken Out from the market. Both standards on building insulation and air conditioning will have a huge impact on cutting electricity consumption since buildings (residential, commercial and government) account for up to 70% of the total electricity consumed. In addition, efficiency was targeted at the generation level by upgrading old plant and increasing the share of more efficient combined cycle (CC) turbines. The number of CC unit went from 74 in 2014 to 122 in 2017 and the share of CC increased from 8.3% in 2010 to 30.8% in 2018. The improved efficiency was directly translated into more fuel saving, so oil consumption went from 2.01 barrels per MW in 2009 to 1.71 barrels in 2017 [9].

On the restructuring side, the transmission business was separated from the Saudi Electricity Company (SEC) and named National Grid SA beginning 2012. The single buyer company was created as Saudi Company for Energy Procurement on 31/5/2017. Both companies are still 100% owned by the government. So with WERA acting as an independent regulator and the new single buyer, the electricity market became more open for competition. This resulted in IPPs share in generation increasing to about 30%. For example, Table 1 shows the list of power projects planned for 2018-20, totaling 17 GW, SEC share is only 7.5GW (44%) while the remaining 9.5GW (56%) by other companies [10]. IPPs relieve budget pressure by providing upfront capital. In addition, they are more cost competitive due to bidding process and can deliver projects faster.

Table 1. Major Projects 2018-20 [10]

Project	Capacity MW	Fuel	Startup date
PP13 (SEC)	1800	Gas	2018
Shuqaiq (SEC)	2650	Oil	2018
Waad Alshamal	1390	Gas/Solar	2018
Jizan IGCC (Aramco)	4000	-	2018
Duba 1 (SEC IPP)	550	Gas/Solar	2019
Fadhili IPP (SEC/ Aramco)	1500	Gas	2019
Sakaka 1 (REPDO IPP)	300	Solar PV	2019
PP14 (SEC)	1640	Gas	2020
Yanbu 3 (SWCC)	3100	Oil	2020
Total	16,930		

Tariff Reforms
The electricity tariff was heavily subsidized for a long period which encouraged overconsumption and inefficiency. According to WERA 2014 report, SEC collected an average of 0.138SR/ kWh, while its subsidized average cost was 0.154 SR/kWh and real unsubsidized cost was 0.80SR/kWh. This means that customers pay only 17% of the real cost on average. Table 2 shows the details of the two tariff increases beginning 2016 and 2018 for the residential sector only. The first increase (2016) only affected customers with consumption higher than 4000kWh. This increase did not bring significant income to the power companies as shown by a detailed analysis by WERA [11],

77.2 % of the residential consumption is less than 6000 kWh. For commercial sector 65.1% consume more than 6000 kWh, for government and industrial sectors 90.7% and 99.82% consume more than 8000 kWh. This explains why for the second tariff reform in 2018 the focus was on the majority consuming less than 6000 kWh to increase the income for the power companies. According to SEC 2018 report [9], the income from residential and commercial increased from about SR4.83 Billion for 2017 to SR11.8 Billion for 2018.

Table 2. Electricity Tariff Reform (Residential)

Consumption Categories (kWh)	Pre- January 2016	Jan 2016- Dec 2017	January 2018
1- 2000	5	5	18
2001- 4000	10	10	
4001- 6000	12	20	
6001-7000	15	30	30
7001-8000	20		
8001-9000	22		
9001-10000	24		
More than 10000	26		

** Tariff is in Saudi halala, 1SR=100 halala=$0.267

Impact of Reforms on Electricity Demand

Since the electricity sector reforms are not complete yet and the tariff major reform is less than two years old, it will be very hard to make good conclusions. Therefore, only the initial facts will be mentioned until more information becomes available.

For the impact of reforms on electricity demand, Figs. 1 and 2 clearly show that the growth started slowing after 2016 and went below 1% for the last two years. Figure 3 shows the change kWh per customer between 2009 and 2021 which supports the same trend from Figs. 1 and 2, that there is a decrease in demand growth. If we compare the change by sector, Figure 4 and 5 show the residential and commercial sectors. Again a clear reduction in growth, in particular from 2017 to 2019, the residential consumption decreased by 9.1% and commercial by 3.1%. This indicates that the tariff reform beginning 2018 had a clearer impact on residential consumers. Does this decrease in demand come from consumer reaction to tariff increase or from improved efficiency? most probably both until further information becomes available.

Figure 1. Total Energy Sold in TWh from 2009 to 2018 [12].

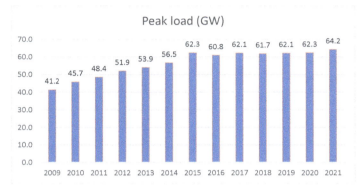

Figure 2. Peak Load in GW between 2009 and 2018 [12].

Figure 3. Consumption kWh per customer from 2009 to 2018 [12].

Figure 4. Residential consumption from 2009 to 2018 [12].

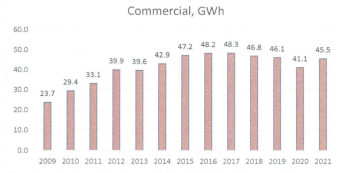

Figure 5. Commercial consumption from 2009 to 2018 [12].

Impact of reforms on Renewable Energy Program

Saudi Arabia enjoys perennial clear skies with approximately 3,000 hours of sunshine per year and annual insolation levels reaching 2450 kWh/m^2. In addition, KSA has empty stretches of desert that can host solar arrays and vast deposits of sand that can be used in the manufacture of silicon photovoltaic cells. As a comparison, Germany has insolation levels that barely reach 1700 kWh/m2 in Freiburg and is one of the top five global leaders. As mentioned earlier, since the initial KACARE plan was announced in 2010, it experienced repeated delays and revisions. Several companies were setup in preparation for getting a share of the expected large number of solar energy projects. Some of these companies invested in factories to make photovoltaic (PV) solar cells and panels while other companies focused on solar energy project installation. Table 3 lists the main players in PV modules fabrication and project implementation.

Table 3. List of KSA Renewable Energy Companies

No.	Company	Date
1	KACST: R&D Labs	2010
2	Taqnia : Eng Solar, Wind	2014
3	AlAfandi Group: Solar (Jeddah)	2015
4	Desert Technologies (Jeddah): Solar PV Modules	2016
5	ACWA Power: Eng Projects (Riyadh)	2008

Another important strength is the establishment of research centers such as the KACST Water Energy Research Institute (WERI) and the KACST R&D Labs. The KACST WERI cooperated with the US National Renewable Energy Laboratory (NREL) to establish a network of 12 solar radiation stations spread over all KSA regions [11]. The radiation stations have been used by KACARE to produce the Renewable Resource Atlas of Saudi Arabia in support of achievement of a sustainable energy mix. The KACST R&D Labs provide another important resource that specializes in PV module testing and certification based on International Electrotechnical Commission (IEC) and newly developed desert certification procedures by KACST [13].

KSA had an early start on Solar energy (1983) compared to Gulf and MENA, but fell behind until the recent pickup in pace. After the recent tariff, regulation reforms and the decline in renewable cost, ACWA was able to complete the first large scale PV project at Sakaka (300MW) at record LCOE of 2.32 c/kWh which was completed mid November 2019 [14]. The project is a good example of IPP with a 25 years contract of power purchase. Another renewable energy

400MW wind project was awarded in July 2018 for a record LCOE of 2.13c/kWh at Dumat AlJandal (near Sakaka PV project), different renewables completed projects shown in Table 4. Figure 6 shows the revised renewable energy program for the next 5 years and 12 years [15].

Table 4. List of KSA Completed Renewable Energy Projects

No.	Project	Power	Com Date
1	KACST (Uyaynah)	350kW	1983
2	KAUST Solar Park	2 MW	2010
3	Pilot project	500 kW	2011
4	Saudi Aramco Solar Car Park	10.5 MW	2012
5	Princess Noura University (Thermal)	25 MWth	2012
6	King Abdullah Financial District	200 KW	2012
7	King Abdulaziz Int. Airport Dev Proj.	5.4 MW	2013
8	Al-Khafji PV plant (RO Desalination)	15 MW	2017
9	PV Plant at Al-Aflaj (Taqnia)	10 MW	2019
10	PV Plant at Sakaka (ACWA)	300 MW	2019
11	Dumat Al Jandal Wind Farm	400 MW	2020
12	Al Rajaf Wind Farm	400 MW	2021
13	Waad Al Shamal Solar Park	600 MW	2022
14	Sudair Solar Power Plant	1,500 MW	2022
15	Red Sea Project Solar Plant	100 MW	2022
16	Domat Al Jandal North Wind Farm	600 MW	2023
17	Al Kharsaah Solar Power Plant	1,200 MW	2023

Figure 6. National Renewable Energy Program [11].

Conclusion

The research presented a review of the KSA electricity sector challenges and the government initiatives and reforms undertaken to address these challenges. Even though the reforms are in their early stages, the data shows that the reforms had a positive impact on the electricity demand growth and acceleration of renewable energy program implementation. However, more time and data are needed to have a clearer idea about the reforms impact.

References

[1] Saudi Geological Survey. Kingdom of Saudi Arabia Numbers and Facts. Saudi Geographical Survey. 1st ed. (2012), Riyadh. KSA.

[2] World Bank. Global Economic Prospects of the Middle East and North Africa. June 2015. https://www.worldbank.org/content/dam/Worldbank/GEP/GEP2015b/Global-Economic-Prospects-June-2015-Middle-East-and-North-Africa-analysis.pdf>

[3] World Bank. "A Water Sector Assessment Report on Countries of the Cooperation Council of the Arab State of the Gulf". Report No. 32539-MNA, (2005).

[4] General Authority for Statistics, https://www.stats.gov.sa/en/13

[5] WERA. Water & Electricity Regulatory Authority, https://wera.gov.sa/en/DataAndStatistics/NRList/ElectricityGeneration/Pages/default.aspx

[6] Ouda, Omar, KACARE (2012) presentation, "Renewable Energy- Waste to Energy, A pillar of the sustainable energy kingdom", First International Environment Conference, King Fahd Civic Centre, Yanbu Al Sinaiyah, KSA, 20-21 November, 2012.

[7] Royal Decree establishing King Abdullah City for Atomic and Renewable Energy (KACARE) No. A 35, 17 April 2010, https://www.energy.gov.sa/ar/Pages/default.aspx

[8] Al Harbi F." Saudi Arabia's Electricity: Energy Supply and Demand Future Challenges",1st Global power, energy and communication conference (IEEE GPECOM2019) , June 2019, Turkey, https://doi.org/10.1109/GPECOM.2019.8778554

[9] Energy Research: Saudi Energy price reform getting serious," APICORP Group, Saudi Arabia, https://www.apicorp.org/

[10] Saudi Electricity Company, " Consumption Tariffs", https://www.se.com.sa/en/Ourservices/ColumnC/Bills-and-Consumption/ConsumptionTariffs

[11] National Renewable Energy Program (NREP). (https://powersaudiarabia.com.sa/web/index.html

[12] Saudi electricity company (SEC), Annual Reports. https://www.se.com.sa/en/Investors/Reports-and-Presentations/Annual-Reports

[13] KACST R&D and Centers of Excellence, https://kacst.gov.sa/

[14] ACWA Connects the Sakaka PV 300MW Project, https://www.acwapower.com/news/acwa-power-connects-the-first-renewable-energy-project-in-the-kingdom-sakaka-pv-ipp-to-the-national-electricity-grid-commencing-initial-production/,

Optimising solar: A techno-economic assessment and government facility compensation framework power generation

Navaid ALI[1,a]*, Faheem Ullah SHAIKH[2,b]* and Laveet KUMAR[3,c]*

[1]Directorate of Postgraduate Studies, Mehran UET, Jamshoro, Pakistan

[2]Department of Electrical Engineering, Mehran UET, Jamshoro, Pakistan

[3]Department of Mechanical Engineering, Mehran UET, Jamshoro, Pakistan

[a]engrnavaid08@hotmail.com, [b]faheemullah.shaikh@faculty.muet.edu.pk, [c]laveet.kumar@faculty.muet.edu.pk

Keywords: Centralized Solar PV System, CO_2 Emissions Reduction, Government Facilities, HESCO, PVsyst

Abstract. In light of the necessity to meet growing energy demands while minimising expenses, this article examines two solar power system designs, considering peak and average load situations. The peak system, designed to handle a peak load of 190 MW with 260,568 PV panels installed, generates a significant excess of approximately 49,911 MWh annually. On the other hand, the average load design suggests a little deficit in energy. The monetary analysis reveals considerable capital expenditures for the total installation costs, totalling $58.77 million for an average system and $66.71 million for the peak system. However, the Levelized Cost of Energy (LCOE) rates are relatively competitive, at 0.0835 USD/kWh and 0.0736 USD/kWh respectively. Notably, the study highlights that the environmental impact analysis demonstrates a significant decrease in CO_2 emissions, with the peak system achieving a reduction of up to 3,601,588 tonnes per year. This research has explicitly validated the capabilities of centralised solar power systems in addressing the current and future energy difficulties faced by the Government of Sindh in a sustainable and economically viable manner.

Introduction

On a global scale, the energy industry faces substantial obstacles, such as the exhaustion of natural resources, environmental issues, and rising expenses, which necessitate an urgent transition to sustainable energy alternatives. Pakistan is experiencing a serious energy crisis characterised by severe power shortages, heavy dependence on non-renewable energy sources (61% thermal, 24% hydropower, 12% nuclear as of June 2023), and inadequate infrastructure [1]. Pakistan actively seeks solutions by adopting renewable energy and developing infrastructure to promote economic growth. Crucial organisations facilitate this shift, including Generation Companies (GENCOs), the Water and Power Development Authority (WAPDA), the Private Power Infrastructure Board (PPIB), National Transmission and Dispatch Company (NTDC), Distribution Companies (DISCOs), and National Electric Power Regulatory Authority (NEPRA) [2]. Following the 2010 amendment to the 1973 constitution, provinces were granted the power to generate, transmit, and distribute electricity at the provincial level. The provincial government of Sindh utilised it by implementing the Sindh Transmission and Dispatch Company (STDC) and the 100MW Nooriabad power project [3,4]. These initiatives aim to ease financial burdens by lowering power expenses. Solar and wind energy are more cost-effective than hydropower, making them good options for Pakistan's economic and operating problems. They offer a way to grow the economy and protect the environment [5]. The financial reports covering May 2021 to May 2022 reveal that Sindh's power expenditures amounted to PKR 8,704 million ($ 31.1M) [6]. The Institute for Energy Economics and Financial Analysis (IEEFA) conducted a financial analysis of Sindh's power expenditures, revealing that solar and wind energy is the most economically efficient, with photovoltaic power generation being the best renewable option due to affordability and low maintenance costs[7-8].

A feasibility analysis of a 100 MWp solar power facility in Pakistan using PVSOL software was done [9]. A study assessed the facility's financial viability, encompassing revenue, operations, maintenance, interest payments, net profit, and payback period. The facility may generate 180,000 GWh/year and eliminate 90,225 tons/year of CO_2, and the payback period is 3.125 years. Based on feasibility, design, and execution, reference [10] evaluated a grid-tied 150 MW_p solar PV trial project in Karachi, Pakistan, for energy yield. In the PVsyst simulation, the average annual power generation, capacity factor, and performance ratio were 232,518 MWh/year, 17.7%, and 74.73%. Reference [11] evaluated the performance and design of a 20 MW Malaysian airport solar PV system. Power generation, performance ratio, and capacity utilisation factor (CUF) were 26,304 MWh, 76.88%, and 15.22%. A one MW PV grid-tied system in Oman was the subject of a techno-economic analysis by [12], who concluded that the system's annual yield factor was 1875 kWh/kWp. The system indicated a capacity factor of 21.7% and an electricity production cost of roughly USD 0.2258 per kWh.

A 3 MW_p grid-connected solar PV system was tested in Karnataka, India. The annual average performance ratio was 0.7, and the power generation was 1372 kWh/kWp [13]. An Indian grid-connected 10 MW solar PV project was assessed. The annual performance ratio, CUF, and power generation were 86.12%, 17.68%, and 15,798.19 MWh, respectively [14]. However, the techno-economic viability of Ghana's grid-connected 50 MW solar PV was examined [15]. The scientists discovered that monocrystalline, polycrystalline, and thin-film solar cells cost USD 0.124/kWh, USD 0.123/kWh, and USD 0.109/kWh, respectively, to produce power. Mono-Si, poly-Si, and CdTe systems had 75.5%, 75.7%, and 77% performance ratios. Reference [16] investigated a 12 MWp solar-powered airport in Cochin, India. The average annual performance ratio was 86.56%, and CUF was 20.12%. This plant could produce 50,000 kWh per day. During its lifetime, the facility would reduce 12,134.26 metric tons of CO_2.

Previous studies have explored the techno-economic elements of small-scale solar PV systems, but there is a lack of research on expansive systems, especially in government facilities. This study aims to examine the economic and operational benefits of implementing 155 MW average and 190 MW peak grid-connected photovoltaic (PV) systems in Sindh, Pakistan, to address existing shortcomings such as lower energy and operational costs, energy independence, improved safety, and promote sustainability.

Methodology
A well-established technique has been employed to evaluate the feasibility of the proposed centralised solar-powered site, illustrating the sequential procedures involved in conducting the research.

i. Data Collection
The operational zone of Hyderabad Electric Supply Company (HESCO) was analysed. This analysis used HESCO and Sindh's Energy Department's Electricity Monitoring and Reconciliation Cell (EM&RC) data. Field visits to multiple places supplemented the Management Information System (MIS) data accuracy and reliability of HESCO and the energy department's electricity consumption statistics. The study examined power use data from January to December 2022 to determine energy consumption patterns—advanced data processing methods like cleaning, mining, filtering, and estimating ensured data accuracy.

Table 3 shows Sindh government-controlled power use by 75 provincials, affiliated, autonomous, and special institutions.

ii. Climatic conditions of the understudy locations

After analysing the Table 1 parameters, Manjhand, Sindh, is the best place for a solar PV installation. The location meets technical standards and supports the project's clean energy, efficiency, and sustainability goals.

iii. Simulation Tool

Solar energy companies widely use PVsyst for simulations. Solar power plants can be efficiently designed using this software for various climates, different PV modules and inverters, meteorological irradiation data, manual import data from other databases, appropriate sizes, and related components [17].

iv. Design of grid-connected large-scale solar PV system

a) Peak Load Design:

$$P_{annual} = P_{max,monthly} \times 12 \tag{1}$$

P_{annual} represents the solar PV system's projected annual peak load, measured in kilowatt-hours (kWh) and $P_{max,\ monthly}$ represents the highest electricity usage recorded in a month over a year.

b) Average Load Design:

$$L_{avg,monthly} = \frac{E_{total,annual}}{12} \tag{2}$$

$L_{avg,\ monthly}$ represents the average monthly load of the solar PV system, measured in kilowatt-hours (kWh) and $E_{total,\ annual}$ value represents the total yearly electricity consumption of the building (kWh).

v. Panel Generation Factor

$$PGF = \frac{\text{daily solar radiation at } the \text{ site per day}}{\text{STC irradiance}} \tag{3}$$

The PGF values for Hyderabad, Thatta, Nawabshah, Manjhand, and Tharparkar are calculated as 5.69, 5.558, 5.65, 5.668 and 5.53, respectively, using the solar irradiance values.

vi. PV module selection

AE Solar's 132-AE solar modules are ideal for average and peak PV systems, offering a typical panel efficiency of 22.56% and a power rating of 700 W_p. These modules are resistant to extreme weather conditions and elevated temperatures. Table 2 of the Monocrystalline 132-AE Solar PV Module shows the technical specifications [18].

vii. Inverter Size Optimisation

It is a crucial optimisation strategy in PV systems, determining the optimal size of a solar inverter by dividing the installed DC power capacity by the AC power output rating [19]. P_{inv} represents the power rating of the inverter's AC output.

$$ILR = \frac{W_{T,peak}}{P_{inv}} \tag{4}$$

viii. Inverter selection

The chosen PVS800 ABB central inverter has a maximum system voltage rating of 1100 V_{dc}, which precisely matches the maximum input voltage rating of the inverter. The selected inverters have a frequency range of 50/60 Hz and a rated maximum power output of 1200 kW_{ac} [20].

Table 1 Site selection parameters

Location	Annual Global Irradiation [kWh/m²]	Annual PV output [kWh/kWp]	Grid Connectivity [distance, existing/new]	Annual Climatic Conditions	Infrastructure and Access
Nawabshah	2,251.4	1737.9	Approx. 250 km, new grid	35.6 [°C]	It needs to be built.
Hyderabad	2,266.5	1,754	Approx. 62 km, new grid	35 [°C]	Road access, utilities, and proximity to resources.
Tharparkar	2,217	1642.1	Approx. 270 km, new grid	35 [°C]	It needs to be built.
Thatta	2221.7	1724.6	Approx. 110 km, new grid	31 [°C]	It needs to be built.
Manjhand	2100	1744.9	Approx. 80 km, existing grid	35.1 [°C]	Road access, utilities, and proximity to resources.

Table 2 Technical specifications solar PV module

Parameters	Value (unit)
Manufacturer	AE Solar Germany
Model	AE700TME 132BDS
Rated maximum power capacity	700 [W]
Max voltage (V_{mp})	42.10 [V]
Max power current (I_{mp})	16.63 [A]
Open circuit voltage (V_{oc})	50.13 [V]
Short circuit current I_{sc}	17.43 [A]
Efficiency	22.56 [%]

Performance parameters for PVsyst software

In PVsyst, the required input data is reduced, calculation times are sped up, and accuracy is maintained.

1. **Technical Parameters**

Table 3 Performance technical parameters for PVsyst software

Parameter	Description	Equation	Unit	Reference
PV Array Yield	The array yield is the ratio of the energy output from a photovoltaic (PV) array for a specific time (such as a day, month, or year) divided by its rated power.	$Y_a = \dfrac{E_{DC,array}}{P_{PV,rated}}$	[kWh/kW/day]	[21]
Reference Yield	Reference yield is the ratio of solar radiation H_t (kWh/m²) absorbed by the solar module plane to G_o (1kW/m²). It represents a solar plant's daily peak sun hours in any location.	$Y_r = \dfrac{H_t}{G_o}$	[kWh/kW/day]	[21]
Specific Production	Annual energy production per kW_p is "specific production". It evaluates the plant's financial value and compares technologies and systems' operational performance.	Specific production $= \dfrac{\text{produced energy}}{P_o}$	[kWh/W_p]	[21]
Performance Ratio	The ratio between the final and reference yields represents the performance ratio (PR). The PR compares installed PV systems at different places by percentage.	$PR = \dfrac{Y_f}{Y_r} * 100$	[%]	[21]
System Losses	The system loss L_s corresponds to the dissipation of energy on the AC side, which includes the irregular functioning of the inverter, AC transformer, and wiring.	$L_S = Y_a - Y_f$	[hour/day]	[21]
Final Yield	The final yield Y_f is the total system useable AC energy E_{grid} (kWh) over a specified period divided by the installed plant's nominal power P_o (kW_p).	$Y_f = \dfrac{E_{grid}}{P_o}$	[kWh/kW/day]	[21]

Parameter	Description	Equation		Reference
PV Module Efficiency (η_{PV})	The solar module's energy conversion as a percentage of available radiation is called module efficiency.	$\eta_{PV} = \dfrac{E_{AC}}{E_{DC}} * 100$	[%]	[21]
Inverter Efficiency (η_{inv})	The inverter generates the AC power P_{ac} fed into the grid, while the PV array produces the DC power P_{dc}.	$\eta_{inv} = S * \dfrac{E_{DC}}{H_t} * 100$	[%]	[21]
PV System Efficiency	The instantaneous efficiency of a photovoltaic (PV) system is determined by multiplying the efficiency of its PV module efficiency (η_{PV}) by the efficiency of the inverter efficiency (η_{inv}).	$\eta_{sys} = \eta_{PV} * \eta_{inv}$	[%]	[21]
Capacity Utilisation factor	The ratio between a PV plant's actual annual energy output and the amount of energy it would generate at nominal capacity for one year. The capacity utilisation factor of a solar PV power plant is determined by three main factors: the size of the inverter, the tracking ability, and the quality of the resource.	$CF = \dfrac{Y_f}{8760}$	[%]	[21]

2. Key Economics Parameters

Table 4 Key Economics Parameters for the PV Systems

Parameter	Description	Equation	Reference
Net Present Value	A project's net present value (NPV) is the difference between the present value of cash flows over the project's lifetime and the initial capital investment cost.	$NPV = \sum_{n=0}^{N} \dfrac{\text{Revenue}}{(1+r)^n} - \text{initial cost}$	[22]
Internal rate of return	In financial terms, the internal rate of return (IRR) is the discount rate at which a project's net present value (NPV) becomes zero.	$NPV = \sum_{n=0}^{N} \dfrac{C_n}{(1+IRR)^n}$	[22]
Simple Payback Period	The simple payback period is the duration, measured in years, required for the cash flow (minus loan payments) to equal the total investment, including debt and equity.	$SPB = \dfrac{\text{initial investment}}{\text{annual savings}}$	[23]
Levelised cost of energy	The levelised cost of electricity, also known as energy production cost, is the cost per unit of electricity required to achieve zero net present value.	$LCOE = \dfrac{\text{yearly cost} + O\&M}{\text{yearly energy system produces}}$	[24]

3. Environmental Analysis

Greenhouse gases, primarily absorbed by infrared light, are a significant contributor to global warming, causing distress to millions worldwide [25]—equation (5), which is considered to be consistent for the chosen site.

$$Produced\ emissions = Annual\ generation * CO_2/kWh \qquad (5)$$

Results and Discussion

I. Technical viability under peak and average load scenario

Both peak and average simulation results reveal the techno-economics of solar PV systems. The peak system with 260,568 PV modules produces 339,721 MWh annually, or 1,863 kWh/kWp. The average system with 221,430 PV modules generates 155 MW and 288,687 MWh yearly. For both suggested systems, the performance ratio is 0.872, with an inverter output of 87.2% and a PV array output loss of 11.3%. Calculations used a 30/180° tilt angle. The peak system's excess units may be cheaper for the facility's 289,809 MWh yearly energy needs.

Table 5 shows that PVsyst software provides the facility with the required and generated units from the average and peak PV systems. Compared to the needed units, the average solar PV system produces 1,122 MWh less yearly. During the same period, the peak solar system produced 49,911 MWh, exceeding the facility's energy needs. The analysis shows how well the solar systems perform over time and meet the facility's energy needs. Additionally, Table 6 shows the technical specifications calculated by the PVsyst software.

II. Financial analysis

Table 7 shows the average and peak PV solar system financial performance. The average PV system installation costs are 58,770,750 USD, with annual operating costs of 21,066,800 USD. The system has a 4.8-year Simple Payback Period and an LCOE of 0.0835 USD/kWh. This yields 211.3% ROI. Its IRR is 17.17%. On the other hand, Peak PV solar systems cost 66,714,400 USD to install and 21,485,900 USD to operate. Its Levelized Cost of Energy (LCOE) is 0.0736 USD/kWh and has a shorter Simple Payback Period of 6.5 years. NPV for the peak system yields 320.6% ROI and 26.10% IRR. Peak PV solar systems offer a longer payback period, higher returns, and a more favourable electricity cost than average PV systems. This study provides valuable insights for decision-makers involved in large-scale solar PV systems.

I. Environmental Analysis

Energy from solar PV facilities is not emission-free. Tables 8 and 9 break down the life cycle emissions for the system's primary components to compute a solar plant's CO_2 emissions. The average and peak PV solar systems are assessed for greenhouse gas (GHG) emissions. An average PV solar system produces 288,687.29 MWh annually, lowering CO_2 emissions by 3,853,975.3 tons. This shows its significant carbon emission reduction. Meanwhile, the system generates 333,511.75 tonnes of CO_2, totalling 3,010,449.3 tons.

In contrast, the peak PV solar system, which generates 339,720.83 MWh of electricity annually, effectively offsets 535,273.1 tonnes of CO_2. Although the system produces 333,511.75 tonnes of CO_2, it achieves a net reduction, resulting in a total CO_2 balance of 3,601,588. These findings highlight the complex interaction between emissions reduction and generation within each system, offering essential insights for sustainable energy decisions.

Table 5 Comparison of generated and required units (MWh) for government facilities

Month	Units required [MWh] by the government sites	Units generated [MWh] by the average solar system	Units generated [MWh] by peak solar system
January	19,018	24,026	28,274
February	18,533	22,478	26,452
March	19,694	27,052	31,836
April	24,213	24,182	28,456
May	29,954	24,420	28,736
June	29,215	24,413	28,729
July	25,685	22,748	26,768
August	24,013	22,345	26,294
September	22,203	24,309	28,606
October	23,992	25,548	30,065
November	29,395	25,231	29,694
December	23,895	21,935	25,812
Total	289,809	288,687	339,721

Table 6 Technical specifications of the simulation's PV panels and inverters

Sr no.	Parameters	Calculated values (Average)	Calculated values (Peak)
1	Unit Nominal Power	700 [W_p]	700 [W_p]
2	No. of PV modules	221,430 units	260,568 units
3	Cell area	644,494 [m^2]	758,409 [m^2]
4	Module area	687,023 [m^2]	808,455 [m^2]
5	No. of inverters	150 units	182 units
6	Operating Voltage	700-1500 [] V	700-1500 [V]
7	Nominal Power (STC)	155 [MW_p]	182.4 [MW_p]
8	Modules	10,065 string x 22 In series	10,857 string x 24 In series
9	Inverter rating	1,000 [kWac] (150 units)	1,000 [kWac] (182 units)
10	Inverter power	1,000 [kW_{ac}]	1,000 [kW_{ac}]
11	CO_2 emissions cut	3,010,449.323 [tCO_2/year]	3601580.433 [tCO_2/year]

Table 7 Financial parameters of the proposed PV solar systems

System parameters	Average PV Solar System	Peak PV Solar System
Total Installation	$58.8 M	$66.7 M
Operating Cost	$21.1M [$/year]	$21.5 M [$/year]
Energy Generated	288,687 [MWh/year]	339,721 [MWh/year]
LCOE	0.0835 [$/kWh]	0.0736 [$/kWh]
Payback Period	4.8 [years]	6.5 [years]
NPV	$ 124.0 M	$213.5 M
ROI	211.3 [%]	320.6 [%]
IRR	17.17 [%]	26.10 [%]

Table 8 Peak PV Systems' Lifecycle Emissions

Item	LCE	Quantity	Subtotal [kgCO$_2$]
Modules	1713 [kgCO$_2$/kW$_p$]	189,998 [kW$_p$]	325,412,518
Supports	2.97 [kgCO$_2$/kg]	2,714,250 [kg]	8,056,301
Inverters	294 [kgCO$_2$/Units]	182 units	53,508

Table 9 Average PV Systems' Lifecycle Emissions

Item	LCE	Quantity	Subtotal [kgCO$_2$]
Modules	1713 [kgCO$_2$/kW$_p$]	144,999 [kW$_p$]	248,383,287
Supports	2.97 [kgCO$_2$/kg]	2,714,250 [kg]	8,056,301
Inverters	294 [kgCO$_2$/Units]	146 units	42,924

References

[1] D. D. A. C. Goodman Tom Prater, Joe, "The Carbon Brief Profile: Pakistan," Carbon Brief, May 26, 2023. https://interactive.carbonbrief.org/the-carbon-brief-profile-pakistan/

[2] S. A. Khatri et al., "An Overview of the Current Energy Situation of Pakistan and the Way Forward towards Green Energy Implementation," Energies, vol. 16, no. 1, p. 423, Dec. 2022. https://doi.org/10.3390/en16010423

[3] W. Rafique, "Electricity As A Subject After The 18th Amendment," Courting The Law, Aug. 26, 2016. https://courtingthelaw.com/2016/08/23/commentary/electricity-as-a-subject-after-the-18th-amendment/

[4] Information on "Nooriabad Power Company." https://www.snpc.com.pk/

[5] U. Qazi and M. Jahanzaib, "An integrated sectoral framework for the development of sustainable power sector in Pakistan," Energy Reports, vol. 4, pp. 376–392, Nov. 2018. https://doi.org/10.1016/j.egyr.2018.06.001

[6] Information on "EM&RC – SINDH ENERGY DEPARTMENT." [Online]. Available: https://sindhenergy.gov.pk/emrc/

[7] "Year Book 2022-23," Ministry of Energy (Power Division), Nov. 28, 2023. https://power.gov.pk/SiteImage/Publication/YearBook2022-23.pdf (accessed Feb. 21, 2024).

[8] M. Irfan, Z.-Y. Zhao, M. Ahmad, and M. Mukeshimana, "Solar Energy Development in Pakistan: Barriers and Policy Recommendations," Sustainability, vol. 11, no. 4, p. 1206, Feb. 2019. https://doi.org/10.3390/su11041206

[9] N. Abas, S. Rauf, M. S. Saleem, M. Irfan, and S. A. Hameed, "Techno-Economic Feasibility Analysis of 100 MW Solar Photovoltaic Power Plant in Pakistan," Technology and Economics of Smart Grids and Sustainable Energy, vol. 7, no. 1, Apr. 2022. https://doi.org/10.1007/s40866-022-00139-w

[10] Jamil, J. Zhao, L. Zhang, R. Jamil, and S. F. Rafique, "Evaluation of Energy Production and Energy Yield Assessment Based on Feasibility, Design, and Execution of 3 × 50 MW Grid-Connected Solar PV Pilot Project in Nooriabad," International Journal of Photoenergy, vol. 2017, pp. 1–18, 2017. https://doi.org/10.1155/2017/6429581

[11] S. Sreenath, K. Sudhakar, Y. A.F., E. Solomin, and I. M. Kirpichnikova, "Solar PV energy system in Malaysian airport: Glare analysis, general design and performance assessment," Energy Reports, vol. 6, pp. 698–712, Nov. 2020. https://doi.org/10.1016/j.egyr.2020.03.015

[12] A. Kazem, M. H. Albadi, A. H. A. Al-Waeli, A. H. Al-Busaidi, and M. T. Chaichan, "Techno-economic feasibility analysis of 1 MW photovoltaic grid-connected system in Oman," Case Studies in Thermal Engineering, vol. 10, pp. 131–141, Sep. 2017. https://doi.org/10.1016/j.csite.2017.05.008

[13] K. Padmavathi and S. A. Daniel, "Performance analysis of a 3MWp grid-connected solar photovoltaic power plant in India," Energy for Sustainable Development, vol. 17, no. 6, pp. 615–625, Dec. 2013. https://doi.org/10.1016/j.esd.2013.09.002

[14] B. Shiva Kumar and K. Sudhakar, "Performance evaluation of 10 MW grid-connected solar photovoltaic power plant in India," Energy Reports, vol. 1, pp. 184–192, Nov. 2015. https://doi.org/10.1016/j.egyr.2015.10.001

[15] M. Obeng, S. Gyamfi, N. S. Derkyi, A. T. Kabo-bah, and F. Peprah, "Technical and economic feasibility of a 50 MW grid-connected solar PV at UENR Nsoatre Campus," Journal of Cleaner Production, vol. 247, p. 119159, Feb. 2020. https://doi.org/10.1016/j.jclepro.2019.119159

[16] S. Sukumaran and K. Sudhakar, "Performance analysis of solar powered airport based on energy and exergy analysis," Energy, vol. 149, pp. 1000–1009, Apr. 2018. https://doi.org/10.1016/j.energy.2018.02.095

[17] M. A. Anrizal Akbar, A. M. S. Yunus, and J. Tangko, "PVSYST-Based Solar Power Plant Planning," INTEK: Jurnal Penelitian, vol. 9, no. 1, p. 89, Apr. 2022. https://doi.org/10.31963/intek.v9i1.3789

[18] "AE700TME-132BDS - AESOLAR," AESOLAR, Jan. 09, 2024. https://aesolar.com/products/ae700-tme-132bds/

[19] T. E. K. Zidane, S. M. Zali, M. R. Adzman, M. F. N. Tajuddin, and A. Durusu, "PV array and inverter optimum sizing for grid-connected photovoltaic power plants using optimisation design," Journal of Physics: Conference Series, vol. 1878, no. 1, p. 012015, May 2021. https://doi.org/10.1088/1742-6596/1878/1/012015

[20] "Abb Solar Inverter 1000kw 1mw Pvs 800 Central Inverter Ip42 Ip65," indiamart.com. https://www.indiamart.com/proddetail/abb-solar-inverter-1000kw-1mw-pvs-800-central-inverter-ip42-ip65-23097354773.html

[21] N. Ahmed et al., "Techno-economic potential assessment of mega-scale grid-connected PV power plant in five climate zones of Pakistan," Energy Conversion and Management, vol. 237, p. 114097, Jun. 2021. https://doi.org/10.1016/j.enconman.2021.114097

[22] S. Abdelhady, "Performance and cost evaluation of solar dish power plant: sensitivity analysis of levelized cost of electricity (LCOE) and net present value (NPV)," Renewable Energy, vol. 168, pp. 332–342, May 2021. https://doi.org/10.1016/j.renene.2020.12.074

[23] M. M. Rafique and H. M. S. Bahaidarah, "Thermo-economic and environmental feasibility of a solar power plant as a renewable and green source of electrification," International Journal of Green Energy, vol. 16, no. 15, pp. 1577–1590, Oct. 2019. https://doi.org/10.1080/15435075.2019.1677237

[24] W. Shen et al., "A comprehensive review of variable renewable energy levelized cost of electricity," Renewable and Sustainable Energy Reviews, vol. 133, p. 110301, Nov. 2020. https://doi.org/10.1016/j.rser.2020.110301

[25] "Climate Change Indicators: Greenhouse Gases | US EPA," US EPA, Feb. 09, 2024. https://www.epa.gov/climate-indicators/greenhouse-gases

A developed system design for blue energy generation

Omar D. MOHAMMED[1,a,*], Saud ALHARBI[1,b], Shoja ALHARBI[1,c], Waleed ALDHUWAIHI[1,d], and Nasser BINI HAMEEM[1,e]

[1] Mechanical Engineering Department - Prince Mohammad bin Fahd University PMU, Khobar, Saudi Arabia

[a]osily@pmu.edu.sa, [b]201801600@pmu.edu.sa, [c]202002618@pmu.edu.sa, [d]201901074@pmu.edu.sa, [e]201502982@pmu.edu.sa

Keywords. Wave Energy, Linear Generator, Power Generation, Renewable Energy

Abstract. Sustainability is an important factor in energy generation. Sea or ocean waves can be used as a sustainable source of energy generation. A technique of using buoyant floats submerged in water to drive a magnetic generator for energy conversion can be applied and studied. In the current article, a developed system design is presented for harvesting wave energy (blue energy). The wave power and generator power are calculated based on the studied model. The presented design enhances the concept of wave energy harvesting, as a renewable energy source, and shows the possibility of applying it. Applying wave energy harvesting should get more attention as a sustainable and clean source of power electricity.

Introduction

Sea or ocean wave energy, which is an abundant and untapped source of energy, can be used to reduce the use of fossil fuels. Wave energy, which can be called "blue energy", is a clean renewable source of energy and supports sustainability. This method reduces greenhouse gas emissions and protects marine ecosystems, unlike traditional power generation. Innovation in wave energy technology boosts energy security. Waves can help coastal regions become more energy self-sufficient and less vulnerable to energy supply disruptions. Exploring and refining a linear wave energy generator advances wave energy conversion technology, potentially paving the way for future developments and commercial applications [1,2]. Electricity generation in remote or island regions often relies on imported fossil fuels. In such areas wave energy generators provide a more sustainable and independent energy source than expensive and environmentally harmful fuel imports [3]. These generators use ocean waves to provide clean renewable energy that supports the global transition to a carbon-neutral future [4]. The generator design, materials and operation, which is driven by the need for reliable and efficient wave energy converters, was discussed in [5,6]. The environmental assessments were examined to ensure sustainable ocean resource use. It helps evaluate existing technologies, identify gaps and guide this project toward a more efficient and sustainable wave energy generator [6].

A comprehensive literature review of wave energy was presented in [7] to discuss the concept development since the 1970s, highlighting its growing global importance and promise. It thoroughly examines sea wave energy understanding and use. The paper focused on wave energy resource characterization and the theoretical foundations for wave energy absorption and control hydrodynamics. The review emphasizes the complexity of wave dynamics and the theoretical frameworks needed to extract energy from ocean waves.

Wave energy using linear generator system was presented and discussed in [8,9]. Wave energy conversion and the variety of generator systems were discussed. This variety highlights the different ways to design wave energy-efficient systems. The authors discussed modern linear generator systems, including the Archimedes Wave Swing (AWS) and Uppsala systems. These systems demonstrate linear wave energy conversion technology's progress. The authors also suggested investigating air-cored machines for integrated electrical-mechanical-structural designs suggesting a holistic generator development approach. The search for cost-effective solutions suggests exploring generator constructions that may be cheaper. This includes cylindrical generators and concentrated coil generators.

Wave energy conversion using linear generators was examined in [10]. The differences between rotary and linear generators application were discussed. The fact that linear generators convert wave energy efficiently is important in energy generation. Linear generators can adjust wave speeds unlike rotary generators. Flexible motion matches wave variations resulting in different voltages, currents and phase sequences. The peak to average power ratio increases due to this variability improving efficiency. The ocean wave energy fundamental calculations were presented in [11]. The ongoing research on wave energy at Oregon State University was sumarasied.

Despite the benefits of the wave energy technique, its applicability is still challenging. The device durability in harsh marine environments, cost-effectiveness, energy production and grid integration complexity still require more investigations. The previous works involved immersing the generator in the water. In the current work a new system design is developed for power generation using the wave energy. The presented system is designed to be attached to a structure above the water on the coastal wave breaker. The design concept is presented and discussed in terms of manufacturability, productivity and applicability.

System Design
In the current work a system design is developed for generating power using the wave energy. The system consists of a magnetic shaft and electric coil directly connected to a float movement. As the magnetic shaft moves inside the coil, an electrical current is generated. Electromagnetic induction generates electricity from the float's wave induced kinetic energy. Linear motion of the magnetic shaft inside the coil changes the magnetic field generating an electric current. The heart of the wave energy generator system is converting mechanical motion into electrical energy.

Fig. 1 shows the developed system model and Fig.2 shows the coil and armature used in the system. The designed model is provided by a platform to be attached to a structure above the water on the coastal wave breaker. This will provide the feasibility to locate number of the generator units for more energy generation.

The design meets practical deployment size and spatial constraints. A small footprint and streamlined design allow for efficient integration into existing marine infrastructures maximizing system effectiveness without taking up too much space.

Sustainability guides material selection and operational strategies, aiming to minimize the environmental footprint. Using eco-friendly materials and energy-efficient components creates a renewable energy and environmental preservation system. The system minimizes marine ecosystem damage. The system's deployment and operation are designed to minimize noise pollution and aquatic life disturbances to preserve marine environments. Beyond technology, by considering social impacts, the generator design concept encourages community for sustainable energy initiatives.

Efficiency and affordability are balanced by cost effectiveness. To encourage adoption, strategies optimize system efficiency while lowering manufacturing and operational costs. The system design prioritizes readily available components and construction methods for large-scale manufacturing, streamlining production for efficient and cost-effective deployment. The system's strict safety protocols and design features reduce risks for maintenance personnel and the marine environment.

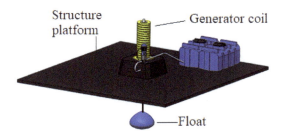

Figure 1 the developed generator system design

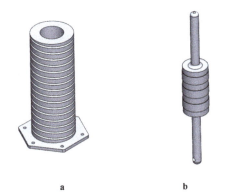

Figure 2 the generator parts, a. the generator coil, b. the magnetic armature

Theoretical Calculations

The wave motion can be caracterized by the wave length or the time period (T_{wave}), and the wave height or the wave amplitude (H_{wave}). The input power to the generator is P_{wave}, which is the wave power per one meter of the wave length, can be calculated as follows [11,12,13].

$$P_{wave} = \frac{(\rho \times g^2 \times H_{wave}^2 \times T_{wave})}{32\pi} \tag{1}$$
$$= 7662 \ W/m$$

Where,
The water density, $\rho = 1000 \ kg/m^3$
The gravity constant, $g = 9.81 \ m/s^2$
The wave amplitude, $H_{wave} = 1 \ m$
The wave time period, $T_{wave} = 8 \ s$
The power output of the linear *generator* (*Pgenerator*) can be calculated as follow.
$$P_{generator} = \eta \times P_{wave}$$
$$= 0.85 \times 7662 = 6512.8 \ W/m$$

Where,

The efficiency of the generator, $\eta = 0.85$

These calculations reveal wave power, linear generator output and system energy conversion efficiency.

Discussion

The system is designed to provide the feasibility to be located and attached to the wave breakers. A number of the generator units can be used for power generation and at the same time attaching the units to the wave breaker structure will give more strength to the units and will make the maintenance easier.

The buoyant float should be made of a sturdy material for hostile sea settings. Material weight, corrosion and wave resistance matter. The linear generator component selection is crucial to the subsystem. Effective electromagnetic induction is achieved by the magnetic shaft, coil and materials. Optimizing wave motion to electricity conversion demands high quality materials and precision engineering. Primary capture systems should use oscillating water columns or point absorbers to efficiently catch wave energy. In diverse wave situations, subsystem components must maximize energy absorption. Frame, bearings and anchoring systems are crucial converter components. For converter structural integrity and lifetime, durability, stability and seawater resistance are used. Controlling and converting electricity requires inverters, transformers and control units. Critical components must work well under different electrical loads and environments. These components should be chosen for performance, durability, pricing and system compatibility. The goal is an integrated system that efficiently converts wave energy into electrical power in marine conditions while being reliable and durable. Designers must choose components that enhance efficiency, safety and sustainability. To ensure alignment and functionality, the system model parts have to be fabricated, assembeled and tested.

Conclusions

The wave energy converter is caracterized by the aspects of innovation, perseverance and sustainablity. This current work advances renewable energy and shows a shared commitment to energy issues. This work demonstrates how engineering and real-world application can use the sustainable wave energy (blue energy) for power electricity generation. A developed design is presented to use the wave energy and promote the growth of renewable energy dependency. In an age of clean energy and environmental concerns energy generation can be more sustainable. It symbolizes the commitment to fighting climate change, conserving resources and boosting economic resilience through technology. To reduce design complexity, a linear generator is applied in the project to avoid the use of extra mechnism for converting the linear motion to a rotary motion to adapt the rotary geneartor. Developing wave energy harvesting techniques and improving the efficiency should get more attention as a sustainable and clean source of power electricity.

References

[1] O. Edenhofer et al. Renewable Energy Sources and Climate Change Mitigation: Special Report of the Intergovernmental Panel on Climate Change. Cambridge University Press, 2012. https://doi.org/10.1017/CBO9781139151153

[2] M. Okur Dinçsoy and H. Can. Optimizing Energy Efficiency During a Global Energy Crisis. IGI Global, 2023. https://doi.org/10.4018/979-8-3693-0400-6

[3] E. Hossain and S. Petrovic, Renewable Energy Crash Course: A Concise Introduction. Springer International Publishing, 2021. https://doi.org/10.1007/978-3-030-70049-2

[4] F. R. Spellman and R. M. Bieber, The Science of Renewable Energy. CRC Press, 2016. https://doi.org/10.1201/b13592

[5] A. Pecher and J. P. Kofoed, Handbook of Ocean Wave Energy. Springer International Publishing, 2016. https://doi.org/10.1007/978-3-319-39889-1

[6] L. Peppas. Ocean, Tidal and Wave Energy: Power from the Sea. Crabtree Publishing Company, 2008.

[7] A. Falcao and s. e. reviews, Wave energy utilization: A review of the technologies, vol. 14, no. 3, pp. 899-918, 2010. https://doi.org/10.1016/j.rser.2009.11.003

[8] H. Polinder, M. Mueller, M. Scuotto, and M. G. de Sousa Prado. Linear generator systems for wave energy conversion, in Proceedings of the 7th European Wave and Tidal Energy Conference, Porto, Portugal, 2007, pp. 11-14.

[9] P. Khatri, X. Wang. Comprehensive review of a linear electrical generator for ocean wave energy conversion. The Institution of Engineering and Technology, 2020, vol. 14, no. 6, pp. 949-958. https://doi.org/10.1049/iet-rpg.2019.0624

[10] A. A. Faiad and I. Gowaid, Linear generator technologies for wave energy conversion applications: A review in 2018 53rd International Universities Power Engineering Conference (UPEC), 2018, pp. 1-6: IEEE.

[10] T. Brekken, A. von Jouanne, Hai Yue Han. Ocean Wave Energy Overview and Research at Oregon State University. 2009 IEEE Power Electronics and Machines in Wind Applications. https://doi.org/10.1109/PEMWA.2009.5208333

[11] R. Parthasarathy. Linear PM generator for wave energy conversion. Louisiana State University and Agricultural and Mechanical College, Master thesis, 2012.

[12] A. Khaligh, O. Onar. Energy Harvesting: Solar, Wind, and Ocean Energy Conversion Systems. Taylor and Francis Group, LLC, 2010.

A review of the renewable energy technologies and innovations in geotechnical engineering

Eman J. Bani ISMAEEL[1,a], Samer RABABAH[1,b], Mohammad Ali KHASAWNEH[2,c], Nayeemuddin MOHAMMED[2,d*], Danish AHMED[2,e]

[1] Civil Engineering Department, College of Engineering, Jordan University of Science &Technology, Irbid 22110, Jordan

[2] Department of Civil Engineering, Prince Mohammad Bin Fahd University, Al Khobar, Kingdom of Saudi Arabia

[a]egbaniissmaeel20@eng.just.edu.jo, [b]srrababah@just.edu.jo, [c]mkhasawneh@pmu.edu.sa, [d]mnayeemuddin@pmu.edu.sa, [e]dahmed@pmu.edu.sa

Keywords: Sustainability, Renewable Energy, Recent Trends, Fossil fuels, Climate Change, CO_2 emission

Abstract. The recent trend in the energy sector is seeking eco-friendly and cost-effective solutions. In the last few years, the globe has given attention to renewable energy resources to overcome the depletion problems of fossil fuels and take advantage of abundant natural resources such as solar, wind, and natural gas. Renewable energy resources provide a long-term efficiency solution in different applications. This review paper comprehensively reviews recent advancements in renewable energy technologies and innovations, focusing on solar energy, wind energy development, smart grid technologies, and energy storage solutions for a cleaner and more sustainable future.

Introduction

Renewable energy sources have become increasingly important in recent years [1]. They offer a solution to combat the depletion of fuels and address environmental issues like global warming [2]. Generally, energy sources can be classified into two categories: fossil fuels and renewable energy [3]. The European Union has pledged to cut its emissions by 20% in the second phase of the Kyoto Protocol. The intermittent renewable energy sources solar and wind must be transmitted and stored as effectively as feasible in order to meet this objective. Renewable energy, known for its nature, includes clean energy options [4]. With the increasing global demand for energy consumption and environmental challenges such as climate change, carbon dioxide (CO_2) emissions, and greenhouse gas (GHG) emissions, authorities are increasingly encouraged to transition from conventional energy sources, particularly fossil fuels, to clean and secure renewable energy alternative [5]. Renewable energy resources offer energy security, environmental protection, economic advantages, and a pollution-free environment. Switching from relying on fuels for energy consumption to adopting a zero-carbon emission approach marks a shift in the energy sector. The transition to renewable energy will be facilitated through the integration of smart technologies, information technology, and the implementation of clean energy policy frameworks.

The adoption of renewable energy sources has prompted the development of numerous technologies and innovations that boost progress within the energy industry. These technologies provide environmentally friendly substitutes for energy derived from fossil fuels and establish the foundation for a future distinguished by carbon-neutral energy. Recent developments in renewable energy technologies and innovations, such as advancements in solar energy, energy storage solutions, wind energy development, and smart grid technologies, are examined in detail in this article. Moreover, this review will address the challenges and opportunities of transitioning to renewable energy.

Solar Energy Advancement
Solar energy is generated abundantly by the sun. Solar energy is considered a vital renewable resource, intercepting around 1.8×1014 kW on Earth. Its ubiquity, zero cost, and sustainability make it a promising solution to global energy demands. Solar energy is applicable across diverse sectors and is crucial for addressing the continuously increasing demand for energy resources. It is one of the renewable energy resources that has gained significant importance in the recent development of the energy sector [6]. Solar energy can be converted into different forms through photovoltaic (PV) systems. PV systems provide benefits such as minimal environmental impact, low maintenance expenses, and the lack of moving parts despite their typically 18-23% lower efficiency. However, numerous variables, including temperature fluctuations and solar insolation, influence the output of PV systems, resulting in power generation fluctuations. Various approaches, including maximum power point tracking (MPPT) methods and sun trackers, can increase the efficiency of a PV system [7].

The Integration of artificial intelligence (AI) into the utilization of solar energy resulted in the development of advanced solar panels optimized in structure, performance, and efficiency. [8] examined the AI can transform PV technology in the solar energy sector. According to the study, AI can potentially improve solar energy system efficiency, power grid integration, and the transition to sustainable energy. Researchers used machine learning models to accelerate the discovery of high-performance solar cell materials. These algorithms predict the material properties and searched large data sets quickly. [9] developed a solar energy tracking prototype to optimize solar energy collection using Arduino technology. The prototype could automatically adjust the position of the solar panel by utilizing servo motors and Light-Dependent Resistors (LDRs). An 18% increase in energy output compared to static panels indicates that these panels may be useful in various situations.

Advanced battery management algorithms and Neural Maximum Power Point Tracking (MPPT) control to optimize photovoltaic solar systems. This work optimized solar energy conversion to electricity by dynamically modifying the MPP and managing battery charging and discharging [10]. Thorough MATLAB/Simulink simulations showed that neural MPPT control outperforms other methods even in variable sunlight conditions. [11] investigated whether tiny machine learning (TinyML) could predict solar energy yield in real time for microcontrollers and other resource-constrained edge IoT devices. Four popular machine learning models: unidirectional long short-term memory (LSTM), bidirectional gated recurrent unit (BiGRU), bidirectional long short-term memory (BiLSTM), and simple bidirectional recurrent neural network (BiRNN) are extensively evaluated for predicting solar farm energy yield. This study adds to the body of knowledge on cost-effective IoT solutions and emphasizes edge device limitations in ML architecture selection. This makes the energy landscape more sustainable and efficient, benefiting residential and industrial sectors.

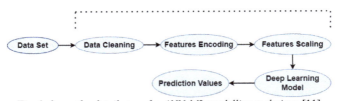

Fig. 1 shows the detail steps for ANN-ML modelling technique [11]

Wind Energy Development

Renewable energy sources, such as wind energy, offer a promising way to reduce fossil fuel dependency while mitigating environmental impact. Wind, considered an indirect derivative of solar energy, is continually replenished by natural processes driven by the sun's energy [12]. Wind energy arises from the differential heating of the Earth's surface, considered an abundant and sustainable energy resource. Estimates suggest that more than 10 million MW of energy is continuously available in the Earth's wind, significantly contributing to global energy needs [13]. The advancement in wind turbines leads to stronger, lighter, and more efficient turbine blades, which means enhancements in annual energy output and reductions in turbine weight, emission, and noise over recent years [14]. Several measurements are considered to harness wind energy's potential further, such as establishing more wind monitoring stations, enhancing turbine maintenance techniques, carefully selecting wind farm sites, adapting high-capacity machines, and using advanced design techniques. These procedures lead to optimizing wind energy generation by maximizing the turbine's efficiency, increasing machine availability, and expanding operational capacity [15].

A recent study employs computational strategies for converting wind energy into hot water. This study optimized wind farm layout, modeling and simulation systems, energy management strategy, wind resource assessment, turbine design and size, power conversion control, and layout engineering [16]. An evaluation of the semicircular wind tower blasting and painting system's potential to revolutionize conventional wind tower coating methods was conducted by [17]. Contributing to advancing the wind energy industry toward a more sustainable future, the findings of this research demonstrated that the semicircular system presents notable benefits compared to robotic arms, including a substantial enhancement in efficiency, reduced costs, and an environmentally friendly solution. In a study conducted by [18], the objective was to investigate the feasibility of utilizing wind towers as natural ventilation and cooling systems in residential buildings. The primary focus of this research was the integration of wind towers with energy storage systems and renewable energy sources to maximize their efficiency and effectively reduce greenhouse gas emissions.

Smart Grid Technologies

In general, smart grids are advanced electricity networks engineered to utilize information and communication technologies to ensure improved dependability and productivity in power delivery [19]. Promoting energy conservation and sustainability, these technologies facilitate the seamless integration and administration of renewable energy sources. Storage solutions are crucial in mitigating fluctuations in energy output, especially when considering the decentralized nature of energy production facilitated by renewable resources. In order to prevent blackouts, smart grids attempt to maintain a balance between energy supply and demand by addressing a number of issues, such as rising pressures and peak consumption [20]. Smart Grids use cutting-edge technologies like intelligent control, communication, and self-healing to advance electricity networks. Demand side management (DSM) and demand response initiatives in these power grids give consumers more control over their energy use. Meters, microgeneration, and smart appliances enable this control. Smart grids use various generator and storage technologies, including distributed generation (DG) and renewable energy sources, to reduce the environmental impact of electricity provision. They optimize asset management and delivery independently, maximizing resource use. Smart grids also ensure reliable energy supply during disasters by strengthening resilience to physical and cyber threats. Smart Grids increase supply aggregation and transmission capacity to improve power supply and market access [21].

Recent developments in smart grid technologies have been instrumental in addressing the changing demands of the power distribution sector. These developments optimize grid operations, enhance resilience to cyber and physical threats, and facilitate the integration of renewable energy sources and energy storage solutions by utilizing cutting-edge innovations. In addition, the continuous advancement and enhancement of smart grid technologies possess tremendous potential to fortify and transform the distribution system, facilitating the transition to a future characterized by enhanced energy efficiency, dependability, and sustainability [22]. The smart grid technologies are driving significant progress toward enhancing modern power systems' reliability, sustainability, and efficiency. One notable area of advancement lies in the development of energy management systems, which play a crucial role in ensuring the seamless integration and coordination of various grid components from generation to consumption. Initiatives such as the Smart Grid Interoperability (SGIP) standards, initiated by the National Institute of Standards and Technology (NIST), have promoted interoperability among grid components and facilitated efficient planning and implementation [23].

Moreover, integrating Internet of Things (IoT) solutions has revolutionized grid communication and automation, marking an important advance in smart grid technology. IoT technologies provide greater connection and automation, allowing grid components to interact more efficiently and reliably. However, using IoT in smart grids introduces additional issues, notably regarding security and privacy. Researchers are currently addressing these problems by establishing strong security mechanisms and authentication methods to preserve the integrity and confidentiality of grid data [24]. Furthermore, big data analytics transforms smart grid operations by giving essential insights into grid efficiency and management. However, difficulties such as data storage, processing, integration, and security continue to be important emphasis areas for researchers looking to fully realize the potential of big data in smart grid technologies [25].

Energy Storage Solutions
Among the most recent developments in the renewable energy field, energy storage solutions are critical to assuring the reliability and scalability of renewable energy sources. Furthermore, this technology considerably increases energy consumption capacity. The increasing demand for sustainable and clean energy continues to rise, driving the development of new imperative solutions for energy storage [26]. This section aims to provide insights into the role of recent advancements in energy storage solutions in facilitating the transition towards a more sustainable energy landscape. The Energy PLAN modeling to find Finland's cheapest 100% renewable energy scenario by 2050. In the study, electricity and heat storage meet 15% of end-user demand, while thermal storage discharge meets 4%. High renewable energy integration requires electrical storage devices at 50% variable renewable energy penetration and seasonal storage devices at over 80% renewable energy (RE) penetration. Energy storage is crucial in Finland's 100% renewable energy system [27]. [28] analyzed the challenges during the remote Arctic region's transition from diesel and fossil fuels to renewable energy sources. Various energy storage options, including battery storage, underground solar power/storage, and hydrogen storage, are explored to achieve energy self-sufficiency at Flatey's. These energy storage options are summarized in Table 1.

Table 1: The suggested energy storage options for remote areas [28]

Storage Solution	Description
Storing Solar Power in Battery Banks	- Utilizes battery storage in PV systems. - Flexible connection of batteries in series for larger facilities. - It requires less space than hydrogen storage.
Underground Solar Power Storage	- Utilizes borehole thermal energy storage (BTES). - Ground source heat exchangers (BHE). - Efficiency depends on geological factors.
Storing Power by Using Hydrogen	- Green hydrogen production through electrolysis - Challenges in storing hydrogen. - Can be stored under high pressure or in liquid form. - Utilized in fuel cells for high-efficiency electricity production. - Economic evaluation required.

Renewable Energy in geotechnical engineering applications

Geotechnical engineering is a crucial in providing good foundation and anchor systems for MRE devices. Marine renewable energy (MRE) systems on a commercial scale will consist of a variety of devices secured to the bottom by foundations or anchors [29]. The renewable energy is a substitute resource to address the high demand for conventional hydrocarbon energy, reduce the impact on the environment, and ensure sustainability for many years. The review of geotechnical engineering concerns and their connection to renewable energy engineering is intended to motivate geotechnical engineers to participate in the developing area of energy research [30]. Owing to the geo-dependent nature of renewable energy, geo technology can help optimize the effective use of renewable resources Environmental impact assessments and offshore wind turbine engineering designs both depend heavily on seabed characterization. The investigation into screw piles for the development of offshore renewable energy, significant upscaling of the currently in use onshore piles was necessary [31].

The main benefit is that it is a clean, renewable energy source with no adverse effects on the environment. In the past, Croatia's experience with geothermal energy extraction has mostly focused on the potential for deep geothermal resource extraction. "Investment Valuation Model for Renewable Energy Systems in Buildings" outlines the real options model that is being offered and identifies the special characteristics that set it apart from other valuation models [32]. The needs of a profession that will increasingly be involved in sustainable design, energy geo technology, waste management, underground utilization, enhanced/more efficient use of natural resources, and alternative/renewable energy sources must be addressed in the geotechnical engineering curriculum, from undergraduate education through continuing professional education [33]. Every technology was found to have a very wide range in terms of electricity costs, greenhouse gas emissions, and generation efficiency. This is mostly because each renewable energy source has a different geographical reliance and a different range of technological alternatives [34].

Table 2: The efficiency and cost aspect to renewable energy [35]

Aspect	Energy Production	Energy Storage	Energy Harvesting
Efficiency	-Solar Photovoltaics: 15-20% -Wind Turbines: >40%	-Lithium-ion Batteries: 80-90% -Lead-acid Batteries: 70-85%	-Photovoltaic Cells: >20% -Thermoelectric Generators: 5-10%
Cost	-Solar Photovoltaics: Decreasing, currently competitive -Wind Turbines: Decreasing.	-Lithium-ion Batteries: Decreasing, still significant cost -Lead-acid Batteries: Relatively low cost.	-Photovoltaic Cells: Decreasing, currently cost-competitive -Thermoelectric Generators: Costly,

Summary

Reviewing recent advancements in renewable energy technologies highlights the significant progress towards a sustainable energy landscape. Solar energy innovations, propelled by AI, promise enhanced efficiency and performance, while wind energy developments offer efficient and eco-friendly solutions for wind tower coatings and ventilation systems. Smart grid technologies facilitate seamless energy distribution and management, bolstering grid reliability and efficiency. Furthermore, energy storage solutions, such as thermal storage and PtG concepts, contribute to energy scalability and sustainability. These advancements underscore the critical role of renewable energy in mitigating environmental challenges and transitioning towards a greener future. Continued research and innovation in renewable energy technologies are essential to drive further progress and accelerate the global transition to clean and sustainable energy sources.

References

[1] F. Rizzi, N. J. van Eck, and M. Frey, "The production of scientific knowledge on renewable energies: Worldwide trends, dynamics and challenges and implications for management," *Renewable Energy*, vol. 62, pp. 657–671, 2014. https://doi.org/10.1016/j.renene.2013.08.030.

[2] D. C. Momete, "Analysis of the Potential of Clean Energy Deployment in the European Union," *IEEE Access*, vol. 6, pp. 54811–54822, 2018. https://doi.org/10.1109/access.2018.2872786.

[3] N. A. Ludin *et al.*, "Prospects of life cycle assessment of renewable energy from solar photovoltaic technologies: A review," *Renewable and Sustainable Energy Reviews*, vol. 96, pp. 11–28, 2018. https://doi.org/10.1016/j.rser.2018.07.048.

[4] P. Trop and D. Goricanec, "Comparisons between energy carriers' productions for exploiting renewable energy sources," *Energy*, vol. 108, pp. 155–161, 2016. https://doi.org/10.1016/j.energy.2015.07.033.

[5] C. Bhowmik, S. Bhowmik, A. Ray, and K. M. Pandey, "Optimal green energy planning for sustainable development: A review," *Renewable and Sustainable Energy Reviews*, vol. 71, pp. 796–813, 2017. https://doi.org/10.1016/j.rser.2016.12.105.

[6] N. Kannan and D. Vakeesan, "Solar energy for future world: - A review," *Renewable and Sustainable Energy Reviews*, vol. 62, pp. 1092–1105, 2016. https://doi.org/10.1016/j.rser.2016.05.022.

[7] R. Rajesh and M. Carolin Mabel, "A comprehensive review of photovoltaic systems," *Renewable and Sustainable Energy Reviews*, vol. 51, pp. 231–248, 2015. https://doi.org/10.1016/j.rser.2015.06.006.

[8] A. Mohammad and F. Mahjabeen, "Revolutionizing solar energy with ai-driven enhancements in photovoltaic technology," *BULLET: Jurnal Multidisiplin Ilmu*, vol. 2, no. 4, pp. 1174–1187, 2023.

[9] A. H. Soomro, S. Talani, T. Soomro, F. A. Khushk, and A. A. Bhatti, "Prototype Development for Solar Energy Tracking Based on Arduino in QUEST Campus Larkana," *Sir Syed University Research Journal of Engineering & Technology*, vol. 13, no. 2, 2024. https://doi.org/10.33317/ssurj.579.

[10] F. F. Ahmad, C. Ghenai, and M. Bettayeb, "Maximum power point tracking and photovoltaic energy harvesting for Internet of Things: A comprehensive review," *Sustainable Energy Technologies and Assessments*, vol. 47, p. 101430, 2021. https://doi.org/10.1016/j.seta.2021.101430.

[11] A. M. Hayajneh, F. Alasali, A. Salama, and W. Holderbaum, "Intelligent Solar Forecasts: Modern Machine Learning Models and TinyML Role for Improved Solar Energy Yield Predictions," *IEEE Access*, vol. 12, pp. 10846–10864, 2024. https://doi.org/10.1109/access.2024.3354703.

[12] M. Seif, M. A. Warsame, and W. Kasima, "Wind energy: energy sustainability perspective." Department of Technical and Vocational Education (TVE), Islamic University …, 2013.

[13] G. M. Joselin Herbert, S. Iniyan, E. Sreevalsan, and S. Rajapandian, "A review of wind energy technologies," *Renewable and Sustainable Energy Reviews*, vol. 11, no. 6, pp. 1117–1145, 2007. https://doi.org/10.1016/j.rser.2005.08.004.

[14] J. de A. Y. Lucena, "Recent advances and technology trends of wind turbines," *Recent Advances in Renewable Energy Technologies*. Elsevier, pp. 177–210, 2021. doi: 10.1016/b978-0-323-91093-4.00009-3.

[15] J. Charles Rajesh Kumar, D. Vinod Kumar, D. Baskar, B. Mary Arunsi, R. Jenova, and M. A. Majid, "Offshore wind energy status, challenges, opportunities, environmental impacts, occupational health, and safety management in India," *Energy & Environment*, vol. 32, no. 4, pp. 565–603, 2020. https://doi.org/10.1177/0958305x20946483.

[16] P. Patil, N. Kardekar, R. Pawar, and D. Kamble, "Wind Energy-Based Hot Water Production: Computational Approaches," *International Journal of Early Childhood Special Education*, vol. 14, no. 06, Nov. 2022. https://doi.org/10.48047/INTJECSE/V14I6.417.

[17] S. Nishar, "Enhancing Efficiency and Supply Chain Management in Wind Tower Fabrication through Cellular Manufacturing," *Journal of Logistics Management*, vol. 11, no. 1, pp. 1–5, 2023.

[18] S. Dehghani, U. Daon, and M. Shariatzadeh, "Propelling a Paradigm Shift: Revolutionizing Energy Yield in Solar Photovoltaic Systems," *2023 Middle East and North Africa Solar Conference (MENA-SC)*. IEEE, 2023. doi: 10.1109/mena-sc54044.2023.10374472.

[19] M. E. El-hawary, "The Smart Grid—State-of-the-art and Future Trends," *Electric Power Components and Systems*, vol. 42, no. 3–4, pp. 239–250, 2014. https://doi.org/10.1080/15325008.2013.868558.

[20] I. Alotaibi, M. A. Abido, M. Khalid, and A. V Savkin, "A Comprehensive Review of Recent Advances in Smart Grids: A Sustainable Future with Renewable Energy Resources," *Energies*, vol. 13, no. 23, p. 6269, 2020. https://doi.org/10.3390/en13236269.

[21] G. Dileep, "A survey on smart grid technologies and applications," *Renewable Energy*, vol. 146, pp. 2589–2625, 2020. https://doi.org/10.1016/j.renene.2019.08.092.

[22] O. Majeed Butt, M. Zulqarnain, and T. Majeed Butt, "Recent advancement in smart grid technology: Future prospects in the electrical power network," *Ain Shams Engineering Journal*, vol. 12, no. 1, pp. 687–695, 2021. https://doi.org/10.1016/j.asej.2020.05.004.

[23] V. Gupta, "Non-destructive testing of some Higher Himalayan Rocks in the Satluj Valley," *Bulletin of Engineering Geology and the Environment*, vol. 68, no. 3, pp. 409–416, Aug. 2009. https://doi.org/10.1007/s10064-009-0211-4.

[24] C. Bekara, "Security Issues and Challenges for the IoT-based Smart Grid," *Procedia Computer Science*, vol. 34, pp. 532–537, 2014. https://doi.org/10.1016/j.procs.2014.07.064.

[25] R. C. Qiu and P. Antonik, *Smart Grid using Big Data Analytics*. Wiley, 2017. doi: 10.1002/9781118716779.

[26] A. N. Abdalla *et al.*, "Integration of energy storage system and renewable energy sources based on artificial intelligence: An overview," *Journal of Energy Storage*, vol. 40, p. 102811, 2021. https://doi.org/10.1016/j.est.2021.102811.

[27] M. Child and C. Breyer, "The Role of Energy Storage Solutions in a 100% Renewable Finnish Energy System," *Energy Procedia*, vol. 99, pp. 25–34, 2016. https://doi.org/10.1016/j.egypro.2016.10.094.

[28] M. Hjallar, E. Viðisdóttir, and O. Gudmestad, "Transitioning towards renewable energy and sustainable storage solutions at remote communities in the Arctic, Case study of Flatey, Iceland," *IOP Conference Series: Materials Science and Engineering*, vol. 1294, p. 12035, Dec. 2023. https://doi.org/10.1088/1757-899X/1294/1/012035.

[29] J. E. Heath, R. P. Jensen, S. D. Weller, J. Hardwick, J. D. Roberts, and L. Johanning, "Applicability of geotechnical approaches and constitutive models for foundation analysis of marine renewable energy arrays," *Renewable and Sustainable Energy Reviews*, vol. 72, pp. 191–204, May 2017. https://doi.org/10.1016/j.rser.2017.01.037.

[30] T. S. Yun, J.-S. Lee, S.-C. Lee, Y. J. Kim, and H.-K. Yoon, "Geotechnical issues related to renewable energy," *KSCE J Civ Eng*, vol. 15, no. 4, pp. 635–642, Apr. 2011. https://doi.org/10.1007/s12205-011-0004-8.

[31] M. Coughlan, M. Long, and P. Doherty, "Geological and geotechnical constraints in the Irish Sea for offshore renewable energy," *Journal of Maps*, vol. 16, no. 2, pp. 420–431, Dec. 2020. https://doi.org/10.1080/17445647.2020.1758811.

[32] H. Kashani, B. Ashuri, S. M. Shahandashti, and J. Lu, "Investment Valuation Model for Renewable Energy Systems in Buildings," *Journal of Construction Engineering and Management*, vol. 141, no. 2, p. 04014074, Feb. 2015. https://doi.org/10.1061/(ASCE)CO.1943-7862.0000932.

[33] R. J. Fragaszy *et al.*, "Sustainable development and energy geotechnology — Potential roles for geotechnical engineering," *KSCE J Civ Eng*, vol. 15, no. 4, pp. 611–621, Apr. 2011. https://doi.org/10.1007/s12205-011-0102-7.

[34] K. Li, H. Bian, C. Liu, D. Zhang, and Y. Yang, "Comparison of geothermal with solar and wind power generation systems," *Renewable and Sustainable Energy Reviews*, vol. 42, pp. 1464–1474, Feb. 2015. https://doi.org/10.1016/j.rser.2014.10.049.

[35] N. Yabuuchi, K. Kubota, M. Dahbi, and S. Komaba, "Research Development on Sodium-Ion Batteries," *Chem. Rev.*, vol. 114, no. 23, pp. 11636–11682, Dec. 2014. https://doi.org/10.1021/cr500192f.

Theoretical modeling to analyze the energy and exergy efficiencies of double air-pass solar tunnel dryer with recycled organic waste material

Dagim Kebede GARI[1,a,*], A. Venkata RAMAYYA[1,b], L. Syam SUNDAR[2,c]

[1]Faculty of Mechanical Engineering, Jimma Institute of Technology, Jimma University, Jimma, Ethiopia

[2]Department of Mechanical Engineering, Collge of Engineering, Prince Mohammad Bin Fahd University, Al-Khobar, Saudi Arabia

[a]k.dagim82@gmail.com, [b]dra.venkata@ju.edu.et, [c]sslingala@gmail.com

Keywords: Thermal Insulation Thickness; Recycled Organic Waste; Energy and Exergy; Solar Tunnel Dryer

Abstract. In this study, the effect of recycled organic waste thermal insulation materials and insulation thickness on a double air pass solar tunnel dryer performance for the charcoal briquette drying process was investigated based on the heat transfer model, that was computed with Python code. The result reveals that, an average temperature of bagasse insulated a double air pass solar tunnel dryer was higher than wood, paper, corn cop, and sawdust insulation by 17.5%, 12%, 8.3%, and 6%, and higher than rock wool, mineral wool, and glass fiber by 3.01%, 4%, and 25.41%, respectively. The energy and exergy efficiency were considerably enhanced with bagasse insulation, comparatively. Thus, the assimilation of recycled organic waste thermal insulation materials enhances the evaporation rate besides reducing energy and exergy losses to the ambient while concomitantly minimizing costs and environmental impacts.

Introduction

In recent years solar energy-driven technologies have been widely used in power generation for industrial processes, household appliances, and commercial centers and in thermal energy extraction for air, water, and process heating, drying, and cooling applications. Conventionally, solar drying and air heating systems are the most economically viable ways of solar energy conversion [1]. The significant problem encountered in solar thermal technologies is energy and exergy loss to the ambient through the bottom and edge wall [2]. This phenomenon indicates that the overall thermal and exergetic efficiency of solar thermal systems is considerably affected by heat loss [3]. Solar air heaters were analyzed in terms of exergy loss to the environment due to absorber plate temperature and the result showed that at a higher absorber plate temperature, there was an increment in the exergy and thermal energy loss [4]. In solar air heaters with double glazing, the exergy and heat loss increased as the heater surface area and absorber plate temperature increased [5].

The onion drying system was analyzed using a mathematical model in terms of exergy loss and the result indicates that the rate of exergy loss reached 28.6% of the incoming exergy in the drying chamber at a velocity and temperature of 2 m/s and 80 °C [6]. The exergy destruction of a hybrid-solar drying system for the rosemary drying process was investigated experimentally and the authors documented that the rate of exergy loss varied from 0.009 to 0.028 KW with a variation in air velocity from 1 to 2 m/s and temperature from 40 to 70 °C [7]. A solar tunnel drying system was investigated experimentally for the drying presses of orange peels and the result revealed that the average exergy loss reached 57.99% in the dryer at higher solar radiation [8]. This indicates that exergy and thermal loss increased with the increment of solar intensity. Recently, different research has been conducted to address and improve the effect of thermal insulation materials on the heat resistance capacity in solar thermal and building technologies [9]. The performance of building elements with and without insulation material was investigated experimentally and the authors documented that using insulation materials reduces the indoor air temperature by 8 °C [10].

Besides thermos-physical properties, insulation thickness was one of the major factors that altered the performance of solar thermal systems. The effect of insulation thickness on collector efficiency of solar flat plate collectors was analyzed by ANSYS FLUENT and the result showed that the collector efficiency for 50 mm insulation thickness was higher than 35 and 25 mm by 10% and 18%, respectively [11]. The energy-saving capacity of salinity gradient solar ponds with different insulation thicknesses was analyzed using mathematical modeling and the authors reported that the energy-saving enhanced by 36.7% to 55.2% for the optimum thickness of 62 to 122 mm, respectively compared to those without insulation [12]. The energetic performance of solar water heating systems using flat plate collectors with different insulation thicknesses was analyzed by developing the thermal model and the result indicates that the energy efficiency was enhanced by 3.66% as the insulation thickness increased from 20 to 40 mm [13]. The effect of insulation thickness on energy-saving capacity and cost of the building was analyzed and the authors reported that the selection of proper insulation thickness enhances the average annual energy saving by 33.5% and the total energy saving cost by 4.4 to 53.5 \$/(m^{2}year) through the reduction energy loss factor [14].

The literature mentioned above indicates the energy efficiency, exergy efficiency, and moisture removal rate are reduced in the solar drying technologies due to the heat and exergy loss to the environment through the thermal insulation thickness [15]. Recently, recycled organic waste thermal insulation materials have promising alternatives to organic (synthetic) thermal insulation materials from the perspective of environment and sustainable development [16]. Additionally, recycled organic waste materials are widely available and economically beneficial over organic synthetic thermal insulation materials. Accordingly, this work intends to study the effect of recycled organic waste thermal insulation materials and thickness on the energy, exergy, and evaporation rate performance of a double air pass solar tunnel drying system for the drying of charcoal briquette, to reduce thermal energy loss to the ambient and environmental impacts of thermal insulation materials during the production process. Integrating recycled organic waste insulation material could reduce the material and manufacturing costs. Heat and exergy loss reduction performance and drying rate enhancement by developing heat transfer models, and the solution was computed by using Python code.

Material and methods
Dryer dimensions
This study will be investigated based on the hourly ambient parameters data of the Jimma zone, Oromia region, Ethiopia (7°40'0" N, 36°50'0"E). The hourly ambient conditions vary from 06:00 a.m. to 17:00 for 11 sunshine hours of the day. The effect of recycled organic waste thermal insulation materials and thickness were computed by developing mass and heat transfer models using the principle of exergy, mass, and energy balance. Throughout the heating and drying process, the mass flow rate of the air stream was considered constant at 0.018 and 0.035 Kg/s, at a constant velocity of 2 m/s, respectively. The influence of recycled organic waste thermal insulation materials on the energy outcomes, exergy extraction, and evaporation rate of a double air pass solar tunnel drying system (DAPSTDS) was studied by using five different materials, namely bagasse, paper, corn cop, wood, and sawdust as well as their respective thermophysical properties are listed in **Table 1**. The variation of thermal efficiency, exergy efficiency, and evaporation rate from the surface of a drying product in a DAPSTDS with the change of insulation was investigated by considering three different dimensions 0.05, 0.25, and 0.5 m.

Table 1: The recycled organic waste, and the inorganic insulation material.

	Material	Density (Kg/m³)	Specific heat (J/Kgk)	Thermal conductivity (W/Km)	Emissivity
Recycled organic waste	Bagasse	120	460	0.041	0.88
	Paper	800	1340	0.093	0.93
	Corn cop	282.38	1500	0.16	0.9
	Wood	920	1670	0.4	0.9
	Saw-Dust	415	900	0.062	0.75
Insulating material	Glass Fiber	1857	800	0.36	0.75
	Mineral Wool	200	850	0.046	0.94
	Rock Wool	200	1030	0.043	0.05

The schematic diagram of a double air pass solar tunnel drying system with the drying product and the main components is displayed in **Figure 1.** The drying system is divided into two main chambers. The heating chamber where the air stream enters through the inlet is heated by an absorber plate and the fraction of solar radiation scattered inside the chamber then flows to the next chamber. The drying chamber is where the air stream gains thermal energy from the absorber plate through convection for the second time and simultaneously, the moisture is removed from the surface of the charcoal briquette, and finally, the moist air is discharged to the ambient through, the outlet. The overall dimensions of the drying system are listed in **Table 2**.

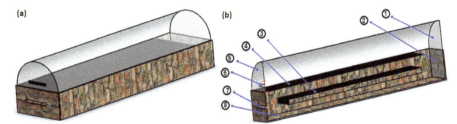

Figure 1: Schematic diagram of a double air pass solar tunnel dryer: (1) heating chamber, (2) drying chamber, (3) charcoal briquette (drying product), (4) absorber plate, (5) cover, (6) inlet, (7) outlet, and (8) floor.

Table 2: The Dimension Specification of DAPSTDS.

Parameters	Dimensions						
	Wet Base				Dry Base		
Mass of briquette	30 Kg				30 Kg		
Initial moisture content	0.5%				100%		
Final moisture content	0.1%				11.111%		
Moisture to be removed	0.4%				88.889		
Mass to be removed	12 Kg				26.667 Kg		
Parameters	Components				Chambers		
	Cover	Plate	Briquette	Floor	Heating	Drying
Length, (m)	2.4	2.25	1.6	2.4	2.4	2.4
Width, (m)	1.885	1.2	0.75	1.2	1.2	1.2
Thickness, (m)	0.0002	0.0004	0.125	0.2	0.6	0.6
Area, (m²)	4.5	2.574	1.2	2.86	0.5652	0.72

Theoretical modeling
Modelling of the dryer
The thermal energy transfer modeling was developed by applying energy balance on the rate of thermal energy exchange between the dryer components (absorber plate, cover, and floor), Charcoal Briquette, and the air stream in both chambers, based on the schematic diagram shown in **Figure 2**. The analysis was performed using the COMSOL Multiphysics 5.2a software.

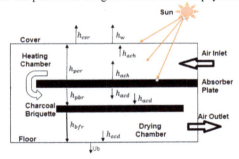

Figure 2: Schematic diagram of heat energy interaction inside the drying system.

Thermal efficiency
Energy is the ability to do work and exergy is the capacity for work extraction, which can be determined from the initial and final temperature with the corresponding time [17].
The rate of input energy ($\dot{Q}_{in}$) is computed as:
$$\dot{Q}_{in} = [A_{pr}I] \tag{1}$$
The amount of useful energy outcome ($\dot{Q}_u$) is evaluated as:
$$\dot{Q}_u = [\dot{m}_a C_{pa}(T_{ao} - T_{ai})] \tag{2}$$
The amount of heat energy loss ($\dot{Q}_{loss}$) is calculated as:
$$\dot{Q}_{loss} = [A_f U_b(T_{ad} - T_{ai})] \tag{3}$$
The thermal efficiency (η_{th}) is determined as:

$$\eta_{th} = \left[\frac{\dot{Q}_u}{\dot{Q}_{in}}\right] \quad (4)$$

Exergy efficiency

The rate of input exergy ($\dot{E}x_{in}$) is computed as,

$$\dot{E}x_{in} = \left[A_{pr}I\left\{1 - \left(\left(\frac{4}{3}\right)*\left(\frac{T_{am}}{T_{sun}}\right)\right) + \left(\left(\frac{1}{3}\right)*\left(\frac{T_{am}}{T_{sun}}\right)^4\right)\right\}\right] \quad (5)$$

The amount of exergy outcome ($\dot{E}x_u$) at a given time is computed as:

$$\dot{E}x_u = \left[\dot{m}_a C_{pa}\left\{(T_{ao} - T_{am}) - \left(T_{am}Log\left(\frac{T_{am}}{T_{sun}}\right)\right)\right\}\right] \quad (6)$$

The exergy efficiency (η_{ex}) is expressed by:

$$\eta_{ex} = \left[\frac{\dot{E}x_u}{\dot{E}x_{in}}\right] \quad (7)$$

Results and discussion

The energy outcomes, exergy extraction, and evaporation rate of the drying systems mainly depend on the ambient conditions. Thus, the mathematical model simulation was computed based on the hourly ambient parameters data of Jimma zone, Ethiopia available on 16th, March, 2023, the maximum solar intensity, temperature, and wind speed were 1051 W/m^2, 28 °C, and 1.61 m/s, respectively.

Air temperature at outlet

The effect of recycled organic waste thermal insulation materials on the temperature was analyzed by applying equations. The hourly variation in the temperature of the drying system under recycled organic waste thermal insulation materials was illustrated in **Figures 3(a-b)**. The average temperature achieved from the analysis in the drying and heating chamber from 09.00 AM to 17:00 PM was 52.2 (40 °C), 48 (38.4 °C), 46.5 (37.9 °C), 44.1 (37.3 °C), and 41.1°C (36.3°C) for bagasse, sawdust, corn cop, paper, and wood, respectively. The lower temperature difference was found in the morning which increased with an increment of solar insolation, then it reached the pick-point in the afternoon at higher insolation, and after that, it declined concerning the amount of solar irradiance.

The average temperature in the dryer with bagasse insulation was higher than wood, paper, corn cop, and sawdust insulation by 17.5%, 12%, 8.3%, and 6%, respectively. This shows that integrating bagasse insulation with a double-air pass solar tunnel drying system could play a role in enhancing the temperature in the heating and drying chamber during the sunshine period than other recycled organic waste insulation materials.

The comparison results of recycled organic waste material (bagasse) with inorganic materials (glass fiber, mineral wool, and rock wool) are illustrated in **Figures 3(c-d)** for similar insulation thickness. The result rivals that the average temperature from 09.00 am to 17:00 was 52.2 (40 °C), 50.63 (39.4 °C), 50.2 (39.3 °C), and 39 (35.6 °C), for bagasse, rock wool, mineral wool, and glass fiber in descending order, respectively. This indicates that the average temperature in the drying system with bagasse insulation was found higher compared to rock wool, mineral wool, and glass fiber by 3.01%, 4%, and 25.41%, respectively. This occurrence was due to bagasse having a better combination of thermos-physical properties compared to rock wool, mineral wool, and glass fiber [18]. Therefore, integrating customized recycled organic waste thermal insulation materials with a solar drying system enhances the temperature.

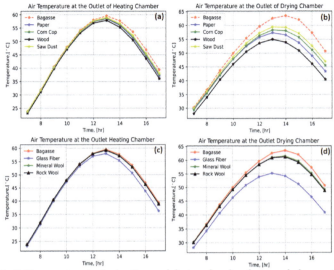

Figure 3: *(a-b) Temperature in the heating and drying chamber for recycled organic waste, and (c-d) Inorganic (synthetic) thermal insulation materials.*

Energy efficiency

The hourly variation of thermal efficiency in the drying system insulated with recycled organic waste thermal insulation material was displayed in **Figure 4(a)**, and the drying system insulated with inorganic thermal insulation materials was presented in **Figure 4(b)**. The result shows that the average thermal efficiency from 07:00 a.m. to 17:00 was 40.8%, 36.93%, 35.24%, 33.93%, and 30.24%for bagasse, sawdust, corn cop, paper, and wood, respectively, and for mineral wool, rock wool, and glass fiber was 39.1%, 38.76%, and 30.65%, respectively. The minimum thermal efficiency was observed at noon, where the solar insolation and the heat loss reached maximum. This phenomenon indicates that the change in thermal efficiency is due to the variation in heat loss caused by solar radiation variation with time. The result indicates that higher thermal efficiency was achieved in a drying system with bagasse than other recycled organic waste and inorganic thermal insulation materials. Because comparatively lower heat loss and higher temperature differences during the daytime were achieved in bagasse-insulated drying systems [19].

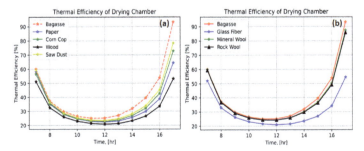

373

Figure 4: *Thermal efficiency of the drying system with (a) recycled organic waste and (b) standard thermal insulation materials.*

Exergy efficiency

The exergetic performance of the drying system integrated with recycled organic waste thermal insulation materials and for comparison with commonly used thermal insulation materials. **Figure 5(a)** and **Figure 5(b)** shows the hourly variation of the exergetic efficiency of the drying system under recycled organic waste thermal insulation material and inorganic thermal insulation materials. The average exergy efficiency from 07:00 a.m. and 17:00 was 45%, 36.76%, 33.71%, 31.35%, and 25.47% for bagasse, sawdust, corn cop, paper, and wood, respectively.

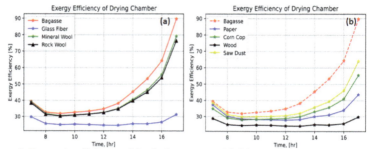

Figure 5: *Exergetic efficiency of the drying system with (a) recycled organic waste and (b) standard thermal insulation materials.*

The average exergy efficiency of mineral wool, rock wool, and glass fiber was 41.17%, 40.53%, and 26.08%, respectively. The useful work extraction efficiency fluctuation was considerably influenced by the input ambient condition variation during the drying period.

Comparatively, at a constant insulation thickness and emissivity of charcoal briquette, the exergetic efficiency of the dryer was higher with bagasse insulation than other recycled organic waste and inorganic thermal insulation materials. Thus, a double air pass solar tunnel drying system with bagasse insulation has a better useful work extraction capacity than other recycled organic waste standard insulation materials, because of their lower heat loss and higher temperature difference.

Energy and exergy efficiency on insulation thickness

Figures 6(a)-(b) shows the effect of insulation thickness on a double air pass solar tunnel drying system thermal efficiency and exergy efficiency investigated. The hourly average thermal efficiency was 38.5%, 40.8%, and 40.31%, and the hourly average exergy efficiency was 40.5%, 45%, and 43.88% for insulation thickness of 0.05, 0.25, and 0.5 m, respectively. The thermal and exergy efficiency enhanced until the thickness reached 0.25m due to the increment of temperature, then reduced as the thickness increased further because temperature also decreased. Thus, the maximum thermal and exergy efficiency was found when the thickness was 0.25 m.

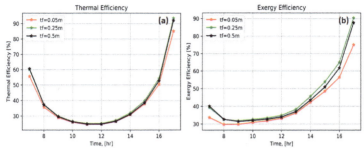

Figure 6: (a) Thermal efficiency and (b) exergy efficiency of drying system for different insulation thickness.

Conclusion

This study presents the numerical analysis of the effect of recycled organic waste thermal insulation materials and thickness on the energetic, exergetic, and evaporation rate performance of a double air pass solar tunnel drying system in terms of heat loss and environmental impact reduction for the drying process of biomass charcoal briquette. A comparative analysis of the recycled organic waste materials with commercially available thermal insulation materials and between different insulation thicknesses was investigated in terms of heat and exergy loss reduction performance and drying rate enhancement by developing mass and heat transfer models, and the solution was computed by establishing the Python code.

Comparatively, a higher temperature, thermal efficiency, and exergy efficiency were obtained in the drying system insulated with bagasse material than other recycled organic waste materials, and the most commonly used thermal insulation. Comparatively, a higher temperature, thermal efficiency, exergy efficiency, and moisture evaporation rate of the charcoal briquette were achieved at an optimum thermal insulation thickness of 0.25 m.

References

[1] P.G. Kumar, K. Balaji, D. Sakthivadivel, V.S. Vigneswaran, M. Meikandan et al. Effect of using low-cost thermal insulation in a solar air heating system with a shot blasted V-corrugated absorber plate, Thermal Science and Engineering Progress 14 (2019) 100405. https://doi.org/10.1016/j.tsep.2019.100403

[2] B.T.H. Huy, P. Nallagownden, R. Kannan, V.N. Dieu, Energetic optimization of solar water heating system with flat plate collector using search group algorithm, J. Advanced Research in Fluid Mechanics and Thermal Sci. 61 (2019) 306-322.

[3] Q. Wu, B. Yan, Y. Gao, X. Meng, Wall adaptability of the phase-change material layer by numerical simulation, Case Studies in Thermal Eng. 41 (2023) 102622. https://doi.org/10.1016/j.csite.2022.102622

[4] M.K. Sahu, R.K. Prasad, Exergy-based performance evaluation of solar air heater with arc shaped wire roughened absorber plate, Renewable Energy 96 (2016) 233–243. https://doi.org/10.1016/j.renene.2016.04.083

[5] M. Hedayatizadeh, F. Sarhaddi, A. Safavinejad, F. Ranjbar, H. Chaji, Exergy loss-based efficiency optimization of a double-pass/glazed V corrugated plate solar air heater, Energy 94 (2016) 799–810. https://doi.org/10.1016/j.energy.2015.11.046

[6] M. Castro, C. Román, M. Echegaray, G. Mazza, R. Rodriguez, Exergy Analyses of Onion Drying by Convection: Influence of Dryer Parameters on Performance, Entropy 20 (2018) 310. https://doi.org/10.3390/e20050310

[7] H. Karami, M. Kaveh, I. Golpour, E. Khalife, R. Rusinek, B. Dobrzanski et al. Thermodynamic evaluation of the forced convective hybrid-solar dryer during drying process of rosemary (Rosmarinus officinalis L.) leaves, Energies 14 (2021) 5835. https://doi.org/10.3390/en14185835

[8] A.K. Karthikeyan, R. Natarajan, Exergy analysis and mathematical modelling of orange peels drying in a mixed mode solar tunnel dryer and under the open sun: a study on performance enhancement, Int. J. Exergy 24 (2017) 235-253. https://doi.org/10.1504/IJEX.2017.087695

[9] G. Deshmukh, P. Birwal, R. Datir, S. Patel, Thermal insulation materials: A tool for energy conservation, J. Food Processing and Technology 8 (2017) 670. https://doi.org/10.4172/2157-7110.1000670

[10] G.K. Abdulsada, T.W. Mohammed-Salih, The impact of efficient insulation on thermal performance of building elements in hot arid region, Renewable Energy and Environmental Sustainability 7 (2021) 2. https://doi.org/10.1051/rees/2021050

[11] M. Mastanaiah, K.R. Hemachandra, Effect of geometric and operating parameters on solar flat plate collector performance-A CFD approach, Int. J. Research in Mechanical Engineering 4 (2016) 6-14.

[12] H. Beiki, E. Soukhtanlou, Determination of optimum insulation thickness for salinity gradient solar pond's bottom wall under different climate conditions, S.N. Applied Science 2 (2020) 1284. https://doi.org/10.1007/s42452-020-3078-4

[13] T.H. Bao-Huy, P. Nallagownden, R. Kannan, V.N. Dieu, Energetic optimization of solar water heating system with flat plate collector using search group algorithm, J. Advanced Research in Fluid Mechanics and Thermal Sci. 61 (2019) 306-322.

[14] M.M. Mohamed, Optimal thermal insulation thickness in isolated air-conditioned buildings and economic analysis. J. Electronics Cooling and Thermal Control 9 (2020) 23-45. https://doi.org/10.4236/jectc.2020.92002

[15] N. Sooriyalakshmi, H.H. Jane, Thermal conductivity of insulation material: An overview, J. Architecture and Civil Engineering 6 (2021) 59-65. https://doi.org/10.9734/bpi/tier/v6/2184A

[16] B.A. Alhabeeb, H.N. Mohammed, S.A. Alhabeeb, Thermal insulators based on waste materials. IOP Conf. Series: Materials Science and Eng. 1067 (2021) 012097. https://doi.org/10.1088/1757-899X/1067/1/012097

[17] J.A. Duffie, W.A. Beckman, Solar engineering of thermal processes, John Wiley & Sons, 1991.

[18] J.H. Watmuff, W.W.S. Charters, D. Proctor, Solar and wind induced external coefficients for solar collectors, Cooperation Mediterraneenne pour l'Energie Solaire, Revue Internationale d'Heliotechnique 2 (1977) 56.

[19] A. Sotoodeh, A.S.A. Hamid, A. Ibrahim, K. Sopian, Experimental investigation on thermal efficiency of indirect forced convection solar tunnel dryer, 7th Int. Conference on Innovation in Science and Technology. 23-25 October, 2020; Amsterdam, Netherlands.

Renewable energy in pavement engineering and its integration with sustainable materials: A review paper

Mohammad Ali KHASAWNEH[1,a], Danish AHMED[1,b*], Fawzyah ALKHAMMAS[1c], Zainab ALMSHAR[1d], Maryam HUSSSAIN[1e], Dalya ALSHALI[1f], Maryam ALKHURAIM[1g], Rana ALDAWOOD[1h], Hidaia ALZAYER[1i], Nayeemuddin MOHAMMED[2,j]

[1]Department of Civil Engineering, Prince Mohammad Bin Fahd University, Al Khobar, Kingdom of Saudi Arabia

[a]mkhasawneh@pmu.edu.sa, [b]dahmed@pmu.edu.sa, [c]202101603@pmu.edu.sa, [d]202001834@pmu.edu.sa, [e]202001160@pmu.edu.sa, [f]202101433@pmu.edu.sa, [g]202000692@pmu.edu.sa, [h]202101700@pmu.edu.sa, [i]202002635@pmu.edu.sa, [j]mnayeemuddin@pmu.edu.sa

Keywords: Renewable Energy, Pavement Engineering, Electromagnetic, Piezoelectric, Thermoelectric, Solar Panels

Abstract. Pavement engineering that incorporates renewable energy sources and uses sustainable materials has emerged as a promising path toward achieving an environmentally friendly and sustainable transportation infrastructure. An inclusive review of renewable energy applications in pavement engineering and its integration with sustainable materials is presented in this paper. The article examines various renewable energy technologies and their potential integration into pavement construction, maintenance, and operation. The findings of this review contribute to advancing sustainable practices in the transportation infrastructure sector and provide valuable insights for researchers, policymakers, and practitioners.

Introduction
Eco-friendly and sustainable transportation infrastructure can be achieved through the incorporation of sustainable materials and the use of renewable energy sources. In pavement engineering, renewable energy refers to the incorporation of energy generation technologies that utilize renewable sources of energy to power various aspects of pavement systems. By integrating renewable energy solutions, pavement engineering can contribute to sustainable and environmentally friendly transportation infrastructure. The applications of renewable energy in pavement engineering include Solar Photovoltaic (PV), Piezoelectric Energy Harvesting, thermoelectric energy harvesting, Kinetic Energy Harvesting and Geothermal Systems.

Whereas, sustainable materials refer to those that are environmentally friendly and have a reduced impact on natural resources throughout their life cycle. These materials are designed to minimize energy consumption, emissions, and waste generation during their production, use, and disposal. Sustainable materials used in pavement engineering include recycled asphalt, porous asphalt, warm mix asphalt etc. The use of sustainable materials in pavement engineering promotes resource conservation, reduces environmental impacts, and contributes to the development of more sustainable transportation infrastructure.

The combination of renewable energy in pavement engineering, along with the integration of sustainable materials, holds great importance in establishing infrastructure systems that are sustainable, resilient, cost-effective, and environmentally friendly. The objective of this study is to provide a detailed investigation of the current state of renewable energy implementation in pavement engineering, as well as its integration with sustainable materials based on existing literature. The paper explores various renewable energy technologies and their potential for integration into pavement construction, maintenance, and operation processes.

Sustainable Materials in Pavement Engineering
Sustainable materials in pavement engineering refer to materials chosen and employed in the construction, maintenance, and rehabilitation of pavements, with the primary objective of minimizing environmental impact, preserving resources, and fostering long-term sustainability.

Yaro et al [1] provides a comprehensive exploration of the utilization of recycled waste materials and technologies in asphalt pavements, with a focus on environmental sustainability and low-carbon roads. The findings emphasize the significant environmental and economic benefits of utilizing recycled materials, including reduced reliance on virgin materials, energy savings, and lower carbon emissions. This study also illustrate that the incorporation of such materials improves the performance characteristics of asphalt pavements.

A study [2] was conducted to evaluate the performance of Asphalt Concrete using 60/70 pen asphalt and a modified binder with resin in hot mix asphalt for road pavement. The research explores the use of renewable resources as alternatives to petroleum-derived materials in bio asphalt production Results revealed that all mixtures showed similar values for optimum bitumen content, but higher percentages of resin resulted in improved stability of the asphalt mixtures.

Praticò et al [3] presents a comprehensive life cycle assessment (LCA) of pavement technologies, specifically hot mix asphalt (HMA) and warm mix asphalt (WMA), with a focus on incorporating recycled materials such as reclaimed asphalt pavements, crumb rubber, and waste plastics. The findings demonstrate the benefits of utilizing WMA and recycled materials in reducing energy consumption and environmental effects by minimizing the use of virgin resources. It also offer cleaner production processes and significant environmental and technical benefits, including reduced energy consumption and greenhouse gas emissions, improved compaction, longer paving periods, and enhanced worker safety.

Renewable Energy Technologies in Pavement Engineering
In pavement engineering, the adoption of renewable energy technologies enhances sustainability and boosts the efficiency of transportation infrastructure by integrating sustainable energy systems into roads. By harnessing renewable energy, transportation infrastructure can become more sustainable and efficient, contributing to a greener and more effective transportation network.

Research [4] focuses on concentrated photovoltaic panels (CPPs) for pavement applications as a clean and renewable energy source in transportation. The study presents a comprehensive analysis of the structural optimization and performance testing of CPPs, demonstrating their feasibility, durability, and economic viability. Mechanical and electrical performance tests reveal the panel's strength, wear resistance, light concentration performance, and power generation capabilities. The CPP system showcases significant economic benefits, with a high return on investment and cost recovery period. Furthermore, it contributes to environmental sustainability by reducing carbon emissions.

The article by Zhou et al. [5] explores the feasibility and performance of a pavement-solar energy system through experimental analysis. The study aims to integrate solar energy collection technology into pavement systems to generate electricity and enhance environmental sustainability (Fig. 1). The experimental results demonstrate that the pavement-solar energy system effectively produces electricity from solar radiation, indicating its potential as a renewable energy source. The study emphasizes the importance of optimizing system design and orientation to improve energy harvesting.

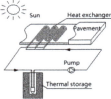

Figure 1 Harnessing solar energy through pavement infrastructure [5]

The study performed by Dunican [6] emphasizes the importance of diversifying solar power generation options in the construction sector to achieve a greener and more sustainable future. It argues that while rooftop solar solutions have become popular, alternative technologies such as transparent solar windows, solar shingles, and solar canopies offer additional opportunities to maximize renewable energy in buildings. Diversification provides benefits such as increased solar density, optimized space utilization, compliance with building codes, and design flexibility. By integrating solar technology into building materials, passive surfaces like windows can be utilized for energy production without compromising functionality.

Ma et al. [7] investigates the use of road pavements as solar energy generators in smart and sustainable cities, focusing on three pavement modules: pavement-integrated photovoltaic (PIPV), pavement-integrated solar thermal (PIST), and pavement-integrated photovoltaic thermal (PIPVT) modules (Fig. 2). PIPVT module achieves slightly higher electricity yield but lower heat yield compared to PIPV and PIST modules, with an average energy efficiency of 37.31%. All modules reduce the maximum asphalt average temperature, with PIPVT having the most significant effect, decreasing it by an average of 10.57°C. Additionally, they contribute to mitigating the urban heat island (UHI) effect, with PIPVT and PIST modules exhibiting the most and least influence, respectively.

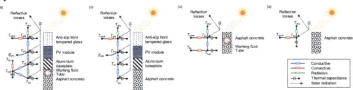

Figure 2 Thermal resistance networks of different solar energy harvesting pavements (a) PIPVT module; (b) PIPV module; (c) PIST module; (d) CP module [7]

Del Serrone et al. [8] explores the concept of utilizing photovoltaic (PV) road pavements to create low-carbon urban infrastructures. It emphasizes the importance of cool pavements in mitigating Urban Heat Islands (UHIs) and discusses the integration of PV panels into road surfaces to generate renewable energy. A case study in Rome exemplifies the potential of photovoltaic road infrastructures (Fig. 3), demonstrating their ability to generate electricity, maintain acceptable temperatures, and provide economic viability. Photovoltaic road pavements offer multiple benefits, including reduced energy consumption, enhanced microclimates, and reduced land use for solar installations.

Figure 3 San Pietro in Vincula Square: (a) top view; (b) site map. [8]

Sumorek & Buczaj [9] performed the study to addresses the technological gap between automotive engineering and road infrastructure development by focusing on energy generation from road pavement. It evaluates the feasibility of capturing solar energy and converting it into usable electricity on the road surface, as well as harvesting mechanical vibrations energy from passing vehicles using piezoelectric transducers. The study confirms the significant potential of solar energy conversion and vibrations energy harvesting, with experimental results supporting the feasibility of converting mechanical vibrations energy into electrical energy (Fig. 4).

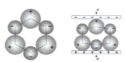

Figure 4 Mechanism of mechanical vibrations energy into electrical energy [9]

The article by Sun et al. [10] provides an in-depth exploration of green technologies for sustainable pavements, focusing on energy harvesting and permeable pavement systems. The article discussed the energy harvesting technologies including piezoelectric, solar, thermoelectric, and geothermal. Permeable pavement systems are highlighted for their ability to facilitate water infiltration, preventing urban flooding and reducing the urban heat island effect. Porous materials such as porous asphalt and concrete are examined, offering benefits like noise reduction and improved hydraulic conductivity.

Al-Qadami et al. [11] provide a systematic analysis of the existing literature on pavement geothermal energy harvesting technologies. Three main sectors are identified: piezoelectric transducer systems, thermoelectric generator systems (TEGs), and solar panel systems (Fig. 5). Piezoelectric transducer systems convert the mechanical stress from moving vehicles into electrical energy using embedded piezoelectric materials. Studies emphasize material selection, pavement design, and traffic characteristics' impact on energy generation, highlighting their potential in high-traffic areas. TEGs harness the temperature difference between the pavement surface and underlying layers to generate electricity. Research focuses on enhancing efficiency through material selection, module configuration, and optimization, indicating their integration into smart pavement systems. Solar panel systems effectively convert solar energy into electricity and can be embedded within or placed on pavement surfaces.

Figure 5 Pavement energy harvesting technologies [11]

Razeman et al. [12] investigated solar pavements as a means of efficiently harnessing solar energy, emphasizing sustainable conductive materials like stainless steel, copper, and aluminum. Numerical simulations are employed to optimize parameters such as pipe material, depth, arrangement, spacing, and flow rate. The results highlight serpentine copper pipe as the most efficient configuration, achieving a heat efficiency rate of 32.22% and an outlet temperature of 327.35K (54.21°C). The study underscores the benefits of solar pavement technology, reduced reliance on fossil fuels, emissions reduction, and reliable electricity generation.

Tahami et al. [13] proposed an innovative approach to harvest energy from asphalt pavements using thermoelectric technology. The findings indicate that the temperature difference in the pavement can produce sufficient electrical energy, especially in areas with intense sunlight exposure and high traffic density. The thermoelectric system demonstrated the ability to generate sustainable energy without compromising pavement performance. The authors discussed the potential applications of this technology in existing infrastructures for renewable energy generation and environmental sustainability. However, challenges such as cost-effectiveness, scaling up, and long-term durability need to be addressed for commercialization and implementation.

Gholikhani et al [14] highlights the potential of utilizing roadway pavements as a renewable energy resource through the electromagnetic speed bump energy harvester (ESE) prototype. This innovative approach aims to capture kinetic energy from passing vehicles while regulating vehicle speed. The ESE prototype absorbs the deflection caused by vehicles passing over a speed bump and converts it into rotational energy using a revolving shaft and embedded generator. Laboratory experiments demonstrate its capability to generate an average power output of up to 3.21 kW under realistic traffic loading conditions. Steel and aluminum are identified as optimal materials for the ESE's top plate due to their favorable properties.

De Fazio et al [15] explored the potential of energy harvesting from various sources on roadways, including mechanical load, solar radiation, heat, and air movement, for the development of self-sustainable smart roads. It examines different technologies such as electromagnetism, piezoelectric and triboelectric harvesters, photovoltaic modules, thermoelectric solutions, and wind turbines optimized for low-speed winds generated by vehicles. The findings emphasize the environmental benefits and potential applications of energy harvesting in the transportation sector, such as autonomous driving, real-time road condition communication, self-powered lighting systems, and security sensors.

The investigation by Saleh et al. [16] focuses on the Hydronic Asphalt Pavement (HAP) system. The HAP system utilizes a network of pipes embedded in the asphalt pavement to remove or reject heat from the pavement using a circulating fluid. The study evaluates the efficiency of the HAP system in reducing pavement surface temperature and enhancing sustainability. The results show that the HAP system successfully lowered the temperature at a depth of 2.5 cm below the surface by approximately 10°C. However, the surface temperature reduction was not significant.

Johnsson [17] study focuses on making the Coastal Highway Route E39 carbon neutral and incorporating energy output facilities. The study explores the use of renewable thermal energy to prevent road surfaces from becoming slippery, considering the warming of sidewalk surfaces and the use of solar energy to melt snow and ice. The study suggests that the proposed hydronic pavement system, when combined with BTES, can enhance winter road maintenance and safety, particularly in regions with milder winters like Scandinavia.

Charlesworth et al. [18] explores the combination of ground source heat (GSH) and pervious paving systems (PPS) to create a sustainable drainage and renewable energy solution. GSH, obtained from the ground, is a renewable energy source that can efficiently heat and cool buildings. Pervious paving systems, such as block pavers, porous asphalt, concrete, and resin, are capable of attenuating storm surges, reducing water quantity, and improving water quality. The integration of GSH collectors with PPS allows for harnessing temperature differences within the ground.

The review paper [19] provides an overview of energy harvesting from roadways and its potential in combating the urban heat island effect. It explores various technologies, including electromagnetic, piezoelectric, thermoelectric, and solar panels, highlighting piezoelectric and thermoelectric systems as promising options. The study emphasizes the importance of hybrid systems that combine multiple energy sources for consistent power supply. Energy harvesting from roadways can reduce greenhouse gas emissions, power roadside applications, and provide electricity to roadside houses, presenting significant opportunities for sustainable energy generation. Table 1 provides a summary of literature references related to the application of renewable energy in pavement engineering.

Table 1 Application of Renewable Energy in Pavement Engineering

S. No	Reference from Literature	Application of Renewable Energy in pavement Engineering	Type of Study
1	Hu et al. [4]	Concentrated (converge sunlight) photovoltaic solar Pavement, (Concentrated photovoltaic panel (CPP) structure for pavement)	Laboratory model test and finite element numerical simulation
2	Zhou et al. [5]	Pavement-solar energy system	Experimental study
3	Ma et al. [7]	Solar energy harvesting pavements	Mathematical modeling and simulation conducted for pavement-integrated photovoltaic (PIPV) module, pavement integrated solar thermal (PIST) module, and pavement-integrated photovoltaic thermal (PIPVT) module.
4	Del Serrone et al. [8]	Utilizing photovoltaic (PV) road pavements is analyzed from the thermal and economic viewpoints	A microclimate simulation of San Pietro in Vincula Square in Rome is conducted using ENVI-Met software.
5	Sumorek & Buczaj [9]	Solar energy and vibrations energy	A combination of theoretical analysis and experimental testing
6	Sun et al. [10]	Energy harvesting technologies (Piezoelectric, solar, thermoelectric, and geothermal technologies) and permeable pavement systems	Literature review
7	Al-Qadami et al. [11]	Harvesting geothermal energy from roadway pavement (piezoelectric transducer systems, thermoelectric generator systems, and solar panel systems)	A systematic review and bibliometric analysis were conducted
8	Randriantsoa et al. [19]	Hybrid energy harvesting systems (combining piezoelectricity and thermoelectricity for pavement applications)	Bibliographic research
9	Razeman et al. [12]	Thermal energy harvesting road pavement	Numerical Simulation using ANSYS Workbench 19.2 (Fluent) and Solidworks 2020 for conducting optimization study.
10	Tahami et al. [13]	Thermoelectric generator system that utilizes the thermal gradients between the pavement surface and the soil below the pavement and converts it to electricity	Prototype testing and finite element analyses were conducted
11	Gholikhani et al. [14]	Electromagnetic speed bump energy harvester (ESE)	laboratory prototype tests and finite element analysis was conducted using software ABAQUS.
12	De Fazio et al. [11]	The study explores different technologies, including electromagnetism, piezoelectric and triboelectric harvesters, photovoltaic modules, thermoelectric solutions, and wind turbines	Literature review

Conclusion

This review paper has explored the significance of renewable energy in pavement engineering and its incorporation with sustainable materials. The utilization of renewable energy sources in pavement engineering has the potential to bring about substantial environmental and economic benefits. By integrating sustainable materials into the design, construction, and maintenance of pavements, the sustainability of the infrastructure can further enhance.

The adoption of renewable energy technologies such as solar, Piezoelectric Energy Harvesting, thermoelectric energy harvesting, Kinetic Energy Harvesting and Geothermal Systems can significantly decrease greenhouse gas emissions associated with energy consumption in pavement construction and operation. This not only helps combat climate change but also reduces reliance on fossil fuels and promotes energy independence.

Moreover, the addition of sustainable materials in pavement engineering has been discussed as a complementary approach to renewable energy utilization. Sustainable materials, including recycled aggregates, reclaimed asphalt pavement, and bio-based binders, offer opportunities to reduce the environmental impact of pavement construction by saving natural resources and reducing waste generation. These materials can also contribute to the circular economy by promoting recycling materials within the pavement engineering.

In conclusion, the integration of renewable energy in pavement engineering, coupled with the use of sustainable materials, holds great promise for achieving a more sustainable and resilient transportation infrastructure. This combination can contribute to a more environmentally friendly and cost-effective pavements.

References

[1] N. S. A. Yaro et al., "A Comprehensive Overview of the Utilization of Recycled Waste Materials and Technologies in Asphalt Pavements: Towards Environmental and Sustainable Low-Carbon Roads," Processes, vol. 11, no. 7, p. 2095, Jul. 2023. https://doi.org/10.3390/pr11072095

[2] Djumari, M. A. D. Yami, M. F. Nasution, and A. Setyawan, "Design and Properties of Renewable Bioasphalt for Flexible Pavement," Procedia Eng., vol. 171, pp. 1413–1420, 2017. https://doi.org/10.1016/j.proeng.2017.01.458

[3] F. G. Praticò, M. Giunta, M. Mistretta, and T. M. Gulotta, "Energy and Environmental Life Cycle Assessment of Sustainable Pavement Materials and Technologies for Urban Roads," Sustainability, vol. 12, no. 2, p. 704, Jan. 2020. https://doi.org/10.3390/su12020704

[4] H. Hu, X. Zha, C. Niu, Z. Wang, and R. Lv, "Structural optimization and performance testing of concentrated photovoltaic panels for pavement," Appl. Energy, vol. 356, p. 122362, Feb. 2024. https://doi.org/10.1016/j.apenergy.2023.122362

[5] Z. Zhou, X. Wang, X. Zhang, G. Chen, J. Zuo, and S. Pullen, "Effectiveness of pavement-solar energy system – An experimental study," Appl. Energy, vol. 138, pp. 1–10, Jan. 2015. https://doi.org/10.1016/j.apenergy.2014.10.045

[6] G. Dunican, "A diversified approach to renewable energy in the construction sector," For Construction Pros, Aug. 11, 2023. https://www.forconstructionpros.com/construction-technology/article/22868964/ubiquitous-energy-a-diversified-approach-to-renewable-energy-in-the-construction-sector

[7] T. Ma, S. Li, W. Gu, S. Weng, J. Peng, and G. Xiao, "Solar energy harvesting pavements on the road: comparative study and performance assessment," Sustain. Cities Soc., vol. 81, p. 103868, Jun. 2022. https://doi.org/10.1016/j.scs.2022.103868

[8] G. Del Serrone, P. Peluso, and L. Moretti, "Photovoltaic road pavements as a strategy for low-carbon urban infrastructures," Heliyon, vol. 9, no. 9, p. e19977, Sep. 2023. https://doi.org/10.1016/j.heliyon.2023.e19977

[9] A. Sumorek and M. Buczaj, "New technologies using renewable energy in road construction," ECONTECHMOD: an international quarterly journal on economics of technology and modelling processes, Jan. 2017.

[10] W. Sun et al., "The State of the Art: Application of Green Technology in Sustainable Pavement," Adv. Mater. Sci. Eng., vol. 2018, pp. 1–19, Jun. 2018. https://doi.org/10.1155/2018/9760464

[11] E. H. H. Al-Qadami, Z. Mustaffa, and M. E. Al-Atroush, "Evaluation of the Pavement Geothermal Energy Harvesting Technologies towards Sustainability and Renewable Energy," Energies, vol. 15, no. 3, p. 1201, Feb. 2022. https://doi.org/10.3390/en15031201

[12] N. A. Razeman et al., "Optimization of Thermal Energy Harvesting Road Pavement using Sustainable Conductive Material in Malaysia by Numerical Simulation," Engineering, preprint, Jul. 2023. doi: 10.20944/preprints202307.0151.v1

[13] S. A. Tahami, M. Gholikhani, R. Nasouri, S. Dessouky, and A. T. Papagiannakis, "Developing a new thermoelectric approach for energy harvesting from asphalt pavements," Appl. Energy, vol. 238, pp. 786–795, Mar. 2019. https://doi.org/10.1016/j.apenergy.2019.01.152

[14] M. Gholikhani, R. Nasouri, S. A. Tahami, S. Legette, S. Dessouky, and A. Montoya, "Harvesting kinetic energy from roadway pavement through an electromagnetic speed bump," Appl. Energy, vol. 250, pp. 503–511, Sep. 2019. https://doi.org/10.1016/j.apenergy.2019.05.060

[15] R. De Fazio, M. De Giorgi, D. Cafagna, C. Del-Valle-Soto, and P. Visconti, "Energy Harvesting Technologies and Devices from Vehicular Transit and Natural Sources on Roads for a Sustainable Transport: State-of-the-Art Analysis and Commercial Solutions," Energies, vol. 16, no. 7, p. 3016, Mar. 2023. https://doi.org/10.3390/en16073016

[16] N. F. Saleh, A. A. Zalghout, S. A. Sari Ad Din, G. R. Chehab, and G. A. Saad, "Design, construction, and evaluation of energy-harvesting asphalt pavement systems," Road Mater. Pavement Des., vol. 21, no. 6, pp. 1647–1674, Aug. 2020. https://doi.org/10.1080/14680629.2018.1564352

[17] J. Johnsson, "Winter Road Maintenance Using Renewable Thermal Energy" - ProQuest. In ProQuest. Chalmers University of Technology. Department of Civil and Environmental Engineering, (2017).

[18] S. M. Charlesworth, A. S. Faraj-Llyod, and S. J. Coupe, "Renewable energy combined with sustainable drainage: Ground source heat and pervious paving," Renew. Sustain. Energy Rev., vol. 68, pp. 912–919, Feb. 2017. https://doi.org/10.1016/j.rser.2016.02.019

[19] A. N. A. Randriantsoa, D. A. H. Fakra, L. Rakotondrajaona, and W. J. Van Der Merwe Steyn, "Recent Advances in Hybrid Energy Harvesting Technologies Using Roadway Pavements: A Review of the Technical Possibility of Using Piezo-thermoelectrical Combinations," Int. J. Pavement Res. Technol., vol. 16, no. 4, pp. 796–821, Jul. 2023. https://doi.org/10.1007/s42947-022-00164-z

Keyword Index

2nd Life Application	223
Absorber Plate	254
Adaptive Cooling	73
Additive Manufacturing	164
Aerodynamics	44
AI Tool	238
ANFIS	188
Apparent Solar Time	254
Array Yield	277
Artificial Intelligence (AI)	96, 104, 148, 188, 230, 269
Augmentation	21
Batteries	51, 299
Battery Charging	324
Battery Degradation	197
Battery Management System (BMS)	332
Bidding Optimization	214
Bidding	179
Capacity Factor	277
Capacity Fade	197
Carbon Emissions	307
Centralized Solar PV System	345
Cleaning Methods	316
Climate Change	360
Climate	60
CNN	261
CO2 Emission	360
CO2 Emissions Reduction	345
Complexing Agents	1
Composite	31
Construction Industry	66
Construction	31
Convolutional Neural Network	205
Coordination	179
Copper Indium Diselenide	1
Critical Thinking	238
Current Controllers	246
Current Output	96
Cyclic Voltammetry	1
Date Fruit Type Classification	205
Day Ahead Market	179
Daylight Time	254
DC Bus Regulation	246
Deep Convolution Neural Network	13
Deep Learning	156, 230
Demand Growth	337
Desalination	104, 172, 179
Design and Development	197
Distillate	104
Dust Accumulation	316
EDM	124
Electric Vehicle Charging	299
Electrical Efficiency	290
Electricity Market	179
Electricity Reforms	337
Electrode Wear Rate	124
Electrodeposition	1
Electrolyzer	214
Electromagnetic	377
Energy and Exergy	368
Energy Consumption	140
Energy Dispersive X-Ray Spectroscopy	1
Energy Generation	324
Energy Performance Index	277
Energy Storage Systems (ESS)	223
Energy, Artificial Neural Network	156
Environment	60
Environmental Impact	172

a

EV	51
Exergy Efficiency	290
Experimental Investigation	316
Failure Analysis	197
Final Yield	277
Finite Element Method	164
Flat Plate Collector	21, 254
Flexibility	179
Fossil Fuels	360
Full Factorial	124
Future	60
Generation	179
Genetic Algorithm	96
Glazed	254
Global Solar Radiation	88
Government Facilities	345
Green Buildings	66
Green Hydrogen	36
Grid Integration	246
Headers	254
Healthcare	140
Heat Transfer Rate	254
HESCO	345
Hybrid Nanofluids	21
Hydrogen Production	214
Hydrogen	36
IEEE Smart Village (ISV)	332
Image Classification	13
Impacts of Renewable Energy on Power Systems	112
Internet of Things (IoT)	148
IoMT	140
IoT	140, 261

KSA 2030 Vision	277
KSA	337
Levelized Cost of Hydrogen (LCOH)	36
Li-ion Battery	223
Linear Generator	355
LSTM	88
Machine Learning	269
Martian Architecture	238
Mass Flow	254
Maximizing Profits	179
Maximum Power Point Tracking (MPPT)	332
Mechanisms	324
Minimum Chip Thickness	132
Nano-Powder Mixed Electro-Discharge Machining	124
Natural Binder	31
Object Detection	230
Optimization	179
Orchid Seeding	261
Pavement Engineering	377
Performance Ratio	277
Perturb and Observe (P&O)	332
Phase Locked Loop Point Absorber	246
Photocatalytic	104
Photovoltaic (PV)	285, 307
Photovoltaic Energy	36
Photovoltaic Solar Panels	13
Photovoltaic Systems	73
Photovoltaic-Thermal Applications	73
Photovoltaic-Thermal	290
Piezoelectric	324
Piezoelectric	377
Policies	51

Power Generation	355
Power Harvesting	324
Power Systems Stability	112
Precision Farming	269
Prediction Models	156
Pretrained Network	205
Production	179
PV Panel Thermal Inspection	230
PV Systems	188, 316
PV	179
Pvsyst	345
Raisers	254
Random Variables	82
Recent Trends	360
Recycled Organic Waste	368
Recycling	51
Regression	156
Regulations	51
Reject Brine	172
Reliability Analysis	82
Renewable Energy	13, 36, 44, 60, 66, 82, 112, 179, 299, 337, 355, 360
Renewable	307
Resistive Loading	246
Resource Management	148
Resource Utilization	172
Retired Batteries	223
Revenue	179
RHINO/Grasshopper	238
RO	179
Scanning Electron Microscopy	1
Size Effect	132
Sliding Mode Control	246
Smart Agriculture	269
Smart Health	140
Solar Energy	118

Solar Energy	21, 73,188. 230. 285, 299
Solar Irradiation	254
Solar Panels	377
Solar Still	104
Solar Tunnel Dryer	368
Solar	307
Specific Heat	254
Squeezenet	205
State of Health (SOH)	223
Statistical	156
Strategy	179
Surface Roughness	132
Sustainability	60, 66, 214, 360
Sustainable Agriculture	269
Sustainable Construction	172
Sustainable Material	31
Sustainable Micro-Milling	132
System Efficiency	277
Temperature Control	285
Temperature Regulation	73
Temporal Convolutional Network (TCN)	88
Test Matrix Design	197
Testing and Characterization	197
Thermal Conductivity	254
Thermal Efficiency	21, 290
Thermal Insulation Thickness	368
Thermoelectric	377
Thin-Film Solar Cells	1
Transfer Learning	205
Unglazed	254
Vibration Analysis	164
Vibration	324

Water and Electricity Management	148
Water Power Nexus	179
Wave Energy	36, 96, 246, 355
Wind Speed	96
Wind Turbine	44, 82, 96, 307
Windmill	96
X-Ray Diffraction	1
YOLOv7	230

Milton Keynes UK
Ingram Content Group UK Ltd.
UKHW021133040824
446423UK00008B/15